# *Modeling Axisymmetric Flows*

## *Dynamics of Films, Jets, and Drops*

# Modeling Axisymmetric Flows

## Dynamics of Films, Jets, and Drops

**Stanley Middleman**
Department of Applied Mechanics and Engineering Sciences
University of California, San Diego

**Academic Press**
San Diego New York Boston
London Sydney Tokyo Toronto

Academic Press, Inc.
A Division of Harcourt Brace & Company
525 B Street, Suite 1900, San Diego, California 92101-4495

*United Kingdom Edition published by*
Academic Press Limited
24-28 Oval Road, London NW1 7DX

Library of Congress Cataloging-in-Publication Data

Middleman, Stanley.
Modeling axisymmetric flows : dynamics of films, jets, and drops / Stanley Middleman.
p. cm.
Includes index.
ISBN 0-12-388694-5 (Paperback)
1. Fluid dynamics--Mathematical models. 2. Boundary element methods. I. Title.
TA357.M53 1995
620.1'06'0113--dc20 95-2196
CIP

PRINTED IN THE UNITED STATES OF AMERICA
95 96 97 98 99 00 EB 9 8 7 6 5 4 3 2 1

*This book is dedicated to*
*my granddaughter,*
*Ilana Shira Jaffe.*

# Contents

# 9 Ultrathin Films: Disjoining Pressure



# 10 The Interaction of a Deformable Drop with a Surface

A vertical laminar jet of a viscous Newtonian liquid issues from a long capillary far upstream (above) the frame of this photograph. In this image we see the effect of an axisymmetric surface tension-driven instability which leads to the breakup of the jet into a series of drops of varying sizes. In Chapter 4 we review various attempts, largely successful, at predicting the onset of this instability and the drop sizes produced. Photo by Mario Errico, and reproduced with his kind permission.

# Preface

One afternoon, while trying to find some logic in my professional life, I came to the realization that much of my research efforts had centered around the fluid dynamics of what seemed to be a very restricted class of flows: low Reynolds number flows into, through, and from a capillary tube, and the creation and subsequent behavior of drops formed from capillaries. Was this, I wondered, the result of some dementia? Had a childhood trauma led me inexorably along this path? Or could one argue, instead, that such flows are interesting from both a fundamental and a practical perspective?

This "slim volume" is an attempt to demonstrate that one can teach an interesting and enjoyable second course in fluid dynamics around—by and large—axisymmetric flows from capillaries, and the dynamics of drops and bubbles and films. In the end, very little of my own work appears, testimony to the reality that so much more and better has been done on these flows since I began my studies in the laboratory of Jerry Gavis, in the moist and reverberant basement of Maryland Hall at The Johns Hopkins University. At The Hopkins I had the great privilege of studying fluid dynamics in the classrooms of Stanley Corrsin and Owen Phillips. In the laboratory, Gavis was a wise mentor, letting me follow my intuition as I tried to find something interesting to make into a dissertation. An indispensable addendum to my education came from the happy fact that I shared the laboratory (and, one summer, digs on the Left Bank of the Raritan River) with my friend and classmate Simon Goren. Simon painted honey onto wires, I squirted liquids out of long tubes, and we both tried, with very little success, to explain to our friends and relatives what it was we were doing, and harder yet, why.

There is a richness in the topics of this small book that transcends the space they occupy. This is especially true of free boundary problems, where surface tension plays an essential role in the dynamics. Most classic texts in fluid dynamics—and, sadly then, most courses—fail to give attention to this class of flow. Hence one of my goals is to provide a vehicle for the continuing education of the student of fluid dynamics into the realm of free boundary flows. Another is to demonstrate the art of mathematical modeling—the iterative building of a theory, or at least a predictive model, from knowledge, intuition, and observation. Hence I try to provide throughout the book an abundant variety of examples of comparisons of experimental data with models of the flows described. In some cases this has necessitated that I convince a student to disappear into my own laboratories, therein to generate the required evidence, *pro* or *con*, of the utility of some simple mathematical model of a complex flow. In the end, this is the goal: to make the complex simple, without losing the essence—the virtue—of the complexity.

*Stanley Middleman*

# Acknowledgments

This book is built upon the accomplishments of hundreds of investigators whose papers I studied, and whose work I have attempted to present in a coherent manner. Several people were kind enough to answer specific questions I had when I was uncertain of my understanding of some key point in their presentation. Others kindly provided original photos or artwork for reproduction here. Their generosity is greatly appreciated, and there are too many such people to thank individually.

There are several people to whom I must express my gratitude individually. Professor Douglas Bousfield of the University of Maine read some early sections of this manuscript and offered several suggestions for improving the presentation. At a number of points in the preparation of this manuscript I found that the literature was lacking the experimental basis for evaluation of the models developed here. On many of these occasions I convinced an undergraduate student to spend a quarter or two in my laboratory producing the relevant body of data. I am grateful to Humphrey Chow, Shelley W. Chien, Jainie Mandrusov, Saul Ovalle and Lisa Strong for their efforts and accomplishments that produced the results presented in various places throughout the book.

Three of my former Ph.D. students have contributed to the book, each in a very distinctive style. The excellent experimental work of Liz Mickailey appears in Chapter 2. Andrew Yeckel made contributions cited in several places, none of which has anything to do with his Ph.D. thesis topic. My great fortune lies in the fact that Andrew has an abundant curiosity, and would often come back to me with a thoroughly worked out analysis of a problem that I happened to mention in passing as challenging but difficult. Mario Errico is an enormously skilled photographer with a wonderful eye for the aesthetics of fluid dynamics. His photographs grace these pages in several places, and I thank him for permitting their reproduction here.

The people at Academic Press have been a pleasure to work with, and I owe special thanks to Chuck Arthur, who encouraged me to write a book on an interesting topic, and Heidi Cruz, who managed the production of this book so skillfully.

Finally, I thank my wife Jo-Ann, herself an accomplished writer, for her patience and understanding when the writing was hard for me.

# 1 Laminar Axisymmetric Flows

*The choice of* axisymmetric flow fields has been made in order to provide a specific focus for this book. Although axial symmetry is clearly a limitation that will restrict us to the study of a specific class of flows, we will find that the range of challenging problems available to our study is enormous. As a second restriction, we will consider only laminar flows, although not necessarily flows at such small Reynolds numbers that inertial effects vanish. Again, this restriction still leaves us with such a variety of fluid dynamics problems that we will cover a great deal of ground as we pursue the methods of developing mathematical models of these flows. In trade for the simplification that attends the restriction to axisymmetric flows, we take on a significant complicating issue in a large fraction of the problems discussed herein. It is the complication that arises when we consider "free boundary" flows—flows which have an interface with another fluid, and for which the shape and position of the interface are unknown *a priori*. Free boundary flows include as examples the coating of a liquid onto a cylindrical wire that is steadily withdrawn from a liquid bath, the breakup of a liquid jet into droplets, and the spreading, under the action of gravity, of a viscous drop placed on a rigid impermeable horizontal surface.

One of the primary goals of this book is to describe how one may develop mathematical models of fairly complex flows, using relatively simple tools in many cases. At every opportunity we will compare the predictions of these models with experimental observations. A second goal is to broaden the exposure of the student of fluid dynamics into the realm of free boundary flows, a topic barely mentioned in most introductory courses in fluid dynamics and often neglected in advanced courses as well.

## *1-1 BASIC DYNAMIC AND KINEMATIC EQUATIONS*

We can begin by writing the mathematical statements of conservation of mass: the continuity equation for axisymmetric flow and conservation of momentum: the Navier–Stokes equations for axisymmetric flow of a Newtonian fluid under isothermal conditions. With these restrictions, these equations take these forms, in cylindrical coordinates $(r,\theta,z)$:

**Continuity:**

$$\frac{\partial u_z}{\partial z} + \frac{1}{r}\frac{\partial (r\,u_r)}{\partial r} = 0. \tag{1-1.1}$$

**Navier–Stokes:**

$r$ component

$$\rho\left(\frac{\partial u_r}{\partial t} + u_r\frac{\partial u_r}{\partial r} - \frac{u_\theta^2}{r} + u_z\frac{\partial u_r}{\partial z}\right) = -\frac{\partial p}{\partial r} + \mu\left\{\frac{\partial}{\partial r}\left[\frac{1}{r}\frac{\partial}{\partial r}(r\,u_r)\right] + \frac{\partial^2 u_r}{\partial z^2}\right\}. \tag{1-1.2}$$

$z$ component

$$\rho\left(\frac{\partial u_z}{\partial t} + u_r\frac{\partial u_z}{\partial r} + u_z\frac{\partial u_z}{\partial z}\right) = -\frac{\partial p}{\partial z} + \mu\left[\frac{1}{r}\frac{\partial}{\partial r}\left(r\frac{\partial u_z}{\partial r}\right) + \frac{\partial^2 u_z}{\partial z^2}\right] + \rho g_z. \tag{1-1.3}$$

$\vartheta$ component

$$\rho\left(\frac{\partial u_\theta}{\partial t} + u_r\frac{\partial u_\theta}{\partial r} + \frac{u_r u_\theta}{r} + u_z\frac{\partial u_\theta}{\partial z}\right) = \mu\left\{\frac{\partial}{\partial r}\left[\frac{1}{r}\frac{\partial (r u_\theta)}{\partial r}\right] + \frac{\partial^2 u_\theta}{\partial z^2}\right\}. \tag{1-1.4}$$

Implicit in writing the Navier–Stokes equations in these forms is the assumption that the viscosity $\mu$ is a constant. We have also assumed that if there is a body

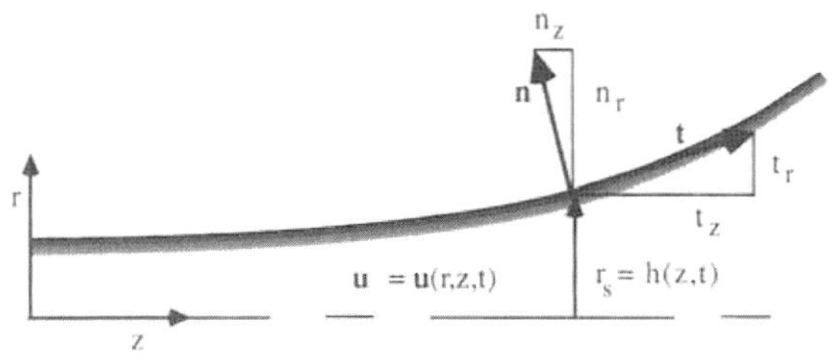


*FIGURE 1-1.1* Geometry at an axisymmetric interface.

force, it acts only along the axial direction $z$. Of course, for axisymmetric flows all derivatives $\partial/\partial\theta$ are set to zero. All of the flows of interest to us will satisfy these restrictions.

These equations must be solved subject to *initial* conditions, in the case of time-varying flows, and *boundary* conditions in all cases. The simplest and most familiar boundary condition occurs when the fluid is constrained in some region by a solid impermeable surface. This is the usual "no-slip" condition that matches the fluid velocity vector to that of the surface:

$$\mathbf{u} = \mathbf{u}_S \quad \text{on solid boundaries.} \tag{1-1.5}$$

Of course, if the surface is fixed we obtain

$$\mathbf{u} = \mathbf{0} \quad \text{on fixed solid boundaries.} \tag{1-1.6}$$

When there is a free boundary the situation is more complex.

With reference to Fig. 1-1.1, suppose that we consider the flow of a liquid bounded in some region by an axisymmetric free surface that separates the liquid from a gas phase. The position of the free surface is defined as

$$r_S = h(z,t). \tag{1-1.7}$$

At some point on the free surface, where the radius is $h$, the surface has normal and tangential vectors **n** and **t**. In terms of the geometry the components of the normal vector, in the cylindrical coordinate system, are

$$\begin{aligned} n_z &= -h' \, (1 + h'^2)^{-1/2}, \\ n_r &= (1 + h'^2)^{-1/2}, \end{aligned} \tag{1-1.8}$$

where $h' = \partial h/\partial z$.

Our primary interest here is in the boundary conditions at a free surface with curvature. A purely *kinematic* condition is the assertion that fluid does not cross the free surface. This implies that

$$\frac{\partial h}{\partial t} - \alpha(\mathbf{u}\cdot\mathbf{n}) = 0 = \frac{\partial h}{\partial t} - \alpha\left(u_r\, n_r + u_z\, n_z\right) \quad \text{along}\, h(z,t), \tag{1-1.9}$$[1]

[1]This follows from the fact that **u·n** is the component of velocity normal to the surface. Multiplication by $\alpha$ gives the radial component of velocity, which is itself simply $\partial h/\partial t$.

where

$$\alpha \equiv \left(1 + h'^2\right)^{1/2} \tag{1-1.10}$$

from which it follows that

$$u_r = \frac{\partial h}{\partial t} + h' u_z \quad \text{at } y = h(z,t). \tag{1-1.11}$$

The *dynamic* boundary conditions are much more difficult to write and explain. Physically, one boundary condition states that (for the cases we will be considering of gas–liquid interfaces and low-speed relative motion) there is no shear stress at (tangential to) the interface. The other dynamic condition states that the stress within the liquid, normal to the interface, just inside the surface, must be balanced by a stress acting normal to the surface arising from surface tension, and described by the Young–Laplace equation.

Let's set about writing these stress boundary conditions. We will suppose that the state of stress at some point on the free surface is determined by the stress tensor (within the liquid phase) **T**. We assume that the external fluid (taken to be a gas, here) is stress free.[2] The stress vector normal to the free surface is given by **T·n**. The $z$- and $r$-directed components of **T·n** are

$$(\mathbf{T}\cdot\mathbf{n})_z = T_{rz}\, n_r + T_{zz}\, n_z$$

and

$$(\mathbf{T}\cdot\mathbf{n})_r = T_{rr}\, n_r + T_{rz}\, n_z. \tag{1-1.12}$$

The stress boundary conditions on the tangential and normal stresses are now written in the following forms:

$$(\mathbf{T}\cdot\mathbf{n})\cdot\mathbf{t} = 0, \tag{1-1.13}$$

which is the condition that the shear stress vanishes at the interface, and

$$-(\mathbf{T}\cdot\mathbf{n})\cdot\mathbf{n} = \sigma\left(\frac{1}{R_{zr}} + \frac{1}{R_{n\theta}}\right), \tag{1-1.14}$$

which is the Young–Laplace equation. We denote the surface tension of the liquid–gas interface by $\sigma$, and the two $R$'s that appear are the so-called principal radii of curvature of the surface, which are defined on Fig. 1-1.2. Note that the radius of the axisymmetric surface, in the cylindrical coordinate system, is $h(z,t)$. At any point on the surface the curvature of the surface in the $zr$ plane may be described by the radius $R_{zr}$. The second principal radius of curvature is not $h(z,t)$—the radius of the cross section of the surface in the $r\theta$ plane—because

[2]Actually, we are assuming that in the external fluid there is an isotropic uniform stress—a pressure. For an incompressible fluid the isotropic stress can be found only to within an arbitrary constant. In other words, by adding any constant pressure to the solution of the equations for kinematics and dynamics (Eqs. 1-1.1 through 1-1.4), we do not change the solution. Hence we may take the external pressure to be zero.

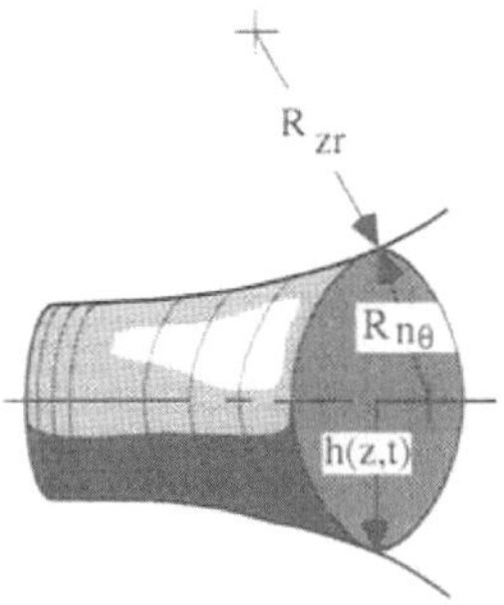


***FIGURE 1-1.2*** The principal radii of curvature of an axisymmetric surface.

$h(z,t)$ is not normal to the surface. The second principal radius of curvature is $R_{n\theta}$, a radius connecting the axis of symmetry with the point on the surface, such that the line $R_{n\theta}$ is normal to the surface.

From the geometry shown in Fig. 1-1.2 we can find (using Eq. 1-1.10 for $\alpha$)

$$R_{n\theta} = \alpha h \tag{1-1.15}$$

and

$$R_{zr} = \frac{\alpha^3}{-\partial^2 h/\partial z^2} = \frac{\alpha^3}{-h''}. \tag{1-1.16}$$

From Fig. 1-1.1 we can note that the tangential and normal unit vectors are related by

$$t_z = n_r \quad \text{and} \quad t_r = -n_z. \tag{1-1.17}$$

Now Eqs. 1-1.13 and 14 may be written as

$$\begin{aligned}(\mathbf{T\cdot n})\cdot\mathbf{t} &= (\mathbf{T\cdot n})_{\mathbf{z}}\cdot t_z + (\mathbf{T\cdot n})_{\mathbf{r}}\cdot t_r = 0\\ &= T_{rz}n_r^2 + T_{zz}n_r n_z - T_{rr}n_r n_z - T_{rz}n_z^2\end{aligned} \tag{1-1.18}$$

and

$$\begin{aligned}(\mathbf{T\cdot n})\cdot\mathbf{n} &= (\mathbf{T\cdot n})_{\mathbf{z}}\cdot n_z + (\mathbf{T\cdot n})_{\mathbf{r}}\cdot n_r = -\sigma\left(\frac{1}{R_{zr}} + \frac{1}{R_{n\theta}}\right)\\ &= T_{rz}n_r n_z + T_{zz}n_z^2 + T_{rr}n_r^2 + T_{rz}n_z n_r.\end{aligned} \tag{1-1.19}$$

We introduce the Newtonian model for the stress components:

$$T_{rr} = -p + 2\mu\frac{\partial u_r}{\partial r}, \qquad T_{zz} = -p + 2\mu\frac{\partial u_z}{\partial z},$$

$$T_{zr} = T_{rz} = \mu\left(\frac{\partial u_z}{\partial r} + \frac{\partial u_r}{\partial z}\right), \tag{1-1.20}$$ [3]

[3] Although there are axisymmetric flows in which there is a θ-component of velocity, we will not consider any in this book. Hence the equations that follow are for the case $u_\theta = 0$.

and after quite a bit of algebra we can write the stress boundary conditions in the following forms. For the boundary condition on the vanishing of the shear stress at the free surface (Eq. 1-1.18) we find

$$\mu\left(\frac{\partial u_z}{\partial r}+\frac{\partial u_r}{\partial z}\right)\left(1-h'^2\right) - 2\mu\left(\frac{\partial u_z}{\partial z}-\frac{\partial u_r}{\partial r}\right)h' = 0. \tag{1-1.21}$$

For the condition on the normal stress (Eq. 1-1.19) the result is

$$\frac{2\mu}{\left(1+h'^2\right)}\left[h'\left(\frac{\partial u_z}{\partial r}+\frac{\partial u_r}{\partial z}\right) - \frac{\partial u_r}{\partial r} - \frac{\partial u_z}{\partial z}h'^2\right] + p = \frac{-\sigma h''}{\left(1+h'^2\right)^{3/2}}. \tag{1-1.22}$$

These boundary conditions are obviously very complex, and they are a major source of the difficulty in solving free boundary problems, since they include terms involving the boundary shape, they are evaluated at that boundary, and the boundary is unknown—it is part of the solution. Nevertheless, at this stage we have written the dynamic and kinematic equations that an axisymmetric flow must obey.

Instead of proceeding directly with this set of equations we illustrate an alternative formulation of the dynamic equations for an axisymmetric flow. In Eqs. 1-1.2 and 3 we have three independent variables—the two velocity components and pressure. These are referred to as "primitive" variables, perhaps because they seem the most natural set of variables with which to describe the flow. There is an alternative, however, which introduces another set of independent variables that contain all of the information about the fluid dynamics. This is the vorticity–stream function formulation. We define the stream function $\psi$ as

$$u_r = \frac{1}{r}\frac{\partial\psi}{\partial z} \quad \text{and} \quad u_z = -\frac{1}{r}\frac{\partial\psi}{\partial r}. \tag{1-1.23a,b}$$

If we now replace the two velocity components in the Navier–Stokes equations through Eqs. 1-1.23a and b, and then eliminate pressure from the equations by differentiating Eq. 1-1.2 with respect to $z$ and 1-1.3 with respect to $r$, followed by subtraction of one equation from the other, the result is a fourth-order partial differential equation in $\psi$. The algebra is a bit messy, but in the case of a steady-state flow, and if we can neglect inertial terms, the result is

$$0 = \nu D^4\psi \tag{1-1.24}$$

where

$$D^2\psi = \frac{\partial^2\psi}{\partial r^2} - \frac{1}{r}\frac{\partial\psi}{\partial r} + \frac{\partial^2\psi}{\partial z^2}. \tag{1-1.25}$$

Thus there is a stream function formulation of the Navier–Stokes equations which represents three second-order partial differential equations by a single fourth-order equation.

By contrast, in the *vorticity*–stream function formulation, we define the so-called vorticity for the flow:

$$\omega \equiv \frac{\partial u_r}{\partial z} - \frac{\partial u_z}{\partial r}. \tag{1-1.26}$$

Vorticity is obviously related to the rate of deformation of the fluid. When Eqs. 1-1.23a and b are introduced into the definition of vorticity, we find the following simple relationship between the stream function and the vorticity:

$$D^2\psi = r\omega = \frac{\partial^2\psi}{\partial r^2} - \frac{1}{r}\frac{\partial\psi}{\partial r} + \frac{\partial^2\psi}{\partial z^2}. \tag{1-1.27}$$

This permits us to write Eq. 1-1.24 in the form

$$0 = \nu\, D^2 r\omega$$

which may be written as

$$0 = \frac{\partial^2\omega}{\partial r^2} + \frac{1}{r}\frac{\partial\omega}{\partial r} - \frac{\omega}{r^2} + \frac{\partial^2\omega}{\partial z^2} = \nabla^2\omega - \frac{\omega}{r^2}. \tag{1-1.28}$$

This is similar to, but not exactly, a steady-state diffusion equation for vorticity, and when we describe the fluid dynamics in terms of the vorticity formulation we often speak of the physics as involving the "diffusion" of vorticity. (In Cartesian coordinates one does, in fact, get a diffusion equation.) Equations 1-1.27 and 28 constitute a pair of second-order linear partial differential equations in the variables $\omega$ and $\psi$—hence this is called the "vorticity–stream function formulation." In computational fluid dynamics it is easier to solve this pair of equations than those of the fourth-order stream function formulation.

As we examine a variety of interesting physical problems, including free boundary problems, we will find that it is possible to introduce various assumptions (approximations) that permit us to simplify the dynamic and kinematic equations to the point that, at least in some cases, an analytical solution is possible. In preparation for that we first review a classical problem in fluid dynamics—steady laminar flow in a long tube of circular cross section, or Poiseuille flow. Then we will examine two "perturbations" on this problem as a way of introducing some notions regarding approximations.

## 1-2 STEADY FULLY DEVELOPED (POISEUILLE) FLOW

By definition, this flow is one for which there is only an axial component of velocity, i.e., $u_\theta = u_r = 0$. With this assumption the continuity equation immediately reduces to the form

$$\frac{\partial u_z}{\partial z} = 0. \tag{1-2.1}$$

We call a flow that satisfies this constraint a "fully developed flow," since it is unchanging along the direction of flow. Thus we see that *uniaxial* flow (flow for which there is only the nonzero velocity component $u_z$) is equivalent to fully developed flow. Since $u_r$ vanishes (by assumption), Eq. 1-1.2 is trivially satisfied by

$$0 = \frac{\partial p}{\partial r} \tag{1-2.2}$$

or

$$p = p(z) \text{ only.} \tag{1-2.3}$$

Then Eq. 1-1.3 takes the form

$$0 = -\frac{dp}{dz} + \frac{\mu}{r}\frac{\partial}{\partial r}\left(r\frac{\partial u_z}{\partial r}\right) + \rho g_z. \tag{1-2.4}$$

We may now define a modified pressure function as

$$\wp = p - \rho g_z z, \tag{1-2.5}$$

as a consequence of which Eq. 1-2.4 may be written in the form

$$\frac{d\wp}{dz} = \frac{\mu}{r}\frac{d}{dr}\left(r\frac{du_z}{dr}\right). \tag{1-2.6}$$

Note the use of the ordinary derivative $d/dr$, since Eq. 1-2.1 makes it clear that $u_z$ is not a function of $z$. But $\wp$ is not a function of the $r$-coordinate, by virtue of Eqs. 1-2.3 and 1-2.5. Hence the left-hand side of Eq. 1-2.6 is not a function of $r$, but the right-hand side is a function *only* of $r$. The only possibility, other than the trivial solution that there is no motion, is that each side is equal to a constant $C$, so

$$\frac{d\wp}{dz} = \frac{\mu}{r}\frac{d}{dr}\left(r\frac{du_z}{dr}\right) = C. \tag{1-2.7}$$

Equation 1-2.7 really represents two ordinary differential equations, the solutions of which are immediately found, upon integration, to be

$$\wp = Cz + A, \tag{1-2.8}$$

where $A$ is a constant of integration, and

$$u_z = \frac{Cr^2}{4\mu} + B \ln r + D. \tag{1-2.9}$$

Note that we have four unknown constants appearing in the solutions for $u_z$ and $\wp$. The constants $A$, $B$, $C$, and $D$ must be found through the imposition of boundary conditions that reflect the physics of this flow. We may take the following four conditions as part of the assumptions that define the flow field of interest to us:

1. The fluid does not slip at the inner surface of the tube, i.e.,

$$u_z = 0 \quad \text{at } r = R. \tag{1-2.10}$$

2. The flow is bounded along the axis of the tube; i.e., the velocity is finite along $r = 0$. Upon inspection of Eq. 1-2.9 we see that this can be satisfied if, and only if, the coefficient $B$ vanishes. These two statements lead to the conclusion that $D$ satisfies

$$D = -\frac{CR^2}{4\mu}. \tag{1-2.11}$$

We still have the coefficients $A$ and $C$ as unknowns. We may write them in terms of boundary conditions on the pressure:

3. The pressure is some known value, say

$$\wp = 0 \quad \text{at the downstream end of the tube, } z = L. \tag{1-2.12}$$

4. The pressure is some known value, say

$$\wp = \Delta\wp \quad \text{at the upstream end of the tube, } z = 0. \tag{1-2.13}$$

From Eqs. 1-2.8 and 1-2.13 we find

$$A = \Delta\wp \tag{1-2.14}$$

and from Eqs. 1-2.8 and 1-2.12 we find

$$C = -\Delta\wp/L. \tag{1-2.15}$$

With these results we may write the velocity field in the form

$$u_z = \frac{\Delta\wp R^2}{4\mu L}\left[1 - \left(\frac{r}{R}\right)^2\right]. \tag{1-2.16}$$

This is the well-known parabolic solution for the velocity profile. The classical form of Poiseuille's law follows upon integrating Eq. 1-2.16 to find the volumetric flow rate:

$$Q = \int_0^R 2\pi r u_z \, dr = \frac{\Delta\wp\pi R^4}{8\mu L}. \tag{1-2.17}$$

In the remainder of this introductory chapter we will examine a number of perturbations on this classical solution, in which one or more of the assumptions taken above are modified to correspond to other situations of physical interest.

## *1-3 FLOW THROUGH A TUBE WITH A VARIABLE RADIUS— THE LUBRICATION APPROXIMATION*

One perturbation on the classical Poiseuille flow occurs if we relax the assumption that the radius of the tube is constant along the tube axis. Under these conditions the fluid undergoes local acceleration (and/or deceleration) and the inertial terms of the momentum equations must be considered. We can illustrate this case by selecting a specific geometry, as shown in Fig. 1-3.1. The geometry and hence the flow are axisymmetric, but the radius is taken to vary according to the following equation:

$$R = R_0\left\{1 - \varepsilon \exp\left[-\left(z/\lambda R_0\right)^2\right]\right\}. \tag{1-3.1}$$

The dimensionless parameter $\varepsilon$ is a measure of the amplitude of the constriction to the flow, and $\lambda$ is a measure of the width of the constriction. (Because of the algebraic form selected, we will call this a "Gaussian" constriction.)

We must now write the continuity equation and the Navier–Stokes equations (Eqs. 1-1.2 and 1-1.3) for steady axisymmetric flow. No analytical solution is possible for these equations, for the flow geometry under consideration. Even if we simplify the equations (as we will, shortly), we cannot obtain an analytical solution. However, we can simplify the equations in a manner that permits us to obtain an approximate analytical solution. We could, of course, resort to a numerical solution of the full set of equations. The disadvantage of the numerical approach is that we must obtain a solution for each specific geometry (each set of values of $\varepsilon$ and $\lambda$). With an analytical solution we can explore the effect of these parameters without re-solving the equations for each set of parameters.

For steady flow the defining equations take the forms

$$\frac{\partial u_z}{\partial z} + \frac{1}{r}\frac{\partial (r\,u_r)}{\partial r} = 0, \tag{1-3.2}$$

$$\rho\left(u_r \frac{\partial u_r}{\partial r} + u_z \frac{\partial u_r}{\partial z}\right) = -\frac{\partial p}{\partial r} + \mu\left\{\frac{\partial}{\partial r}\left[\frac{1}{r}\frac{\partial}{\partial r}(r\,u_r)\right] + \frac{\partial^2 u_r}{\partial z^2}\right\}, \tag{1-3.3}$$

$$\rho\left(u_r \frac{\partial u_z}{\partial r} + u_z \frac{\partial u_z}{\partial z}\right) = -\frac{\partial p}{\partial z} + \mu\left[\frac{1}{r}\frac{\partial}{\partial r}\left(r\frac{\partial u_z}{\partial r}\right) + \frac{\partial^2 u_z}{\partial z^2}\right] + \rho g_z. \tag{1-3.4}$$

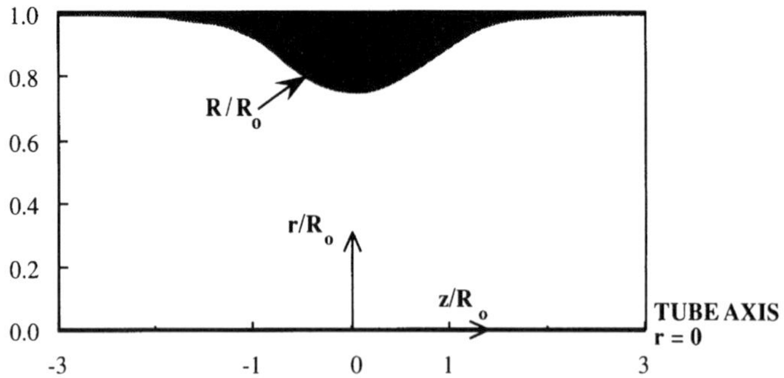


*FIGURE 1-3.1* Axisymmetric tube with a variable radius. (Eq. 1-3.1 with $\varepsilon = 0.25$ and $\lambda = 1$.)

Our further discussion of this problem, and particularly our discussion of approximations to these equations, will be facilitated if we first nondimensionalize the equations. We begin by making the independent variables (the space coordinates, $r$ and $z$) dimensionless. To do so, we first look for any natural length scales in the $r$ and $z$ directions. The magnitude of the radial amplitude is simply $\varepsilon R_0$, so we choose this as a radial length scale and define a dimensionless radial variable as

$$s \equiv \frac{r}{\varepsilon R_0}. \tag{1-3.5}$$

A length scale in the axial direction is $\lambda R_0$, a measure of the distance from the minimum in $R(z)$ required for $R(z)$ to approach $R_0$. Hence we define a dimensionless axial variable as

$$\zeta \equiv \frac{z}{\lambda R_0}. \tag{1-3.6}$$

We suppose that the volumetric flow rate is specified (call it $Q$) and hence a velocity scale in the axial direction may be defined as

$$U_0 \equiv \frac{Q}{\pi R_0^2}. \tag{1-3.7}$$

Then we may define a dimensionless axial velocity $\tilde{u}_z$ as

$$\tilde{u}_z \equiv \frac{u_z}{U_0}. \tag{1-3.8}$$

There seems to be no natural velocity scale in the radial direction. We will make $\tilde{u}_r$ dimensionless by using a yet-to-be defined velocity scale $V_0$:

$$\tilde{u}_r \equiv \frac{u_r}{V_0}. \tag{1-3.9}$$

Finally, we may define a dimensionless pressure as

$$\tilde{p} \equiv \frac{p}{\Pi}, \tag{1-3.10}$$

where a pressure scale $\Pi$ is not yet defined. Now, with these definitions, the continuity equation takes the nondimensional form

$$\frac{U_0}{\lambda}\frac{\partial \tilde{u}_z}{\partial \zeta} + \frac{V_0}{\varepsilon s}\frac{\partial(s\,\tilde{u}_r)}{\partial s} = 0 \tag{1-3.11}$$

or

$$\left(\frac{\varepsilon U_0}{\lambda V_0}\right)\frac{\partial \tilde{u}_z}{\partial \zeta} + \frac{1}{s}\frac{\partial(s\,\tilde{u}_r)}{\partial s} = 0. \tag{1-3.12}$$

Let's look at the coefficient of the first term. There are three possibilities: the coefficient could be of an order of magnitude large compared to unity, or small, or comparable to unity.[4] The derivative terms themselves are each of order unity by virtue of the manner in which the dependent and independent variables were nondimensionalized. Then if, for example, the coefficient of $\partial \tilde{u}_z/\partial \zeta$ is large, it follows that $\partial \tilde{u}_z/\partial \zeta$ itself must be small so that the product of the two, which equals the (negative of the) second term, can be of order unity. But if $\partial \tilde{u}_z/\partial \zeta$ is small, there is no significant variation of $\tilde{u}_z$ in the direction of flow, and this is not consistent with our expectation for this flow. On the other hand, if the coefficient is small, then the product term is small compared to unity, and it follows that the second term is small compared to unity. But this contradicts the expectation that $\tilde{u}_r$ varies in the radial direction. The only thing that makes sense is that the coefficient is of order unity itself. Since the velocity scale $V_0$ has not been defined yet, we conclude that an appropriate choice for $V_0$ is such that

$$\frac{\varepsilon U_0}{\lambda V_0} = 1 \tag{1-3.13}$$

or, specifically

$$V_0 = \frac{\varepsilon U_0}{\lambda}. \tag{1-3.14}$$

If the radius varies to a small extent relative to the axial distance, that is, if

$$\frac{\varepsilon R_0}{\lambda R_0} << 1, \tag{1-3.15}$$

then it follows that

$$\frac{V_0}{U_0} << 1, \tag{1-3.16}$$

as we would expect.

Now let's look at Eq. 1-3.4 in nondimensional form:

$$\rho\left(\frac{U_0 V_0}{\varepsilon R_0}\tilde{u}_r \frac{\partial \tilde{u}_z}{\partial s} + \frac{U_0^2}{\lambda R_0}\tilde{u}_z \frac{\partial \tilde{u}_z}{\partial \zeta}\right)$$
$$= -\frac{\Pi}{\lambda R_0}\frac{\partial \tilde{\wp}}{\partial \zeta} + \frac{\mu U_0}{(\varepsilon R_0)^2}\left[\frac{1}{s}\frac{\partial}{\partial s}\left(s\frac{\partial \tilde{u}_z}{\partial s}\right) + \frac{\varepsilon^2}{\lambda^2}\frac{\partial^2 \tilde{u}_z}{\partial \zeta^2}\right] \tag{1-3.17}$$

or

$$\frac{\varepsilon^2 \rho R_0 U_0}{\mu \lambda}\left(\tilde{u}_r \frac{\partial \tilde{u}_z}{\partial s} + \tilde{u}_z \frac{\partial \tilde{u}_z}{\partial \zeta}\right)$$
$$= -\frac{\varepsilon^2 R_0 \Pi}{\lambda \mu U_0}\frac{\partial \tilde{\wp}}{\partial \zeta} + \frac{1}{s}\frac{\partial}{\partial s}\left(s\frac{\partial \tilde{u}_z}{\partial s}\right) + \frac{\varepsilon^2}{\lambda^2}\frac{\partial^2 \tilde{u}_z}{\partial \zeta^2}. \tag{1-3.18}$$

[4]We follow here the style of argument presented in Chapter 13 of Denn (1980).

(We have used Eq. 1-2.5 to introduce the pressure $\wp$.)

If we again assume that Eq. 1-3.15 holds, we may drop the last term on the right-hand side of Eq. 1-3.18 to yield

$$\frac{\varepsilon^2 \rho R_0 U_0}{\mu\lambda}\left(\tilde{u}_r \frac{\partial \tilde{u}_z}{\partial s} + \tilde{u}_z \frac{\partial \tilde{u}_z}{\partial \zeta}\right)$$
$$= -\frac{\varepsilon^2 R_0 \Pi}{\lambda\mu U_0}\frac{\partial \tilde{\wp}}{\partial \zeta} + \frac{1}{s}\frac{\partial}{\partial s}\left(s\frac{\partial \tilde{u}_z}{\partial s}\right). \tag{1-3.19}$$

We should recognize at this point that a Reynolds number has appeared on the left-hand side of this equation:

$$\mathrm{Re} \equiv \frac{\rho U_0 R_0}{\mu}. \tag{1-3.20}$$

Now there are two cases to consider, depending on the Reynolds number. If the Reynolds number is of order unity or less, then the coefficient of the left-hand side is small (because of the factor $\varepsilon^2/\lambda$), and Eq. 1-3.19 reduces to

$$0 = -\frac{\varepsilon^2 R_0 \Pi}{\lambda\mu U_0}\frac{\partial \tilde{\wp}}{\partial \zeta} + \frac{1}{s}\frac{\partial}{\partial s}\left(s\frac{\partial \tilde{u}_z}{\partial s}\right). \tag{1-3.21}$$

By a similar argument to the one invoked in consideration of the continuity equation, these two terms must be of the same order of magnitude. Otherwise $u_z$ would not be a function of radial coordinate $r$, or the pressure would not vary along the axis. Neither of these possibilities makes sense physically. We conclude that this situation is only avoided by a suitable definition of the scale factor $\Pi$ for pressure as

$$\Pi \equiv \frac{\lambda\mu U_0}{\varepsilon^2 R_0} \tag{1-3.22}$$

and Eq. 1-3.21 becomes

$$0 = -\frac{\partial \tilde{\wp}}{\partial \zeta} + \frac{1}{s}\frac{\partial}{\partial s}\left(s\frac{\partial \tilde{u}_z}{\partial s}\right). \tag{1-3.23}$$

This appears to be the same equation (except in nondimensional format) as Eq. 1-2.6, but this is not the case. In the first place, we no longer have a basis for arguing that each term is constant, independent of $\zeta$ and $s$, since (note Eqs. 1-3.12 and 13 ) we no longer have fully developed flow. In the second place, the boundary of the flow is no longer a tube of uniform radius (see Eq. 1-3.1).

Equation 1-3.23 follows for small Reynolds number. If the Reynolds number of this flow is *not* very small, then the left-hand side (the inertial terms) are larger than order unity and the second term on the right-hand side of Eq. 1-3.19 can be neglected. What remains then is

$$\frac{\rho R_0 U_0}{\mu}\left(\tilde{u}_r \frac{\partial \tilde{u}_z}{\partial s} + \tilde{u}_z \frac{\partial \tilde{u}_z}{\partial \zeta}\right) = -\frac{R_0 \Pi}{\mu U_0}\frac{\partial \tilde{\wp}}{\partial \zeta}. \tag{1-3.24}$$

Since the terms must be of the same order, we conclude that an appropriate definition of $\Pi$ in the large Reynolds number case is

$$\Pi \equiv \rho U_0^2 \tag{1-3.25}$$

and we would solve

$$\tilde{u}_r \frac{\partial \tilde{u}_z}{\partial s} + \tilde{u}_z \frac{\partial \tilde{u}_z}{\partial \zeta} = -\frac{\partial \tilde{\wp}}{\partial \zeta}. \qquad (1\text{-}3.26)^5$$

Finally, let's look at Eq. 1-3.3, the *radial* momentum equation, in nondimensional form:

$$\rho\left(\frac{V_0^2}{\varepsilon R_0}\tilde{u}_r \frac{\partial \tilde{u}_r}{\partial s} + \frac{U_0 V_0}{\lambda R_0}\tilde{u}_z \frac{\partial \tilde{u}_r}{\partial \zeta}\right)$$
$$= -\frac{\Pi}{\varepsilon R_0}\frac{\partial \tilde{\wp}}{\partial s} + \frac{\mu V_0}{\varepsilon^2 R_0^2}\left\{\frac{\partial}{\partial s}\left[\frac{1}{s}\frac{\partial}{\partial s}(s\,\tilde{u}_r)\right] + \frac{\varepsilon^2}{\lambda^2}\frac{\partial^2 \tilde{u}_r}{\partial \zeta^2}\right\} \tag{1-3.27}$$

or

$$\frac{\varepsilon^2 \rho U_0 R_0}{\mu\lambda}\left(\tilde{u}_r \frac{\partial \tilde{u}_r}{\partial s} + \tilde{u}_z \frac{\partial \tilde{u}_r}{\partial \zeta}\right)$$
$$= -\frac{\lambda R_0 \Pi}{\mu U_0}\frac{\partial \tilde{\wp}}{\partial s} + \frac{\partial}{\partial s}\left[\frac{1}{s}\frac{\partial}{\partial s}(s\,\tilde{u}_r)\right] + \frac{\varepsilon^2}{\lambda^2}\frac{\partial^2 \tilde{u}_r}{\partial \zeta^2}. \tag{1-3.28}$$

If we again assume that Eq. 1-3.15 holds, we may drop the last term on the right-hand side of Eq. 1-3.28 to yield

$$\frac{\varepsilon^2 \rho U_0 R_0}{\mu\lambda}\left(\tilde{u}_r \frac{\partial \tilde{u}_r}{\partial s} + \tilde{u}_z \frac{\partial \tilde{u}_r}{\partial \zeta}\right) = -\frac{\lambda R_0 \Pi}{\mu U_0}\frac{\partial \tilde{\wp}}{\partial s} + \frac{\partial}{\partial s}\left[\frac{1}{s}\frac{\partial}{\partial s}(s\tilde{u}_r)\right]. \tag{1-3.29}$$

Again there are two limiting cases (high or low Reynolds number) that correspond to the two definitions of $\Pi$. For low Re the left-hand side is neglected and we have

$$0 = -\frac{\lambda R_0 \Pi}{\mu U_0}\frac{\partial \tilde{\wp}}{\partial s} + \frac{\partial}{\partial s}\left[\frac{1}{s}\frac{\partial}{\partial s}(s\,\tilde{u}_r)\right]. \tag{1-3.30}$$

We have already concluded that Eq. 1-3.22 is appropriate to this case, so this equation becomes

$$0 = -\frac{\partial \tilde{\wp}}{\partial s} + \left(\frac{\varepsilon}{\lambda}\right)^2\left\{\frac{\partial}{\partial s}\left[\frac{1}{s}\frac{\partial}{\partial s}(s\,\tilde{u}_r)\right]\right\} \tag{1-3.31}$$

For the case of interest (Eq. 1-3.15) we conclude that

[5]Since viscosity has disappeared from the problem we no longer could be considering any realistic case of a laminar flow. Hence this is not really an interesting case to consider.

$$0 = -\frac{\partial \tilde{\wp}}{\partial s}. \tag{1-3.32}$$

Hence the pressure is independent of radial position.

In the low Reynolds number case we conclude that the system of equations to be solved is

$$\frac{\partial \tilde{u}_z}{\partial \zeta} + \frac{1}{s}\frac{\partial (s\,\tilde{u}_r)}{\partial s} = 0,$$

$$0 = -\frac{\partial \tilde{\wp}}{\partial \zeta} + \frac{1}{s}\frac{\partial}{\partial s}\left(s\frac{\partial \tilde{u}_z}{\partial s}\right),$$

$$0 = -\frac{\partial \tilde{\wp}}{\partial s}. \tag{1-3.33 a,b,c}$$

Now consider the second case, of high Reynolds number. Equation 1-3.29 becomes

$$\frac{\varepsilon^2 \rho U_0 R_0}{\mu\lambda}\left(\tilde{u}_r \frac{\partial \tilde{u}_r}{\partial s} + \tilde{u}_z \frac{\partial \tilde{u}_r}{\partial \zeta}\right)$$
$$= -\frac{\lambda \rho U_0 R_0}{\mu}\frac{\partial \tilde{\wp}}{\partial s} + \frac{\partial}{\partial s}\left[\frac{1}{s}\frac{\partial}{\partial s}(s\,\tilde{u}_r)\right] \tag{1-3.34}$$

where we have used the definition of $\Pi$ appropriate to high $\mathsf{Re}$, Eq. 1-3.25. We may write this in the form

$$\left(\frac{\varepsilon}{\lambda}\right)^2\left(\tilde{u}_r \frac{\partial \tilde{u}_r}{\partial s} + \tilde{u}_z \frac{\partial \tilde{u}_r}{\partial \zeta}\right) = -\frac{\partial \tilde{\wp}}{\partial s} + \frac{1}{\lambda \mathsf{Re}}\left\{\frac{\partial}{\partial s}\left[\frac{1}{s}\frac{\partial}{\partial s}(s\,\tilde{u}_r)\right]\right\}. \tag{1-3.35}$$

For large $\mathsf{Re}$, $\varepsilon/\lambda \ll 1$, and for $\lambda$ of order unity or greater, we again find Eq. 1-3.32. Hence in the high-$\mathsf{Re}$ case the system of equations to be solved is

$$\frac{\partial \tilde{u}_z}{\partial \zeta} + \frac{1}{s}\frac{\partial (s\tilde{u}_r)}{\partial s} = 0,$$

$$\tilde{u}_r \frac{\partial \tilde{u}_z}{\partial s} + \tilde{u}_z \frac{\partial \tilde{u}_z}{\partial \zeta} = -\frac{\partial \tilde{\wp}}{\partial \zeta},$$

$$0 = -\frac{\partial \tilde{\wp}}{\partial s}. \tag{1-3.36 a,b,c}$$

If we make *only* the geometrical approximation (Eq. 1-3.15) and make no restriction with regard to Reynolds number, then we cannot argue away any of the terms remaining in Eq. 1-3.35. [It appears that we could throw away the left-hand side, because of the factor $(\varepsilon/\lambda)^2$, but we can no longer argue that $\tilde{\wp}$ is of order one, so we cannot compare the terms.] Regardless of the magnitude of the Reynolds number, Eqs. 1-3.33 and 36 imply that the pressure gradient normal to the mean flow is negligible. This is a consequence of the geometrical approximation (Eq. 1-3.15). We often refer to this approximation as the "lubrication approximation" because it arises commonly in lubrication analysis.

In the case of arbitrary Reynolds number (but remember, low enough that the flow is still not turbulent) we must solve Eq. 1-3.35. One method of solution would be through numerical methods. But it is still possible to gain some insight into the behavior of this flow through analytical, albeit approximate, methods. We illustrate that next.

## 1-4 AN INTEGRAL TECHNIQUE FOR SOLUTION OF THE LUBRICATION EQUATIONS

We will approach this problem by using a method similar to the integral boundary layer technique of von Karman and Pohlhausen. In this technique we assume a polynomial form for the velocity function $u_z$ and then integrate the $z$-momentum equation across the radial direction. Although we could continue to work with the dimensionless formulation, it may be a bit easier to follow the logic if we return to the original equations. If we maintain the geometrical approximation (Eq. 1-3.15) then we begin with

$$\rho\left(u_r \frac{\partial u_z}{\partial r} + u_z \frac{\partial u_z}{\partial z}\right) = -\frac{\partial p}{\partial z} + \mu \frac{1}{r}\frac{\partial}{\partial r}\left(r \frac{\partial u_z}{\partial r}\right). \tag{1-4.1}$$

We have dropped the gravitational term. If gravity acts in the direction of the axis of flow, we can just redefine $p$ as $\wp$.

Let's begin with a term-by-term integration of Eq. 1-4.1 across the radius.[6] The first term is integrated to give

$$\begin{aligned}\int_0^{R(z)} r\left(u_r \frac{\partial u_z}{\partial r}\right) dr &= \int_0^{R(z)} r u_r \left(\frac{\partial u_z}{\partial r}\, dr\right) = \int r u_r \left(du_z\right) \\ &= \int d(r u_r u_z) - \int u_z\, d(r u_r) \\ &= \left[r u_r u_z\right]_{r=0}^{r=R} - \int u_z \frac{\partial (r u_r)}{\partial r}\, dr \\ &= \int u_z \left(r \frac{\partial u_z}{\partial z}\right) dr \\ &= \int \frac{r}{2}\left(\frac{\partial u_z^2}{\partial z}\right) dr = \frac{1}{2}\frac{d}{dz}\int_0^{R(z)} r\, u_z^2\, dr.\end{aligned} \tag{1-4.2}$$

[6]Note that these integrals are $\int_0^R (\,)r\,dr$, rather than $\int_0^R (\,)\,dr$, and so we are averaging across the cross-sectional area.

Not all the details of this result are given above. For example, in the third line of Eq. 1-4.2, the term in parentheses $(ru_r u_z)$ is evaluated at the upper and lower limits $r = R(z)$ and $r = 0$. It vanishes at both ends of that range, at the lower limit because $r = 0$ and at the upper limit because $u_z = 0$. The integral at the end of the third line transforms to the integral at the beginning of the fourth line through the use of the continuity equation.

The second term on the left-hand side of Eq. 1-4.1 integrates immediately to

$$\int \left( r\, u_z \frac{\partial u_z}{\partial z} \right) dr = \int \frac{r}{2} \left( \frac{\partial u_z^2}{\partial z} \right) dr = \frac{1}{2} \frac{d}{dz} \int_0^{R(z)} r\, u_z^2 dr. \tag{1-4.3}$$

The terms on the right-hand side of Eq. 1-4.1 are more easily integrated, and the final result is the following:

$$\rho \frac{d}{dz} \int_0^{R(z)} r u_z^2 dr = -\frac{R^2}{2} \frac{dp}{dz} + \mu R \left( \frac{\partial u_z}{\partial r} \right)_{R(z)}. \tag{1-4.4}$$

The approach we take is to assume a polynomial for the velocity component $u_z$, with a set of coefficients that will be determined by forcing this profile to satisfy certain boundary conditions. The profile will also be forced to satisfy the integrated (averaged) form of the $z$ component of the momentum equation, which is Eq. 1-4.4.

Thus we begin by writing the velocity field in the form

$$\frac{u_z}{U(z)} = 1 + a\eta + b\eta^2 + c\eta^3 + d\eta^4, \tag{1-4.5}$$

where

$$\eta = \frac{r}{R(z)}. \tag{1-4.6}$$

$U(z)$ is the velocity along the centerline.

What factors, physical and/or mathematical, have dictated or suggested this form for $u_z$? Equation 1-4.5 is a statement (really, an assumption) that the axial component of velocity at any axial position $z$, and radial position $r$, divided by the centerline or maximum velocity $U(z)$, is a function only of $\eta$, which is the relative distance from the centerline to the wall. In words, one says that we are assuming that the shape of the velocity profile is "similar" with respect to the variable $\eta$. Note that $\eta$ is a function of $r$ and $z$, the latter through the dependence of $R(z)$.

We will require that the no-slip condition be satisfied along $r = R(z)$. Applying this to Eq. 1-4.5, we find that the coefficients must satisfy

$$0 = 1 + a + b + c + d. \tag{1-4.7}$$

A second boundary condition expresses symmetry about the centerline:

$$\frac{\partial u_z}{\partial r} = 0 \quad \text{along} \quad \eta = 0. \tag{1-4.8}$$

This leads to

$$a = 0. \tag{1-4.9}$$

These two boundary conditions are identical to those imposed in the case of a constant tube radius. There are no other natural boundary conditions to be imposed. In order to find two more equations (since we have four coefficients) we will require that the assumed velocity profile satisfy two more conditions. One arises from the expectation that if the amplitude is not too great, the velocity profile should be close to the Poiseuille profile in the central portion of the tube. Hence we will require that

$$\frac{\partial^2 u_z}{\partial r^2} = \frac{-2U(z)}{R^2(z)} \quad \text{along} \quad \eta = 0, \tag{1-4.10}$$

which is exactly true only for the Poiseuille profile. Imposition of this constraint as an approximation leads to the simple result that

$$b = -1. \tag{1-4.11}$$

Hence

$$c = -d. \tag{1-4.12}$$

The final constraint will be imposed along the wall of the tube, where we will require that the profile satisfy the $z$-component momentum equation, that is:

$$0 = -\frac{dp}{dz} + \mu \frac{1}{r}\frac{\partial}{\partial r}\left(r\frac{\partial u_z}{\partial r}\right) \quad \text{along} \quad \eta = 1. \tag{1-4.13}$$

When Eq. 1-4.4 is used in this relationship we find, after some algebraic manipulation, that the coefficient $c$ may be related to the pressure gradient by

$$c = -\frac{\Phi + 4}{7}, \tag{1-4.14}$$

where we simplify the notation by defining a dimensionless pressure gradient as

$$\Phi \equiv \frac{R^2}{\mu U}\frac{dp}{dz}. \tag{1-4.15}$$

At this stage the velocity profile is explicit in the $r$ dependence, but it is expressed in terms of the unknown pressure gradient, and the unknown function $U(z)$, which is the maximum or centerline velocity. Of course, the solution also involves the geometrical shape function $R(z)$.

It is possible to eliminate $U(z)$ in terms of the pressure gradient by using the continuity equation in integral form, which is simply

$$Q = \int_0^{R(z)} 2\pi r u_z(r)\, dr. \tag{1-4.16}$$

From this we find

$$U(z) = \frac{210}{97\pi R^2}\left(Q + \frac{\pi R^4}{105\mu}\frac{dp}{dz}\right). \tag{1-4.17}$$

We have still not used the integral form of the $z$-component momentum equation—Eq. 1-4.4. We may substitute Eq. 1-4.5 into Eq. 1-4.4, using the results for the coefficients $a$, $b$, $c$, and $d$, and the relationship given by Eq. 1-4.17. This would lead to an equation for the pressure gradient as a function of $R(z)$. However, the nonlinearity of the inertial term (the integral on the left-hand side of Eq. 1-4.4) makes it impossible to obtain a solution in explicit analytical form. But we cannot throw away the nonlinearity, since it contains an essential feature of the physics of this flow. The compromise that we make is that we retain the nonlinear term on the left of Eq. 1-4.4, but we approximate it by using the Poiseuille profile for its calculation. Thus we write

$$u_z = 2\,\overline{U}\left(1 - \eta^2\right), \tag{1-4.18}$$

where the average velocity is related to the volume flow rate through

$$Q = \pi R^2 \overline{U}. \tag{1-4.19}$$

Performing the indicated integration, we find that the left-hand term of Eq. 1-4.4 becomes

$$\rho \frac{d}{dz}\int_0^{R(z)} r u_z^2\, dr = \frac{2\rho}{3}\frac{d}{dz}\left(R^2\, \overline{U}^2\right). \tag{1-4.20}$$

The two terms on the right-hand side of Eq. 1-4.4 may be written using Eq. 1-4.5 for $u_z$, 1-4.17 for $U(z)$, and 1-4.19 to eliminate $\overline{U}$, and we find the pressure gradient in the form

$$\frac{dp}{dz} = \frac{5432\,\rho Q^2}{1575\,\pi^2 R^5}\frac{dR}{dz} - \frac{8\mu Q}{\pi R^4}. \tag{1-4.21}$$

With this result we may now go back to the assumed polynomial velocity profile (Eq. 1-4.5) and write it in the following form:

$$\frac{u_z(r,z)}{\overline{U}_0} = 2\left[\frac{R_0}{R(z)}\right]^2 (1 - \eta^2) + \mathrm{Re}_0\left[\frac{R_0}{R(z)}\right]^3 \frac{dR}{dz}\left[\frac{8}{225}(1-\eta^2) - \frac{4}{15}\eta^3 + \frac{4}{15}\eta^4\right]. \tag{1-4.22}$$

With an explicit expression for $R(z)$, such as Eq. 1-3.1, Eq. 1-4.22 is an explicit expression for $u_z(r,z)$. Note that in Eq. 1-4.22 we have introduced a Reynolds number based on the radius and the average velocity of the flow just upstream of the constriction,

$$\mathrm{Re}_0 = \frac{2\rho\overline{U}_0 R_0}{\mu}. \tag{1-4.23}$$

(This definition of $\mathrm{Re}_0$ differs from Eq. 1-3.20 by a factor of 2. It is more common to use Eq. 1-4.23 for tube flow.)

This has been a tedious algebraic exercise, but it has led to two explicit expressions from which we may find the physical characteristics of this flow that are of interest. Let's now examine some of these characteristics.

### *1-4.1 The Excess Pressure Loss across the Constriction*

For flow through a tube of uniform radius the pressure loss across any axial length follows from Poiseuille's law. If we simply take Eq. 1-4.21 and set $dR/dz = 0$ we find the Poiseuille loss.

$$p(z_1) - p(z_2) = \frac{8\mu Q}{\pi R^4}\left(z_2 - z_1\right). \tag{1-4.24}$$

The excess pressure loss associated with the constriction is obtained by integrating the pressure gradient given in Eq. 1-4.21, and subtracting the Poiseuille loss.

$$\Delta P_{\mathrm{ex}} = \int_{z_1}^{z_2} -\frac{dp}{dz}\,dz - \frac{8\mu Q}{\pi R_0^4}\left(z_2 - z_1\right)$$

$$= -\frac{5432\,\rho Q^2}{1575\,\pi^2}\int_{z_1}^{z_2}\frac{1}{R^5}\frac{dR}{dz}\,dz + \int_{z_1}^{z_2}\frac{8\mu Q}{\pi R^4}\,dz - \frac{8\mu Q}{\pi R_0^4}\left(z_2 - z_1\right). \tag{1-4.25}$$

It is to be understood that in these integrals the upper and lower limits are in the upstream and downstream regions where $R(z) = R_0$.

A review of this analysis reveals that nowhere have we assumed anything about the specific shape of the constriction. Hence Eq. 1-4.25, and for that matter Eq. 1-4.23, are general results into which we may substitute the specific shape of interest, such as that given by Eq. 1-3.1. Depending on the complexity of the algebra it may be necessary to perform the indicated integrations in Eq. 1-4.25 numerically. However, one simple result falls out of inspection of Eq. 1-4.25, and it reveals a serious limitation of the technique used in this analysis.

First we nondimensionalize Eq. 1-4.25, and then rewrite it slightly in the form

$$\frac{\pi R_0^3\,\Delta P_{\mathrm{ex}}}{8\mu Q} = -\frac{679}{1575}\,\mathrm{Re}_0\int_{z_1}^{z_2}\left(\frac{R_0}{R}\right)^5 d\left(\frac{R}{R_0}\right)$$

$$+\int_{z_1/R_0}^{z_2/R_0}\left(\frac{R_0}{R}\right)^4 d\left(\frac{z}{R_0}\right) - \frac{(z_2 - z_1)}{R_0}. \tag{1-4.26}$$

If we now understand that the excess pressure drop is to be calculated over a length $z_2 - z_1$ that is symmetric about $z = 0$, we see that the first integral in the expression above vanishes if the shape function is symmetric about $z = 0$. Then the Reynolds number dependence of the left-hand side vanishes, and $\Delta P_{ex}$ is linearly proportional to $Q$. This is characteristic of a purely viscous, inertia-free flow. Hence the approximate method of including inertial effects in this analysis (Eq. 1-4.20) fails to yield a model that accounts for inertial effects on the excess pressure loss across the constriction.

Note that we are not stating that the *pressure distribution* is unaffected by inertia. Equation 1-4.21 has a term that is quadratic in $Q$, and the resulting pressure function $p(z)$ would depend on the Reynolds number. It is the pressure *difference* $\Delta P_{ex}$ that is unaffected by the Reynolds number, according to this approximation. We may write Eq. 1-4.26 in the form

$$\frac{\pi R_0^3 \Delta P_{ex}}{8\mu Q} = \int_{z_1/R_0}^{z_2/R_0} \left[\left(\frac{R_0}{R}\right)^4 - 1\right] d\left(\frac{z}{R_0}\right). \tag{1-4.27}$$

Obviously, the result depends only on the geometry of the constriction. For the shape function selected earlier (Eq. 1-3.1) we may write Eq. 1-4.27 in the form

$$\frac{\pi R_0^3 \Delta P_{ex}}{8\mu Q\lambda} = 2\int_0^{\Lambda} \left[\left(1 - \varepsilon\, e^{-\zeta^2}\right)^{-4} - 1\right] d\zeta. \tag{1-4.28}$$

The upper limit corresponds to a downstream position

$$\zeta = \Lambda = \frac{z_2}{R_0 \lambda} \tag{1-4.29}$$

large enough that $R/R_0$ is essentially unity, i.e., that the tube is of constant radius. With Eq. 1-3.1, $\Lambda = 2$ would be sufficient.

Before we proceed further, we should see if the analysis we have carried out accounts for any inertial phenomena that are important and observable.

### 1-4.2 The Velocity Field through the Constriction

Equation 1-4.22, the expression for the velocity field, has a contribution proportional to the Reynolds number. Let's see if this term accounts for some of the characteristics that we might anticipate for this type of flow. For "creeping flow," i.e., flow at vanishing Reynolds numbers, through a tube that is symmetric about $z = 0$, the velocity field will be symmetric as well. This is easily seen from Eq. 1-4.22, where if we delete the Reynolds number term what we have left is a term that is symmetric about $z = 0$ so long as $R(z)$ exhibits the same symmetry. The Reynolds number term can be seen to be antisymmetric, since for a symmetric tube shape the term $dR/dz$ is antisymmetric. Hence the Reynolds number contribution is additive before the constriction and subtractive after, and as a consequence the velocity field is no longer symmetric at finite Reynolds numbers. This is both expected and observed.

Another observation is that for large enough Reynolds numbers there is a line (really, a surface of rotation) that separates a recirculation region from the main flow. This is suggested in Fig. 1-4.1. We can examine our model to see if it predicts the existence of recirculation regions and the conditions under which they appear.

Physically, the flow separates into two regions at the point on the wall where the shear stress vanishes. To find the conditions for separation we return to Eq. 1-4.22 and differentiate the velocity field so that we can find the axial position where $T_{rz}$ vanishes:

$$T_{rz_{r=R}} = 0 = -\mu\left(\frac{\partial u_z}{\partial r}\right)_{r=R}$$
$$= -4\left[\frac{R_0}{R(z)}\right]^2 + \frac{44}{225}\,\mathsf{Re}_0\left[\frac{R_0}{R(z)}\right]^3\frac{dR}{dz}. \qquad (1\text{-}4.30)$$

This simplifies to

$$\mathsf{Re}_0\left(\frac{R_0}{R(z)}\right)\frac{dR}{dz} = 20.45. \qquad (1\text{-}4.31)$$

The smallest Reynolds number at which separation can occur corresponds to the largest value of the coefficient of $\mathsf{Re}_0$ in Eq. 1-4.31. Thus we must find the value of $z$ such that

$$\frac{d}{dz}\left[\left(\frac{R_0}{R(z)}\right)\frac{dR}{dz}\right] = 0 = \frac{1}{R}\left(\frac{d^2R}{dz^2}\right) - \left(\frac{1}{R}\right)^2\left(\frac{dR}{dz}\right)^2. \qquad (1\text{-}4.32)$$

For the tube shape given by Eq. 1-3.1 we can use Eq. 1-4.32 to find an expression for the value of $z$ that corresponds to the minimum Reynolds number for separation. That expression is (in terms of $\zeta = z/\lambda R_0$)

$$1 - 2\zeta^2 - \varepsilon e^{-\zeta^2} = 0 \qquad (1\text{-}4.33)$$

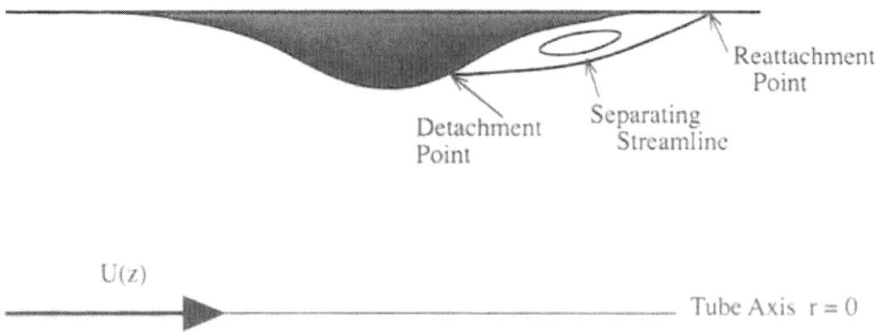


***FIGURE 1-4.1*** The recirculation region downstream of a constriction.

The solution depends upon $\varepsilon$. We then evaluate the terms $R/R_0$ and $dR/dz$ at the particular value of $\zeta$ found from Eq. 1-4.33 and find the corresponding minimum Reynolds number for separation, for the particular geometry. We find, for example, that

$$\text{For } \varepsilon = 0.1,\ \zeta = 0.68,\quad \text{Re}_0 = 223\,\lambda;$$

$$\text{For } \varepsilon = 0.25,\ \zeta = 0.645,\quad \text{Re}_0 = 80\,\lambda;$$

$$\text{For } \varepsilon = 0.5,\ \zeta = 0.563,\quad \text{Re}_0 = 31.6\,\lambda.$$

We should note here that it is unlikely that our model is accurate for $\varepsilon$ as large as 0.5, since we have made several approximations based on the assumption that the flow is perturbed only slightly from Poiseuille flow.

For values of Reynolds numbers exceeding the minima given above, Eq. 1-4.31 has two solutions corresponding to the values of $\zeta$ at which the separating streamline detaches and reattaches. Equation 1-4.31 may be cast in the form

$$\frac{\zeta e^{-\zeta^2}}{1 - \varepsilon e^{-\zeta^2}} = \frac{10.23\,\lambda}{\varepsilon\,\text{Re}_0} \tag{1-4.34}$$

Figure 1-4.2 shows some computations based on this expression.

We have invested quite a bit of time in developing this approach and in presenting some sample computations. It would be appropriate at this point to address the question of the accuracy of the model. We have alluded to this with the occasional *caveat* that the model is good only for small values of $\varepsilon$ and that

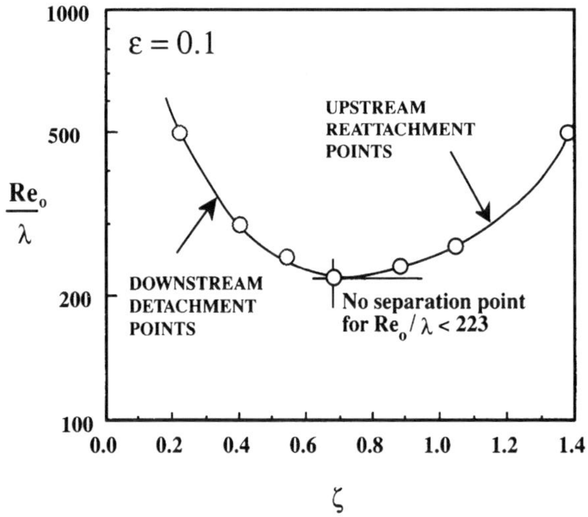


***FIGURE 1-4.2*** Detachment and reattachment points of the separation line.

the approximation to the inertial terms is not expected to hold for large Reynolds numbers. However, we have little idea at this point as to the degree to which the model is accurate, with respect to the parameters ε, λ, and $\mathrm{Re}_0$.

One way to assess this is to solve the Navier–Stokes equations in full, using some numerical method, and compare some features of the model to those of the more exact numerical solution. Another method is to develop an approximation scheme that is more orderly in some sense. A common approach is the so-called perturbation method, where a solution is developed from the case for which an analytical exact solution is available (this would be the Poiseuille solution in the problem at hand). The solution method introduces some parameter that measures the deviation of the approximate solution from the exact one. In the case of the flow considered here, an appropriate parameter would be the amplitude of the constriction, ε. The advantage of this approach [for examples, see Manton (1971), and Chow and Soda (1972)] is that criteria for the accuracy of the solutions can be presented in terms of inequalities that involve parameters such as ε, λ, and $\mathrm{Re}_0$. The disadvantage is that, typically, in order to obtain solutions that are useful in the range of parameters of interest, the solution procedure and its evaluation are computationally intensive.

Examination of the work of Chow and Soda (who applied a perturbation method to a *sinusoidal* constriction) shows that their result for the excess pressure drop across the constriction may be written in the form

$$\frac{\Delta P_{\mathrm{ex}}}{\rho\,\overline{U}_0^2}\,\mathrm{Re}_0 = A + B\,\mathrm{Re}_0^2, \tag{1-4.35}$$

where $A$ and $B$ depend only on the geometry. Our result (Eq. 1-4.28) is of the form

$$\frac{\Delta P_{\mathrm{ex}}}{\rho\,\overline{U}_0^2}\,\mathrm{Re}_0 = A, \tag{1-4.36}$$

which is typical for creeping flows. In the approximation scheme illustrated, leading to Eq. 1-4.28, we lose any effect of inertia on the pressure drop, which would appear as additional terms in Eq. 1-4.36 that depend on Reynolds number. Hence the perturbation scheme of Chow and Soda, carried far enough, will yield the expected additional terms. As noted, however, this is at considerable computational cost.

We have chosen to solve the full Navier–Stokes equations numerically in order to assess the method of solution illustrated here, leading to Eq. 1-4.22 for the velocity field and 1-4.28 for the excess pressure drop. In Fig. 1-4.3 we find the excess pressure drop as a function of Reynolds number from the numerical solutions for two cases, $\varepsilon = 0.1$ and $\varepsilon = 0.25$. (The excess pressure drop, as defined in Eq. 1-4.28, is $A/16$ when $A$ is defined as in Eq. 1-4.36.) For small ε we see that the excess pressure drop, as nondimensionalized here, is independent of Reynolds number up to about $\mathrm{Re}_0 = 20$. For the larger value of ε a Reynolds

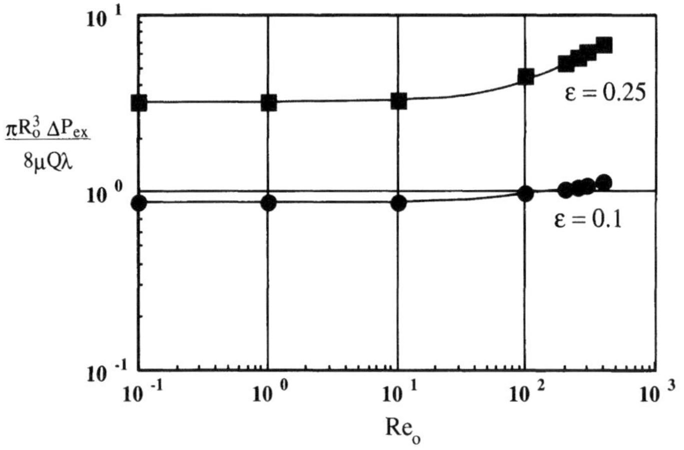


***FIGURE 1-4.3*** Excess pressure drop across a Gaussian constriction. Numerical solution of full Navier-Stokes equations courtesy of Dr. Andrew Yeckel.

number dependence (which means a departure from "creeping flow") sets in at a lower Reynolds number, as we would expect for a larger "disturbance" to the flow.

Although the derivation of Eq. 1-4.27 includes an accounting for the inertial terms, these terms drop out of the final excess pressure calculation. Hence we may regard Eq. 1-4.27 as a "creeping flow" model for the excess pressure drop. In Fig. 1-4.4 we compare the predictions of Eq. 1-4.28 (which is specific to the Gaussian constriction) to the full numerical solution for that shape function, for vanishingly small Reynolds numbers. The agreement is excellent, despite the fact that we have carried the calculation up to an $\varepsilon$ of $\varepsilon = 0.8$. Even though the amplitude of the constriction ($\varepsilon$) gets large, the axial distance over which this constriction is imposed ($\lambda$) is large in the cases illustrated here ($\lambda = 1$), with the result that the flow is quite smooth through this constriction. Hence the integral method, which begins with Eq. 1-4.1 from the *lubrication approximation*, works quite well, when we examine a "global" feature such as the pressure drop.

The lubrication model and the integral method do not produce such good results when we examine the flow field in more detail. The easiest way to visualize the flow field is through the introduction of the stream function $\psi$, defined such that

$$u_r = \frac{1}{r}\frac{\partial \psi}{\partial z} \quad \text{and} \quad u_z = -\frac{1}{r}\frac{\partial \psi}{\partial r}. \tag{1-4.37}$$

Of course, we may find the stream function from the solution for the velocity field. From Eq. 1-4.22 and the second part of Eq. 1-4.37 we can show that the stream function is related to the geometry of the tube by the expression

$$-\frac{\psi}{\overline{U}_0 R_0^2} = \left(\eta^2 - \frac{1}{2}\eta^4\right)\left[\frac{R_0}{R(z)}\right]^2 + \mathrm{Re}_0\left[\frac{R_0}{R(z)}\right]^3 \frac{dR}{dz}\left[\frac{4}{225}\left(\eta^2 - \frac{1}{2}\eta^4\right) - \frac{4}{75}\eta^5 + \frac{2}{45}\eta^6\right]. \tag{1-4.38}$$

This yields a dimensionless stream function that has the value $-0.5$ on the boundary of the tube. It is more common (though not necessary) to define the stream function such that it has the value zero on a no-slip boundary. Hence we can just add 0.5 to the values calculated from the expression above. Then

$$\frac{\psi}{\overline{U}_0 R_0^2} = 0.5 - \left(\eta^2 - \frac{1}{2}\eta^4\right)\left[\frac{R_0}{R(z)}\right]^2 - \mathrm{Re}_0\left[\frac{R_0}{R(z)}\right]^3 \frac{dR}{dz}\left[\frac{4}{225}\left(\eta^2 - \frac{1}{2}\eta^4\right) - \frac{4}{75}\eta^5 + \frac{2}{45}\eta^6\right]. \tag{1-4.39}$$

We may write this in the form

$$\frac{\psi}{\overline{U}_0 R_0^2} = 0.5 - a(\eta)\, F_1\!\left(\frac{z}{R_0}\right) - \mathrm{Re}_0\, b(\eta)\, F_2\!\left(\frac{z}{R_0}\right), \tag{1-4.40}$$

where

$$a(\eta) \equiv \left(\eta^2 - \tfrac{1}{2}\eta^4\right),$$
$$b(\eta) = \left[\frac{4}{225}\left(\eta^2 - \frac{1}{2}\eta^4\right) - \frac{4}{75}\eta^5 + \frac{2}{45}\eta^6\right],$$
$$F_1\!\left(\frac{z}{R_0}\right) = \left[\frac{R_0}{R(z)}\right]^2,$$
$$F_2\!\left(\frac{z}{R_0}\right) = \left[\frac{R_0}{R(z)}\right]^3 \frac{dR}{dz}. \tag{1-4.41}$$

In order to find the separating streamline (see Fig. 1-4.1) we may set $\Psi = 0$ in Eq. 1-4.40. This defines a curve in the $rz$ plane (other than the no-slip boundary $\eta=0$) along which $\Psi = 0$. Alternatively, we can note that the separating streamline defines a recirculation region, and the flow bounded by the recirculating streamline and the solid surface of the tube does not contribute to the volumetric flow rate down the axis of the tube. Hence the recirculating streamline may also be obtained from the expression

$$Q = \pi R_0^2 \overline{U}_0 = \int_0^{R_1} 2\pi r u_z(r,z)\, dr, \tag{1-4.42}$$

where $R_1$ is the radius of the separating streamline as a function of axial position $z$.

For the case of the Gaussian constriction we use

$$R = R_0\left\{1 - \varepsilon \exp\left[-\left(z/\lambda R_0\right)^2\right]\right\} \tag{1-4.43}$$

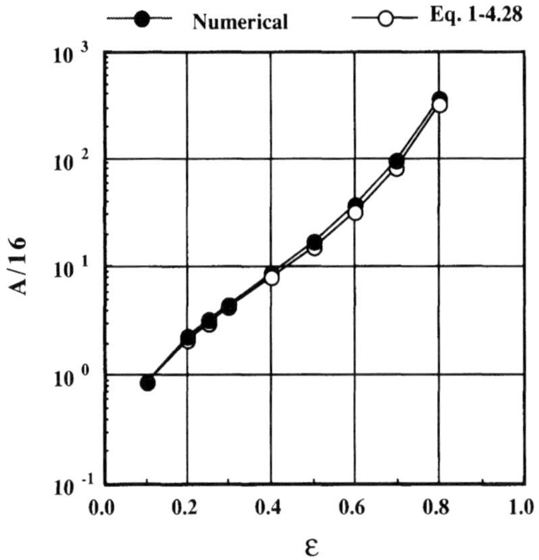


*FIGURE 1-4.4* Excess pressure drop across a Gaussian constriction. $Re_0 = 0$. Numerical solutions courtesy of Dr. Andrew Yeckel.

in the expressions above in order to obtain the separating streamline at various Reynolds numbers. In Fig. 1-4.5 we compare the separating streamline calculated from Eq. 1-4.39 to that based on a numerical solution of the Navier–Stokes equations. At Reynolds numbers where separation occurs the lubrication/integral analysis is very poor. This is not surprising, since the regions of separation and reattachment are very sensitive to the details of the flow field near the boundaries. In averaging the momentum equations across the tube we lose much of this detail.

## *1-4.3 Flow through an Annular Region with a Variable Radius*

We can extend the methods illustrated in this section to another type of

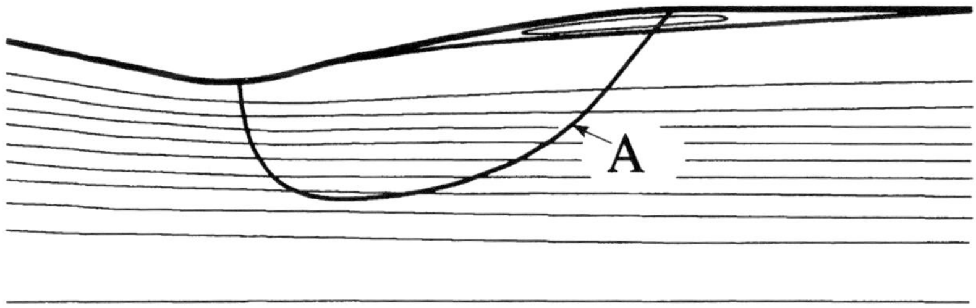


*FIGURE 1-4.5* Separating streamline for the Gaussian constriction. $\varepsilon = 0.25$, $\lambda = 1$, $Re = 400$. Line labeled A is from Eq. 1-4.39.

axisymmetric flow which is of some practical interest—one in which the liquid is confined to the annular region between two coaxial surfaces of revolution. We will do a case of simple geometry (see Fig. 1-4.6) so that the algebra does not get in the way. This is the type of flow field that arises in systems for coating molten polymer onto a wire. The wire, of radius $R_0$, is drawn continuously through a die of conical shape. To simplify the analysis, we assume that the liquid is Newtonian and the flow in the die is isothermal. Further, we assume that the liquid is introduced into the die at some region far upstream from the die entrance, where the pressure is atmospheric. Liquid is then drawn into the die by the motion of the wire. Of particular interest is the flow rate of liquid through the die, since this determines the thickness of the film coated onto the wire. Thus we have a dynamics problem in which the upstream flow is not specified. Instead, the pressures at the two ends of the system are specified, on the assumption that the system is open to the atmosphere at both ends.

Let's define two geometrical parameters, $\lambda$ and $\varepsilon$, such that:

$$L = \lambda R_0 \tag{1-4.44}$$

and

$$H_1 - H_2 = \varepsilon R_0. \tag{1-4.45}$$

We will restrict our attention to the case that

$$\frac{\varepsilon}{\lambda} \ll 1. \tag{1-4.46}$$

The lubrication analysis of Section 1-3 holds, and the pressure is independent of radial position. Equation 1-4.1 is valid, and we integrate it term by term. There is a difference now, because the limits of integration are from $R_0$ to $R_0 + h(z)$, where $h(z)$ is given by

$$h(z) = H_1 + \frac{H_2 - H_1}{L} z. \tag{1-4.47}$$

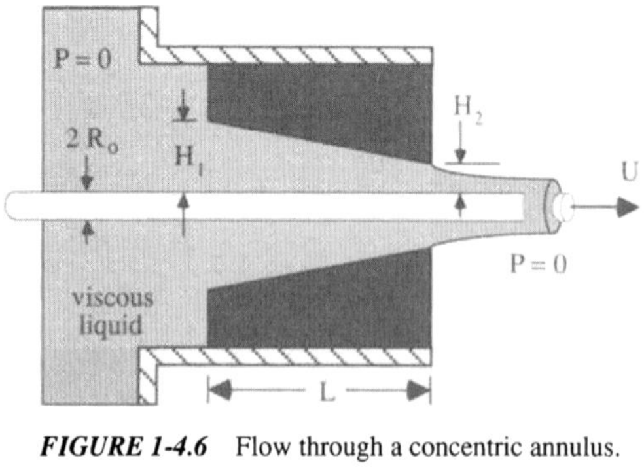


***FIGURE 1-4.6*** Flow through a concentric annulus.

The $z$ coordinate has its origin at the die entrance, and the die exit is at $z = L$.

Equation 1-4.2 is modified accordingly, and we find

$$\int_{R_0}^{R_0+h(z)} r\left(u_r \frac{\partial u_z}{\partial r}\right) dr = \int_{R_0}^{R_0+h(z)} ru_r \left(\frac{\partial u_z}{\partial r}\, dr\right) = \int ru_r \,( du_z)$$

$$= \int d(ru_r u_z) - \int u_z\, d(ru_r)$$

$$= \left[ru_r u_z\right]_{r=R_0}^{r=R_0+h(z)} - \int u_z \frac{\partial(ru_r)}{\partial r}\, dr$$

$$= \int u_z \left(r \frac{\partial u_z}{\partial z}\right) dr = \int \frac{r}{2}\left(\frac{\partial u_z^2}{\partial z}\right) dr$$

$$= \frac{1}{2}\frac{d}{dz}\int_{R_0}^{R_0+h(z)} r\, u_z^2 dr. \qquad (1\text{-}4.48)$$

(Some of the steps involved in writing this equation are described just below Eq. 1-4.2.)

The second term on the left-hand side of Eq. 1-4.1 integrates immediately to

$$\int_{R_0}^{R_0+h(z)} \left(r\, u_z \frac{\partial u_z}{\partial z}\right) dr = \int \frac{r}{2}\left(\frac{\partial u_z^2}{\partial z}\right) dr$$

$$= \frac{1}{2}\frac{d}{dz}\int_{R_0}^{R_0+h(z)} r\, u_z^2\, dr. \qquad (1\text{-}4.49)$$

The terms on the right-hand side of Eq. 1-4.1 are easily integrated, as before, except for minor changes due to the difference in the limits of integration. The final result is the following:

$$\rho \frac{d}{dz}\int_{R_0}^{R_0+h(z)} r\, u_z^2\, dr = -\frac{\left(h^2 + 2hR_o\right)}{2}\frac{dp}{dz}$$

$$+ \mu\left[R_0 + h(z)\right]\left(\frac{\partial u_z}{\partial r}\right)_{R_0+h(z)} - \mu R_0\left(\frac{\partial u_z}{\partial r}\right)_{R_0} \qquad (1\text{-}4.50)$$

We might be tempted to write the velocity field in the form used earlier, Eq. 1-4.5:

$$\frac{u_z}{U} = 1 + a\eta + b\eta^2 + c\eta^3 + d\eta^4, \qquad (1\text{-}4.51)$$

where, now,

$$\eta \equiv \frac{r - R_0}{h(z)} \tag{1-4.52}$$

and $\eta = 0$ is along the moving wire surface, where $u_z = U$, while $\eta = 1$ is along the fixed outer surface of the die. We must be cautious in selection of this velocity field, however, since we no longer can impose some of the conditions used in the earlier example. For example, there is no symmetry about any specific radial position (see Eq. 1-4.8). However, we do have a no-slip condition along $\eta = 1$ :

$$0 = 1 + a + b + c + d. \tag{1-4.53}$$

We might also impose Eq. 1-4.13 along $\eta = 1$:

$$0 = -\frac{dp}{dz} + \mu \frac{1}{r}\frac{\partial}{\partial r}\left(r\frac{\partial u_z}{\partial r}\right) \quad \text{along} \quad \eta = 1. \tag{1-4.54}$$

Of course, we will require that the velocity profile satisfy the averaged momentum equation (Eq. 1-4.50). This gives us three equations from which we might hope to obtain three of the coefficients in Eq. 1-4.53. With that in mind, then, let's drop the fourth-order term in Eq. 1-4.51, eliminating $d$, and see what we find for the coefficients $a$, $b$, and $c$.

In writing Eq. 1-4.54 we need to calculate

$$\frac{\partial u_z}{\partial r} = \frac{\partial u_z}{\partial \eta}\frac{d\eta}{dr} = U\left(a + 2b\eta + 3c\eta^2\right)\frac{1}{h(z)}. \tag{1-4.55}$$

Then Eq. 1-4.54 becomes

$$0 = -p' + \frac{\mu U}{h(z)}\left[\frac{1}{r}\frac{\partial}{\partial r}\left(ar + 2br\eta + 3cr\eta^2\right)\right]_{\eta = 1}, \tag{1-4.56}$$

where $p' = dp/dz$. After performing the operations indicated in the brackets we find

$$\frac{a}{1 + \frac{R_0}{h(z)}} + 2b\left[1 + \frac{1}{1 + \frac{R_0}{h(z)}}\right] + 3c\left[2 + \frac{1}{1 + \frac{R_0}{h(z)}}\right] = \tilde{p}' \tag{1-4.57}$$

A dimensionless pressure gradient has been defined as

$$\tilde{p}' \equiv \frac{p' h^2}{\mu U}. \tag{1-4.58}$$

Equation 1-4.57 is a second equation that we can impose on the coefficients $a$, $b$, and $c$, along with Eq. 1-4.53 (with the $d$ term eliminated):

$$0 = 1 + a + b + c. \tag{1-4.59}$$

The third equation comes from Eq. 1-4.50. We will demonstrate the case of negligible inertia here and drop the left-hand term. At an intermediate stage Eq. 1-4.50 takes the form

$$0 = -\frac{\left(h^2 + 2hR_0\right)}{2}\frac{dp}{dz} + \frac{\mu U\left[R_0 + h(z)\right]}{h(z)}\left(a + 2b + 3c\right) - \frac{\mu U R_0}{h(z)}a, \tag{1-4.60}$$

which may be written in the form

$$a + 2b\left[1 + \frac{R_0}{h(z)}\right] + 3c\left[1 + \frac{R_0}{h(z)}\right] = \frac{\tilde{p}'}{2}\left[1 + \frac{R_0}{h(z)}\right]. \tag{1-4.61}$$

These three equations (Eqs. 1-4.57, 59, and 61) are a set of linear algebraic equations for the coefficients $a$, $b$, and $c$. The solution gives these constants in terms of $\tilde{p}'$ and $h(z)$. We still must find $p(z)$. We have not yet used the continuity equation. We do so in integrated form, so we write another constraint on the velocity field:

$$Q = \int_{R_0}^{R_0 + h(z)} 2\pi\, r\, u_z(r,z)\, dr. \tag{1-4.62}$$

$Q$ is not known at this point, but it is some constant. Integration of Eq. 1-4.62 (after $a$, $b$, and $c$ are written in terms of $\tilde{p}'$) yields an equation for $\tilde{p}'$ as a function of $h(z)$, $Q$, and the parameters of this problem. But this is simply a first-order ordinary differential equation for the pressure gradient, which may be integrated to yield a solution for $p(z)$. Since it comes from a first-order equation, a constant of integration appears. We have a second unknown constant $Q$. These two constants follow from boundary conditions on pressure, which in the simplest case may be written as

$$p = 0 \quad \text{at } z = 0 \tag{1-4.63}$$

and

$$p = 0 \quad \text{at } z = L. \tag{1-4.64}$$

The algebra is laborious, and so we look at an approximation that permits us to write the solutions down in simple forms. We suppose that the radial separation between the two surfaces, $h(z)$, is small compared to $R_0$. This might be the case when we are putting a very thin coating onto a wire whose radius is much greater than the coating thickness. Under this approximation, where we set $R_0/h >> 1$, Eqs. 1-4.57, 59, and 61 reduce to the following:

$$a + b + c = -1, \tag{1-4.65}$$

$$2b + 6c = \tilde{p}', \tag{1-4.66}$$

and

$$2b + 3c = \tilde{p}'. \tag{1-4.67}$$

The only way that Eqs. 1-4.66 and 67 may be satisfied is if

$$c = 0. \tag{1-4.68}$$

Then

$$b = \frac{\tilde{p}'}{2} \quad \text{and} \quad a = -1 - \frac{\tilde{p}'}{2}. \tag{1-4.69}$$

At this point the velocity field is approximated as

$$\frac{u_z}{U} = 1 - \left(1 + \frac{\tilde{p}'}{2}\right)\eta + \frac{\tilde{p}'}{2}\eta^2. \tag{1-4.70}$$

Now we impose Eq. 1-4.62:

$$Q = 2\pi U \int_0^1 \left[\eta h(z) + R_0\right]\left[1 - \left(1 + \frac{\tilde{p}'}{2}\right)\eta + \frac{\tilde{p}'}{2}\eta^2\right] h(z)\, d\eta. \tag{1-4.71}$$

Since $h(z)$ and $p(z)$ are constants with respect to $\eta$, the integration is simple and the result is

$$Q = 2\pi U h(z)\left\{\left[\frac{h(z)}{6} + \frac{R_0}{2}\right] - \frac{\tilde{p}'}{2}\left[\frac{h(z)}{12} + \frac{R_0}{6}\right]\right\}. \tag{1-4.72}$$

Once again we introduce the approximation $R_0/h >> 1$ , and we find

$$Q_w \equiv \frac{Q}{2\pi R_0} = Uh(z)\left(\frac{1}{2} - \frac{\tilde{p}'}{12}\right). \tag{1-4.73}$$

where $Q_w$ is the volumetric flow rate per unit width (perimeter). If $H$ is the film thickness deposited on the wire, then

$$Q_w = UH. \tag{1-4.74}$$

From Eq. 1-4.73 we find the pressure gradient as

$$\tilde{p}' = 6 - \frac{12 Q_w}{Uh(z)} \tag{1-4.75}$$

or

$$\frac{dp}{dz} = \frac{6\mu U}{h^2(z)} - \frac{12\mu Q_w}{h^3(z)}. \tag{1-4.76}$$

The integration is easily performed, and when the boundary conditions (Eqs. 1-4.63 and 64) are imposed, we find

$$p(z) = \frac{6\mu U z}{h^2(z)}\,\frac{h(z) - H_2}{H_1 + H_2}. \tag{1-4.77}$$

Now that we have $p(z)$ we may go back to Eq. 1-4.69 for $a$ and $b$ and then write the velocity field explicitly from Eq. 1-4.51 (with $c = d = 0$). This completes the analysis for the case that inertia is negligible, and $R_0 >> h(z)$. If these approximations are not valid, then the same procedure, but with more effort, yields the solution.

Keep in mind that Eq. 1-4.77 is subject to the approximation that the annular gap around the wire is very thin—specifically that $R_0/h >> 1$. We could derive Eq. 1-4.77 by starting with the Navier–Stokes equations in Cartesian coordinates, introducing the "lubrication approximations" (as in Section 1-3) and solving the resulting equations analytically.

## 1-5 AXISYMMETRIC FLOW THROUGH A LEAKY TUBE

In this section we consider a flow that has a two-dimensional character that arises not from the geometry, as in the previous two examples, but from the allowance of a radial flow at the boundary. Figure 1-5.1 shows the geometry and the axial velocity profile that we anticipate. We assume that the system is at steady state and that the flow along the tube axis is at such a low Reynolds number that we can neglect the inertial terms in the Navier–Stokes equations.

The tube has a uniform inside radius $R$. Downstream from the plane $z = 0$ the tube wall is to be regarded as *uniformly* permeable, so that along the length $L$ there is a continuous *volumetric flux* $j$ (with units of $m^3/m^2$-s) of liquid through the wall. We begin with Eqs. 1-3.2 and 1-3.3, which we rewrite here:

$$0 = -\frac{\partial p}{\partial r} + \mu\left\{\frac{\partial}{\partial r}\left[\frac{1}{r}\frac{\partial}{\partial r}(r\,u_r)\right] + \frac{\partial^2 u_r}{\partial z^2}\right\} \tag{1-5.1}$$

and

$$0 = -\frac{\partial p}{\partial z} + \mu\left[\frac{1}{r}\frac{\partial}{\partial r}\left(r\frac{\partial u_z}{\partial r}\right) + \frac{\partial^2 u_z}{\partial z^2}\right]. \tag{1-5.2}$$

Since there are three unknown functions we need the continuity equation (Eq.1-3.2) to serve as the third equation:

$$0 = \frac{1}{r}\frac{\partial}{\partial r}(r\,u_r) + \frac{\partial u_z}{\partial z}. \tag{1-5.3}$$

The volumetric flux $j$ is really a leakage velocity, and in general $j$ is a function of

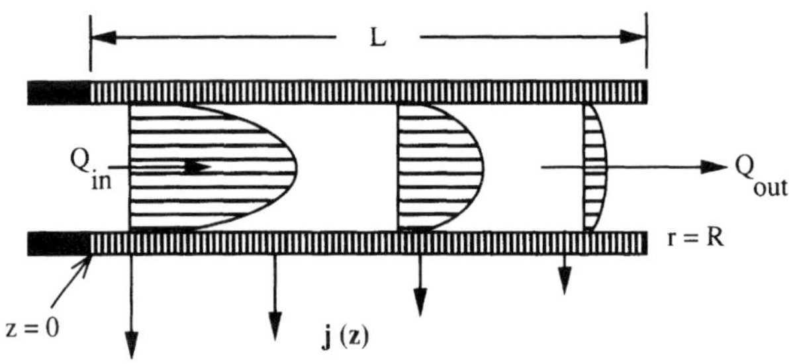


***FIGURE 1-5.1*** Flow through a leaky tube.

position along the axis of the tube. This leakage flow depends upon the internal pressure distribution along the tube axis. We make an arbitrary but reasonable assumption that locally, i.e., at any axial position $z$, the pressure–flow relationship across the wall of the tube is linear. We write this in the form

$$j(z) = K\,[p(z) - p_a], \tag{1-5.4}$$

where $p_a$ is the pressure in the region outside the tube. We may arbitrarily take $p_a = 0$. Equation 1-5.4 defines a *permeability coefficient* $K$ of the tube wall.

The leakage flow across the tube wall creates a *radial* velocity function throughout the flow field. Right at the tube wall this velocity is identical to the volumetric flux, so we can write the leakage flux as

$$j(z) = u_r(R,z) = u_R(z). \tag{1-5.5}$$

This is equivalent to the statement that there is a finite velocity at the tube wall. Hence the fluid does not satisfy a no-slip condition at the wall, although it should be noted that the velocity component we are talking about here is the *radial* velocity. We will continue to assume that the *axial* velocity component is zero at the tube wall.

Now we introduce an important simplifying constraint or assumption. The leakage flow is taken to be small in some sense that we will have to quantify later. Then the flow is almost parallel, except very near the wall where the axial velocity gets small (to satisfy no-slip in the *axial* direction) and the *radial* velocity remains finite.

The following question arises. If the leakage is small enough, is it possible that the axial velocity has a functional dependence on radial position that is nearly the same as in the case of no leakage, but with a minor correction that accounts for leakage? For example, with Eq. 1-2.16 as a basis, we may propose to write the velocity profile for "leaky" Poiseuille flow in the form

$$u_z = 2U(z)\left[1 - \left(\frac{r}{R}\right)^2\right] = 2U_0\left[1 - \left(\frac{r}{R}\right)^2\right] f(z). \tag{1-5.6}$$

Here we use the average velocity at the inlet of the tube, $U_0 = Q_0/\pi R^2$, and introduce an unknown function $f(z)$. Note that we subscript the inlet conditions, since the volume flow rate and the average velocity vary along the tube axis as fluid is lost through the leakage.

We may use the continuity equation to write $u_r$ in terms of $u_z$:

$$u_r = -\frac{1}{r}\int_0^r r\,\frac{\partial u_z}{\partial z}\,dr. \tag{1-5.7}$$

We now differentiate Eq. 1-5.6 with respect to $z$, put the result into the integral above, and perform the integration with respect to $r$. The result is that the radial velocity component $u_r$ must satisfy

$$u_r = -RU_0 f'(z)\left[\left(\frac{r}{R}\right)^2 - \frac{1}{2}\left(\frac{r}{R}\right)^4\right], \tag{1-5.8}$$

where $f' = df/dz$.

At the tube surface ($r = R$) we find

$$u_r(R,z) \equiv u_R(z) = -\frac{RU_0}{2}f'. \tag{1-5.9}$$

Note what we have accomplished here. We have replaced the two unknowns $u_r$ and $u_z$ by the single unknown $f(z)$. But we still have to answer the question: Is this assumed form for $u_z$ given above in Eq. 1-5.6 compatible with the principle of conservation of momentum? To answer this we must substitute $u_z$ in Eq. 1-5.2 and see if it is possible to find a solution for the function $f(z)$. (A key point to remember here is that we are assuming that $f$ is a function only of $z$.) Equation 1-5.2 gives us

$$\frac{\partial p}{\partial z} = \mu\left\{-\frac{8U_0}{R^2}f + 2U_0\left[1-\left(\frac{r}{R}\right)^2\right]f''\right\}. \tag{1-5.10}$$

Between Eqs. 1-5.4, 1-5.5, and 1-5.8 we see that

$$\frac{\partial p}{\partial z} = \frac{1}{K}\frac{dU_R}{dz} = -\frac{RU_0}{2K}f''. \tag{1-5.11}$$

Hence we obtain the following differential equation for the function $f$:

$$f'' = \frac{16K\mu}{R^3}f - \frac{4K\mu}{R}\left[1-\left(\frac{r}{R}\right)^2\right]f''. \tag{1-5.12}$$

Since the radial variable $r$ appears in this differential equation for $f$, it follows that $f$ is a function of both $z$ and $r$, which violates our original assumption that $f = f(z)$ only. Thus the assumed form of the axial velocity distribution, Eq. 1-5.6, does not satisfy conservation of $z$ momentum. *Note, however, that if the dimensionless group $4K\mu/R$ is small compared to unity, the "offending" term can be neglected relative to the second-derivative term on the left-hand side of the equation.* We claim (or at least hope), then, that Eq. 1-5.6 is a good approximation if the leakage is small enough, and specifically if

$$\frac{4K\mu}{R} << 1. \tag{1-5.13}$$

The function $f(z)$ must then be found as the solution to the following ordinary differential equation:

$$f'' - \frac{16K\mu}{R^3}f = 0. \tag{1-5.14}$$

We require two boundary conditions. One states that at the entrance to the leaky section, i.e., at $z = 0$, the flow is fully developed and given by the classical Poiseuille solution. This is equivalent to the condition

$$f = 1 \quad \text{at } z = 0. \tag{1-5.15}$$

The second condition is simply that

$$f \text{ remains finite for } z \to \infty. \tag{1-5.16}$$

Equation 1-5.14 is easily solved, then, and the solution satisfying the boundary conditions may be written as

$$f = \exp\left(-4\,\beta\frac{z}{R}\right), \tag{1-5.17}$$

where a dimensionless permeability parameter β is defined as

$$\beta = \sqrt{\frac{K\mu}{R}}\,. \tag{1-5.18}$$

Now we want to find the axial profile of pressure, $p(z)$.

From Eqs. 1-5.11 and 1-5.14 we have

$$\frac{dp}{dz} = -\frac{8\mu U_0}{R^2}\,f = -\frac{8\mu U_0}{R^2}\exp\left(-4\beta\frac{z}{R}\right). \tag{1-5.19}$$

When we integrate this differential equation with the boundary condition

$$p = \Delta P \quad \text{at } z = 0, \tag{1-5.20}$$

we find that the pressure distribution down the tube axis is

$$p(z) = \Delta P + \frac{2\mu U_0}{\beta R}\left[\exp\left(-4\beta\frac{z}{R}\right) - 1\right]. \tag{1-5.21}$$

Then upon setting $p = 0$ at $z = L$ we find

$$\Delta P = \frac{2\mu U_0}{\beta R}\left[1 - \exp\left(-4\beta\frac{L}{R}\right)\right]. \tag{1-5.22}$$

This permits us to write the pressure distribution in the form

$$p(z) = \frac{2\mu U_0}{\beta R}\left[\exp\left(-4\beta\frac{z}{R}\right) - \exp\left(-4\beta\frac{L}{R}\right)\right]. \tag{1-5.23}$$

We may now find the local leakage flux from Eq. 1-5.4, with the result

$$j = Kp(z) = 2\beta U_0\left[\exp\left(-4\beta\frac{z}{R}\right) - \exp\left(-4\beta\frac{L}{R}\right)\right]. \tag{1-5.24}$$

Over the length $L$ the *total* leakage flow is given by

$$Q_L = \int_0^L 2\pi R\; j(z)\,dz$$

$$= \pi R^2 U_0\left[1 - \left(1 + 4\,\beta\frac{L}{R}\right)\exp\left(-4\,\beta\frac{L}{R}\right)\right]. \tag{1-5.25}$$

The fractional loss of fluid over the length $L$ is

$$\phi_L = \frac{Q_L}{Q_0} = \left[1 - \left(1 + 4\,\beta\frac{L}{R}\right)\exp\left(-\,4\,\beta\frac{L}{R}\right)\right]. \tag{1-5.26}$$

The volumetric flow rate out of the tube, from the end at $z = L$, is given by

$$Q_{\text{out}} = Q_0 - Q_L = Q_L\left(\frac{1}{\phi_L} - 1\right). \tag{1-5.27}$$

This may be written in the form

$$Q_{\text{out}} = \left(\frac{\pi R^4 \Delta P}{8\mu L}\right)\left[\frac{\chi\,(1 + \chi\,)\,e^{-\chi}}{1 - e^{-\chi}}\right], \tag{1-5.28}$$

where we now define

$$\chi = 4\beta\frac{L}{R}. \tag{1-5.29}$$

From Eq. 1-2.17 we can see that the bracketed term in Eq. 1-5.28 may be regarded as the factor by which the leakage flow alters the flow out of the tube. We must keep in mind, in examining this factor, that our analysis is valid only if $\beta^2$ is small (see comments below Eq. 1-5.12). However, there is no such restriction on $\chi$ itself, which would normally be larger than unity since we are interested in long tubes (large $L/R$) for which the assumption of fully developed flow is valid. It is appropriate, then, to examine the behavior of Eq. 1-5.28 over a range of values of $\chi$.

The first thing we discover appears counterintuitive. For small $\chi$ values it is easy to show that

$$Q_{\text{out}} = \left(\frac{\pi R^4 \Delta P}{8\mu L}\right)\left(1 + \frac{1}{2}\chi\right) \quad \text{for } \chi \ll 1. \tag{1-5.30}$$

This result asserts that for a given pressure gradient a leaky tube will exhibit a *higher* outflow than an impermeable tube. How can this be so if the tube is leaking? The answer lies in examining the shear stress exerted by this flow, at the wall of the tube.

To a good approximation the shear stress is given by

$$T_{rz} = -\,\mu\,\Delta_{rz} = -\,\mu\,\frac{\partial u_z}{\partial r}. \tag{1-5.31}$$

From our solution for the velocity field we find

$$T_{rz} = \frac{4\mu U_0}{R}\exp\left(-\,\chi\,\frac{z}{L}\right) \quad \text{at } r = R. \tag{1-5.32}$$

The average shear stress is then

$$\overline{T_{rz}} = \frac{4\mu U_0}{R}\frac{1}{L}\int_0^L \exp\left(-\chi\frac{z}{L}\right)dz$$
$$= \frac{4\mu U_0}{R}\left[\frac{1-e^{-\chi}}{\chi}\right] \quad \text{at} \quad r=R. \tag{1-5.33}$$

For small values of $\chi$ we find

$$\overline{T_{rz}} = \frac{4\mu U_0}{R}\left[1-\frac{1}{2}\chi\right] \quad \text{at } r=R. \tag{1-5.34}$$

Thus we see that for small values of $\chi$ the shear stress averaged along the length of the tube wall is *below* that for Poiseuille flow (nonleaky flow). As a consequence it is possible for the outflow to *exceed* that predicted by Poiseuille's law, for a given pressure gradient, even accounting for some degree of leakage. The situation changes as $\chi$ increases, however. The bracketed term in Eq. 1-5.28 goes to zero in the limit of large values of $\chi$, and so we find that beyond some intermediate value of $\chi$ ( we find that it is for $\chi > 1.793$) the outflow is *reduced* with respect to that predicted from Poiseuille's law. We may also confirm that the enhancement of the flow rate (again, relative to that for the impermeable tube) reaches a maximum factor of 1.175 at a value of $\chi = 0.8$.

## Problems

P1-1 Show that if $(\mathbf{u}, p)$ is a solution to Eqs. 1-1.1 through 1-1.4, then $(\mathbf{u}, p + a)$ is a solution, where $a$ is any constant.

P1-2 Derive Eqs. 1-1.21 and 22.

P1-3 Carry out the vorticity–stream function reformulation of the Navier–Stokes equations for *unsteady* axisymmetric flow. Compare and contrast your result to a diffusion equation for a scalar axisymmetric variable. Do you get a diffusion equation for vorticity? What about in the two-dimensional Cartesian case?

P1-4 Derive Eq. 1-4.14.

P1-5 Derive Eq. 1-4.17.

P1-6 Derive Eq. 1-4.21.

P1-7 Apply the integral method of Section 1-4 to the problem of laminar flow through a tube of circular cross section, for which the radius satisfies

$$R(z) = R_0 \ \text{ for } z < 0, \qquad R(z) = R_1 \ \text{ for } z > L, \tag{P1-7.1}$$

while in the range [0,$L$]

$$R(z) = R_0 - \frac{R_0 - R_1}{L}z. \tag{P1-7.2}$$

In particular, write an expression for $\Delta P/L$ across the region [0,$L$].

P1-8 Show that Eq. 1-4.28 leads to a simple analytical result, valid for small $\varepsilon$:

$$\frac{\pi R_0^3\, \Delta P_{ex}}{8\mu Q} = 4\sqrt{\pi}\, \lambda\varepsilon. \qquad \text{(P1-8.1)}$$

Give the value of the coefficient $A$ in Eq. 1-4.36, for this case.

P1-9 Using Eqs. 1-4.22 and 1-4.37, compute and plot some streamlines for flow through this contraction. Pick $\varepsilon = 0.1$, and $\lambda = 1$, and do Reynolds numbers below and above those expected to produce separation, say $\mathrm{Re}_0 = 200$ and $\mathrm{Re}_0 = 400$. For the case that separation occurs, plot the separating streamline.

P1-10 Give the details of the derivation of Eq. 1-4.71 from Eq. 1-4.70.

P1-11 Show that Eq. 1-4.71 predicts a maximum pressure on the range $[0,L]$ and give an analytical expression for this maximum.

P1-12 The nonlinear (inertial) term in Eq. 1-4.4 was approximated by using the Poiseuille profile (Eq. 1-2.16) to yield Eq.1-4.20 and 1-4.21. Suppose we regard Eq. 1-4.22 as a second approximation to the velocity profile, and calculate the inertial term using that profile. What forms do Eqs. 1-4.20 and 1-4.21 now take?

What do we now obtain for the excess pressure drop $\Delta P_{ex}$, in place of Eq. 1-4.25?

How is Eq. 1-4.31, for the separation points, now changed?

P1-13 Use Eq. 1-4.33 to find a curve for the separation boundaries (as per Fig. 1-4.2) for the case $\varepsilon = 0.25$.

P1-14 Use Eq. 1-4.28 to find the coefficient $A$ in Eq. 1-4.36. Plot $A/\lambda$ vs. $\varepsilon$.

P1-15 Repeat Problem P1-13, but for a tube with a sinusoidally varying radius. Use Eq. 1-4.27, and integrate over one wavelength. Define the shape parameters so that

$$\frac{R}{R_0} = 1 + \varepsilon \sin\left(\frac{2\pi}{\lambda}\right) z. \qquad \text{(P1-15.1)}$$

P1-16 When separation occurs there is a line connecting the detachment–reattachment points along which the axial velocity vanishes. Show that this line is defined by

$$\frac{\eta^3}{1+\eta} = \frac{15\,\lambda}{4\,\varepsilon\,\mathrm{Re}_0}\;\frac{1-\varepsilon\, e^{-\zeta^2}}{\zeta\, e^{-\zeta^2}} + \frac{2}{15} \qquad \text{(P1-16.1)}$$

for the shape function given in Eq. 1-3.1.

P1-17 In writing Eq. 1-4.43 it appears that we are stating that

$$\int_{R_1(z)}^{R_2(z)} \frac{\partial f(r,z)}{\partial z}\, dr = \frac{d}{dz}\int_{R_1(z)}^{R_2(z)} f(r,z)\, dr. \qquad \text{(P1-17.1)}$$

This is not the case when one or both limits on the integral are functions of $z$. In that case we use the Leibniz rule for differentiating an integral:

$$\frac{d}{dz}\int_{R_1(z)}^{R_2(z)} f(r,z)\,dr = \int_{R_1(z)}^{R_2(z)} \frac{\partial f(r,z)}{\partial z}dr$$

$$+ f\left[R_2(z),z\right]\left[\frac{dR_2(z)}{dz}\right] - f\left[R_1(z),z\right]\left[\frac{dR_1(z)}{dz}\right]. \qquad \text{(P1-17.2)}$$

Show that the results given in Eqs. 1-4.2 and 1-4.3, as well as in Eqs. 1-4.42 and 1-4.43, are correct as written.

P1-18 Equation 1-5.6 appears to be an assumption justified on the grounds that it yields a velocity profile that is parabolic, as in Poiseuille flow, but modified by a velocity function that varies along the tube axis. Show that we are led to an equation of this form if we begin with the assumption that the velocity profile satisfies "similarity" with respect to $r/R$, i.e.,

$$\frac{u_z(z,r)}{U(z)} = a + b\left(\frac{r}{R}\right) + c\left(\frac{r}{R}\right)^2 + d\left(\frac{r}{R}\right)^3, \qquad \text{(P1-18.1)}$$

where $a$, $b$, $c$, and $d$ are functions of $z$, determined by imposing the following constraints on velocity:

$$u_z = 0 \quad \text{on } r = R, \qquad \text{(P1-18.2)}$$

$$\frac{\partial u_z}{\partial r} = 0 \quad \text{on } r = 0, \qquad \text{(P1-18.3)}$$

$$\frac{\partial^2 u_z}{\partial r^2} = \frac{-2\,U(z)}{R^2} \quad \text{on } r = 0, \qquad \text{(P1-18.4)}$$

and $U(z)$ is the velocity along the axis.

P1-19 Find the value of $\chi$ (defined in Eq. 1-5.29) at which the outflow from a leaky tube is greatest with respect to that predicted by Poiseuille's law.

P1-20 In calculating the shear stress at the wall of a leaky tube we have used Eq. 1-5.31. This neglects the influence of the term $\partial u_r/\partial z$ in the deformation gradient $\Delta_{rz}$. What parameter determines the extent to which this is a good approximation? Is the neglect of this contribution consistent with approximations already made in the analysis?

P1-21 Find the boundary of the recirculating region using Eq. 1-4.42, and confirm that the result is compatible with the use of Eq. 1-4.40.

P1-22 Confirm that Eq. 1-4.77 follows from an exact lubrication analysis of the equations of motion in Cartesian coordinates. Do the case of vanishing inertial terms. The key assumptions that define the "lubrication" approximations are described in Section 1-3. Apply them to the planar approximation of this problem.

## *References*

Chow, J. C. F., and K. Soda. (1972). Laminar flow in tubes with constriction. *Phys. Fluids* **15**, 1700.

Deiber, J. A., and W. R. Schowalter. (1979). Flow through tubes with sinusoidal axial variations in diameter. *AIChE J.* **25**, 638 .

Denn, M. M. (1980) *Process Fluid Mechanics*. Prentice–Hall, Englewood Cliffs, N.J.

Forrester, J. H., and D. F. Young. (1970). Flow through a converging–diverging tube and its implications in occlusive vascular disease. *J. Biomech.* **3**, 297 (Part I), 307 (Part II).

Manton, M. J. (1970). Low Reynolds number flow in slowly varying axisymmetric tubes. *J. Fluid Mech.* **49**, 451.

Singh, R., and R. L. Laurence. (1979). Influence of slip velocity at a membrane surface on ultrafiltration performance—II. Tube flow system. *Int. J. Heat Mass Transfer* **12**, 731.

# 2 *Laminar Flow through a Tube with a Thin Annular Adherent Liquid Film*

*In this chapter* we return to flow within a uniform tube, but we look at some problems that arise when there is an annular layer of liquid adjacent to the inside wall of the tube, which in turn surrounds an inner or "core" flow of a second fluid. On the assumption that the annular liquid film is stable, we find that this layer has the ability to "lubricate" the tube in the sense that its presence can reduce the pressure drop required to maintain a specified volumetric flow rate of the inner fluid. Since the inner flow can exert a shear stress at the boundary of the lubricating layer, we might expect that the annular layer cannot remain in the tube unless it is replenished at the tube entrance. The rate at which this layer is "sheared out" of the tube is examined, and a simple model for the rate of removal is developed. Next we examine the assumption that the annular film is stable, where we use the word "stable" in the sense that if the film thickness is disturbed, the disturbance will decay in time. We find that if there is an interfacial tension between the two fluids, there is a mechanism through which the annular film develops periodic undulations along the axis. In other words, the annular film forms a periodically "rough" boundary for the inner flow, leading (according to Chapter 1) to increased pressure loss. Then the question of how one might create an annular layer within a tube is examined, and a model is described with which the thickness of the layer may be predicted with respect to the conditions under which it is created. Finally, we drop the assumption that if the annular layer is being sheared out of the tube by the core flow, it is replenished somehow, and a transient model for the removal of this layer is presented and compared to experimental data.

## *2-1 AXISYMMETRIC FLOW THROUGH A LUBRICATED TUBE*

We return to the classical analysis that led us to Poiseuille's law, but we now allow for the presence of a uniform thin layer of a second fluid adjacent to the wall of the tube. Figure 2-1.1 shows the situation we are considering. We assume that there is a well-defined boundary between the two fluids and that this boundary is stable. The liquid film thickness $h$ is assumed to be constant along the axis of the tube. We ignore gravity in this analysis.

If we review the analysis that leads to Eq. 1-2.9 we find that the assumptions introduced hold in both fluids of the case being considered here. Equation 1-2.9 is still the general solution for the velocity profile in each fluid. The boundary condition that the shear stress must be finite along $r = 0$ still holds in fluid I, so we have the following solution for the velocity profile in fluid I:

$$u_z^{\mathrm{I}} = \frac{C^{\mathrm{I}}}{4\mu^{\mathrm{I}}} r^2 + D^{\mathrm{I}} \quad \text{over} \quad 0 \le r \le R - h. \tag{2-1.1}$$

In fluid II, however, the constant $B$ that appears in Eq. 1-2.9 does not vanish since $r = 0$ is not within the domain of fluid II. The general solution in fluid II is

$$u_z^{\mathrm{II}} = \frac{C^{\mathrm{II}} r^2}{4\mu^{\mathrm{II}}} + \frac{B^{\mathrm{II}}}{\mu^{\mathrm{II}}} \ln r + D^{\mathrm{II}} \quad \text{on the region} \quad R - h \le r \le R. \tag{2-1.2}$$

Note that we have five unknown constants in these two equations. These constants will have to come from boundary conditions—further statements about the physics of this assumed flow field.

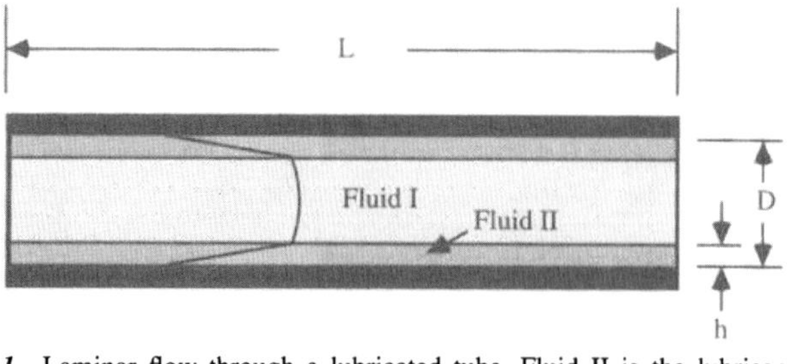


***FIGURE 2-1.1*** Laminar flow through a lubricated tube. Fluid II is the lubricant, with a thickness $h$.

First of all, the statements that led us to Eq. 1-2.15 are still valid in this "two-fluid" problem. Hence we still find

$$C^{\mathrm{I}} = \frac{\Delta P^{\mathrm{I}}}{L} \quad \text{and} \quad C^{\mathrm{II}} = \frac{\Delta P^{\mathrm{II}}}{L}. \tag{2-1.3}$$

A single pressure drop $\Delta P$ is imposed across the ends of the tube, and there is no mechanism whereby the pressure can be different in the two fluids, at a fixed axial position. (We ignore the very small difference in pressure that would arise from surface tension and curvature if the two fluids were immiscible.) Hence it follows that

$$C^{\mathrm{I}} = C^{\mathrm{II}} = \frac{\Delta P}{L}. \tag{2-1.4}$$

The remaining three equations that we need arise from boundary conditions on velocity and shear stress. We still satisfy a no-slip condition in the fluid adjacent to the wall of the tube, so

$$0 = \frac{C^{\mathrm{II}} R^2}{4\mu^{\mathrm{II}}} + \frac{B^{\mathrm{II}}}{\mu^{\mathrm{II}}} \ln R + D^{\mathrm{II}}. \tag{2-1.5}$$

We also satisfy a "no-slip" condition at the interface between the two fluids. In this case, this implies that the velocities in the two fluids are equal at the boundary $r = R - h$, so we write

$$\frac{C^{\mathrm{II}} (R-h)^2}{4\mu^{\mathrm{II}}} + \frac{B^{\mathrm{II}}}{\mu^{\mathrm{II}}} \ln (R-h) + D^{\mathrm{II}} = \frac{C^{\mathrm{I}}}{4\mu^{\mathrm{I}}} (R-h)^2 + D^{\mathrm{I}}. \tag{2-1.6}$$

We need one more boundary condition that reflects something we know (or imply to be true) about the flow. Equation 2-1.6 is really a statement that the velocity field is continuous through the fluids. Another expectation is that the *shear stress* varies continuously as we pass from one fluid into the other, or

$$\mu^{\mathrm{II}} \frac{du_z^{\mathrm{II}}}{dr} = \mu^{\mathrm{I}} \frac{du_z^{\mathrm{I}}}{dr} \quad \text{at } r = R - h. \tag{2-1.7}$$

Applying this to Eqs. 2-1.1 and 2-1.2 we find

$$\frac{C^{\mathrm{I}}(R-h)}{2} = \frac{C^{\mathrm{II}}(R-h)}{2} + \frac{B^{\mathrm{II}}}{(R-h)}. \tag{2-1.8}$$

We now have the required five equations for the five constants of integration in the solutions for the two velocity fields. After completing the algebra the results are

$$\frac{u_z^{\mathrm{I}}}{\Phi} = \left[(1-\eta)^2 + M\eta(2-\eta)\right] - s^2 \tag{2-1.9}$$

and

$$\frac{u_z^{\mathrm{II}}}{\Phi} = M\,(1 - s^2). \tag{2-1.10}$$

To simplify the format we have introduced the following notation:

$$\Phi = \frac{-CR^2}{4\,\mu^{\mathrm{I}}} = \frac{\Delta P\, R^2}{4\,\mu^{\mathrm{I}} L},$$

$$s = \frac{r}{R}, \qquad \eta = \frac{h}{R}, \qquad M = \frac{\mu^{\mathrm{I}}}{\mu^{\mathrm{II}}}. \tag{2-1.11}$$

The volume flow rate of the inner fluid follows from integrating the velocity field across the region occupied by fluid I.

$$Q^{\mathrm{I}} = \int_0^{R-h} 2\pi r u_z^{\mathrm{I}}\, dr. \tag{2-1.12}$$

If there were no lubricating layer ($h = 0$) the volume flow rate would be given by Eq. 1-2.17 (with $\Delta P$ instead of $\Delta\wp$, since we are neglecting gravity). Performing the integration indicated we find (letting $\alpha = 1 - \eta$)

$$\frac{Q^{\mathrm{I}}}{2\pi R^2 \Phi} = \tfrac{1}{4}\alpha^4 + \tfrac{1}{2} M \left(1 - \alpha^2\right) \alpha^2. \tag{2-1.13}$$

The ratio of flow rates with, and without, the lubricating layer is given by

$$\Psi = \alpha^4 + 2\,M \left(1 - \alpha^2\right) \alpha^2. \tag{2-1.14}$$

We show the dependence of $\Psi$ on $M$ and $\alpha$ in Fig. 2-1.2.

The results are consistent with our expectations. Only for $M > 1$, when the lubricating layer is less viscous than the mainstream (core) fluid, is there an enhancement of the flow rate. Further, we see that when $\alpha$ approaches zero (even for large $M$) the flow is reduced, since as the lubricating layer grows thick-

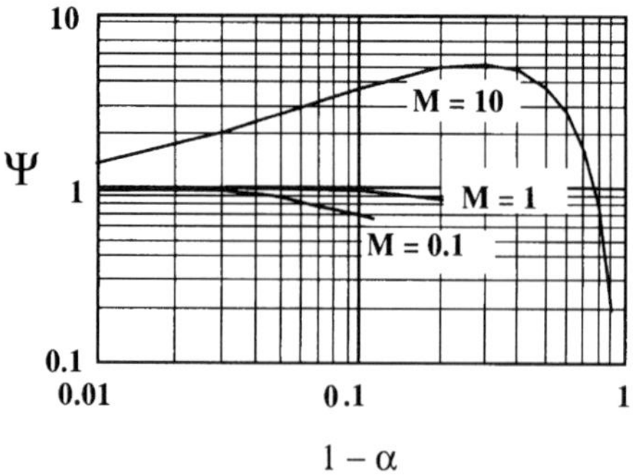


***FIGURE 2-1.2*** Effect of lubricating layer on flow through a tube. (Eq. 2-1.14).

er the region available for flow of fluid I is significantly decreased. For a given value of M there is an optimum film thickness corresponding to an optimum value of $\alpha$.

In this analysis we have assumed that the lubricating layer thickness remains constant along the tube axis, and is independent of time. (This assumption will be dropped in Section 2-6) We might hope to achieve this by providing a means of supplying a thin film of fluid II along the circumference of the tube, at the tube entrance. However, there is another issue here, and that is with respect to the *stability* of a thin liquid film in this type of flow. Without further analysis, or experimental confirmation, we have no guarantee that a thin film of low-viscosity liquid would remain as an annulus of constant thickness in the flow field assumed in our model. This point deserves further attention, and we will deal with the stability problem later.

First, let's see how rapidly the annular film is sheared out of the tube by the action of the inner flow.

## 2-2 REMOVAL OF AN ANNULAR LIQUID LAYER BY SHEARING—A QUASI-STEADY MODEL

The situation of interest to us is depicted in Fig. 2-2.1. On the assumption of fully developed flow in both fluids, there is a constant shear stress exerted by the core fluid and acting along the interface of the two fluids. This shear stress induces a drag flow in the outer (annular) fluid. The flow rate in the annular layer follows from integration of Eq. 2-1.10:

$$Q^{II} = 2\pi R^2 M\Phi \int_{1-\eta}^{1} s(1 - s^2)\, ds = \frac{\pi\, \Delta P\, R^4}{2\, \mu^{II} L} \eta^2 \left(1 - \frac{\eta}{2}\right)^2. \tag{2-2.1}$$

This is the rate at which the annular fluid must be replenished in order to *maintain* the thickness of the annular layer at the initial value of $h$.

Now suppose instead that the annular layer is not replenished, and the layer thickness continually decreases, but in such a way that it remains uniform in

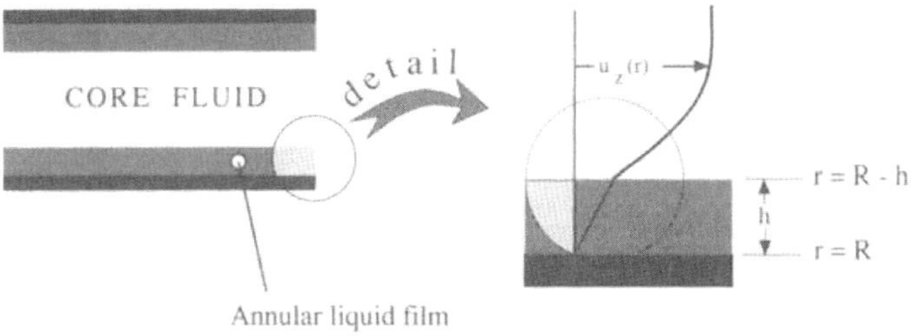


***FIGURE 2-2.1*** An annular layer of liquid is sheared by the internal flow.

thickness. From a simple quasi-steady mass balance we may find the rate of thinning of the annular layer, since

$$Q^{\mathrm{II}} = -\frac{d}{dt}\,\pi R^2\left[1-(1-\eta)^2\right] = \frac{\pi\,\Delta P\,R^4}{2\,\mu^{\mathrm{II}}\,L}\,\eta^2\left(1-\frac{\eta}{2}\right)^2. \tag{2-2.2}$$

From this expression we find $\eta$ as a function of time. The algebra simplifies quite a bit if we confine our attention to films so thin that $h \ll R$ for all time. Under these conditions we may show that

$$\eta(t) = \frac{h(t)}{R} = \left(\frac{R}{h_0} + \frac{\pi\,\Delta P\,R^2}{4\,\mu^{\mathrm{II}}\,L}\,t\right)^{-1}. \tag{2-2.3}$$

This is a very simple result, but one that could be quite useful. For example, Fig. 2-1.2 shows us that for $M = 10$, the ability of the lubricating layer to maintain its effectiveness disappears when $\eta$ falls below a value of about 0.01. Equation 2-2.3 permits us to determine how long a time this would be in some system for which the operating conditions are specified.

The integration of Eq. 2-2.2, leading to Eq. 2-2.3, is based on the approximation that $h \ll R$ (or $\eta \ll 1$). It also assumes that $\Delta P$ is constant. If the inner flow $Q^{\mathrm{I}}$ were held constant, instead, then $\Delta P$ would vary ( and it would depend on $\eta$) and Eq. 2-2.3 would have to be modified. We leave that as an exercise (Problem P2-5).

A key assumption in the analysis of this section is that the flow is steady in time and that the annular film remains an annulus of axially uniform thickness $h(t)$. There is ample evidence that such a steady flow does not ensue under all conditions of interest, even though the annular liquid is continuously replenished at the upstream end of the tube. Quite a sizable literature exists on this topic, and it is too broad and complex to review in any detail. An appropriate starting point for further study is the work of Joseph and coworkers (Bai *et al.*, 1992). We turn instead to a discussion of a relatively simple stability problem—that of an annular layer of liquid within a long tube, in the absence of a core flow.

## *2-3 THE INSTABILITY OF A STATIC ANNULAR FILM OF LIQUID*

We ask the following question. Given an initially uniform and stationary annular film of a viscous liquid, does that film remain of constant thickness $h_0$ or does it evolve toward another configuration $h(z,t)$? The answer is known from experience—there are conditions under which the film evolves toward a more stable configuration which is a periodic series of peaks and valleys. Figure 2-3.1 shows the geometry of interest.

An analysis of the stability of the annular film can be carried out, and we follow the method presented by Goren (1962). Keep in mind that we are assuming that the inner fluid has no effect on the dynamics of the flow in the annular layer.

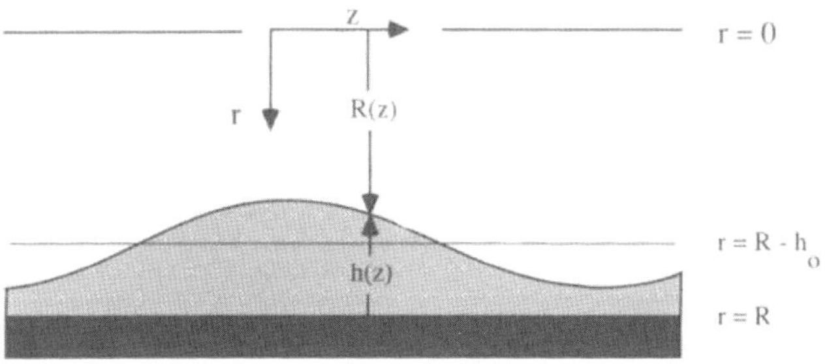


*FIGURE 2-3.1* An annular film of liquid evolving toward a periodic thickness distribution.

We allow for an interfacial tension, but the inner fluid is otherwise incapable of exerting any stresses on the annular layer. We assume that the flow that ensues as the annular layer evolves from its initial state is slow in the sense that second-order inertial effects are negligible in the Navier–Stokes equations.

It is convenient to formulate this problem in terms of the stream function $\psi$, defined so that

$$u_r = \frac{1}{r}\frac{\partial \psi}{\partial z} \quad \text{and} \quad u_z = -\frac{1}{r}\frac{\partial \psi}{\partial r}. \tag{2-3.1}$$

We follow the usual approach of the stream function formulation. We replace the two velocity components in the Navier–Stokes equations through Eqs. 2-3.1, and then eliminate pressure from the equations by differentiating $\partial p/\partial r$ with respect to $z$, and $\partial p/\partial z$ with respect to $r$, followed by subtraction of one equation from the other. The result is a fourth-order partial differential equation which may be written in the format

$$\frac{\partial}{\partial t} D^2\psi = \nu D^4\psi, \tag{2-3.2}$$

where the differential operators are defined by

$$D^2\psi = \frac{\partial^2\psi}{\partial r^2} - \frac{1}{r}\frac{\partial \psi}{\partial r} + \frac{\partial^2\psi}{\partial z^2} \tag{2-3.3}$$

and

$$D^4\psi = \left(\frac{\partial^2}{\partial r^2} - \frac{1}{r}\frac{\partial}{\partial r} + \frac{\partial^2}{\partial z^2}\right)\left(\frac{\partial^2\psi}{\partial r^2} - \frac{1}{r}\frac{\partial \psi}{\partial r} + \frac{\partial^2\psi}{\partial z^2}\right). \tag{2-3.4}$$

The solution of Eq. 2-3.2 may be written in the form

$$\psi = F(r)\exp(\gamma t + ikz), \tag{2-3.5}$$

where $i$ has the usual meaning of $i = \sqrt{-1}$. This form is assumed because we anticipate that the flow field will be spatially periodic. $k$ is the wave number of the periodicity, and the wavelength is $\lambda = 2\pi / k$. The algebra is simplified if we use the complex exponential notation for a periodic function. The parameter $\gamma$ is a growth rate for the surface profile. If $\gamma$ is real and positive the surface is unstable, and disturbances grow exponentially. If $\gamma$ is real and negative, disturbances damp and the surface is stable.

When Eq. 2-3.5 is substituted into Eq. 2-3.2 an ordinary differential equation for $F(r)$ is obtained, and the solution may be written in terms of the modified Bessel functions $I_1$ and $K_1$:

$$\psi = \left\{ \begin{array}{l} A_1 \; r\, I_1(kr) + B_1 \; r\, K_1(kr) \\ + A_2 \; r\, I_1(k_1 r) + B_2 \; r\, K_1(k_1 r) \end{array} \right\} \exp(\gamma t + ikz), \tag{2-3.6}$$

where

$$k_1^2 = k^2 + \gamma / \nu. \tag{2-3.7}$$

The four $A$ and $B$ coefficients must be determined from four boundary conditions on $\psi$. These are

1 and 2. No slip of either velocity component at the solid wall of the tube. Hence

$$u_r = 0 = \left( \frac{1}{r} \frac{\partial \psi}{\partial z} \right)_{r=R} \quad \text{and} \quad u_z = 0 = -\left( \frac{1}{r} \frac{\partial \psi}{\partial z} \right)_{r=R}. \tag{2-3.8}$$

3. No shear stress at the free surface. The shear stress is given by

$$T_{rz} = \mu \left( \frac{\partial u_r}{\partial z} + \frac{\partial u_z}{\partial r} \right). \tag{2-3.9}$$

In terms of the stream function this becomes

$$T_{rz} = \mu \left( \frac{\partial^2 \psi}{\partial r^2} - \frac{1}{r} \frac{\partial \psi}{\partial r} - \frac{\partial^2 \psi}{\partial z^2} \right). \tag{2-3.10}$$

This boundary condition must be applied at the free surface of the liquid, but this is a moving boundary, varying spatially and temporally. To simplify this boundary condition we assume that the deviation of the free surface from its initial position is small, and impose the boundary condition at the *original* surface $r = \alpha R = R - h_0$. Hence we write

$$0 = \mu \left( \frac{\partial^2 \psi}{\partial r^2} - \frac{1}{r} \frac{\partial \psi}{\partial r} - \frac{\partial^2 \psi}{\partial z^2} \right)_{r=\alpha R}. \tag{2-3.11}$$

4. The fourth boundary condition is on the stress *normal* to the free surface. The internal (core) fluid is assumed to exert no normal stress on the interface, other than an isotropic pressure which we set arbitrarily at zero. Because of surface tension, and the curvature of the interface, there is a jump in normal stress across the interface, given by the Young–Laplace equation:

$$\Delta T_{rr} = -\sigma \left( \frac{1}{R_1} + \frac{1}{R_2} \right) = \left( -p + 2\mu \frac{\partial u_r}{\partial r} \right)_{r=\alpha R}. \tag{2-3.12}$$

Note that the internal normal stress is the sum of the pressure and the viscous normal stress. For an axisymmetric surface of radius $R(z)$ the principal radii of curvature of the free surface are given by

$$\frac{1}{R_1} = \frac{-1}{R(z)}\left[1+\left(\frac{\partial R(z)}{\partial z}\right)^2\right]^{-1/2} \tag{2-3.13a}$$

and

$$\frac{1}{R_2} = \frac{\partial^2 R(z)}{\partial z^2}\left[1+\left(\frac{\partial R(z)}{\partial z}\right)^2\right]^{-3/2}. \tag{2-3.14a}$$

If we now change the perspective, so to speak, and look at the film thickness $h(z)$ instead of the radius $R(z)$, these expressions may be written in the forms (see Fig. 2-3.1 for the geometry)

$$\frac{1}{R_1} = \frac{-1}{R-h(z)}\left[1+\left(\frac{\partial h}{\partial z}\right)^2\right]^{-1/2} \approx \frac{-1}{\alpha R}\left(1-\frac{h'}{\alpha R}\right), \tag{2-3.13}$$

where $\alpha = 1 - h/R$ and

$$\frac{1}{R_2} = -\frac{\partial^2 h}{\partial z^2}\left[1+\left(\frac{\partial h}{\partial z}\right)^2\right]^{-3/2} \approx -\frac{\partial^2 h'}{\partial z^2}. \tag{2-3.14}$$

For the disturbance to the free surface we write

$$h(z,t) = h_0 + h' = h_0 + \eta \exp(\gamma t + ikz). \tag{2-3.15}$$

The approximations indicated in Eqs. 2-3.13 and 14 follow from the assumption that the disturbance to the free surface, $h'$, is small compared to $\alpha R$. Note that the curvature of the cross section normal to the $z$ axis is negative.

It is now necessary to express the pressure $p$ and the free surface $h(z)$ in terms of the stream function.

The pressure may be written in the form

$$p = -\frac{\sigma}{\alpha R} + p' \exp(\gamma t + ikz), \tag{2-3.16}$$

where $p'$ is the perturbation to the pressure arising from the motion. Note that in the static case there is a nonzero pressure—the surface tension term. The pressure is related to the stream function through the Navier–Stokes equations, Eq. 1-1.2 and 1-1.3, after writing the velocities in terms of $\psi$. The relationship of $h'$ to the stream function follows (see Eq. 1-1.11) from a kinematic statement:

$$\frac{\partial h'}{\partial t} = -u_{r_{r=h}}. \tag{2-3.17}$$

The ensuing algebra required to convert Eq. 2-3.12 to a relationship involving the stream function is very unpleasant. When it is finished, Eq. 2-3.12 takes the form

$$\left\{ \alpha R \frac{\partial}{\partial r} \left\{ (\alpha R)^2 D^2 \psi - \left[ \frac{\gamma \rho (\alpha R)^2}{\mu} + 2 k^2 (\alpha R)^2 \right] \right\} \psi \right. \\ \left. + 2k^2 (\alpha R)^2 \psi - \frac{\sigma}{\gamma \alpha R \mu} \left[ 1 - k^2 (\alpha R)^2 \right] k^2 (\alpha R)^2 \psi \right\}_{r = \alpha R} \\ = 0. \tag{2-3.18}$$

We finally have all four boundary conditions which yield four linear *homogeneous* equations in the four coefficients $A$ and $B$. We do not solve for these coefficients. Instead we impose the constraint that there be a nontrivial solution for the coefficients, which is simply the requirement that the determinant of the coefficients vanishes. This determinant is made up of Bessel functions and is itself very complicated algebraically. It is easier to consider a simple special case that is a little more transparent, algebraically, and that makes it easier for us to discuss the physics of this stability problem.

## *2-3.1 Stability of an Inviscid Annular Liquid*

We suppose that viscous effects are small and that any ensuing motion within the annular film is controlled by the interaction of inertia (which is still in the analysis through the $\partial/\partial t$ terms) with surface tension. For the case of vanishing viscosity Eq. 2-3.2 reduces to

$$D^2 \psi = 0 \tag{2-3.19}$$

and the solution, maintaining the form of Eq. 2-3.5, is

$$\psi = \left[ A_1 r I_1(kr) + B_1 r K_1(kr) \right] \exp(\gamma t + ikz). \tag{2-3.20}$$

Since Eq. 2-3.19 is only second order, there are only two integration constants. In turn, there can only be two boundary conditions that constrain the solution. In a sense, we lose two boundary conditions when we drop the viscous terms from the Navier–Stokes equations. One condition that we can no longer enforce is the vanishing of the tangential velocity at the inner surface of the tube. An inviscid fluid does not have to satisfy no slip with respect to the tangential velocity. However, the wall is impermeable, and there can still be no radial component of velocity across (at) the wall. Hence we still impose

$$u_r = 0 = \left( \frac{1}{r} \frac{\partial \psi}{\partial z} \right)_{r = R} \tag{2-3.21}$$

as a boundary condition.

The physical argument that leads us to Eq. 2-3.12 is still invoked for an inviscid fluid. The only difference is that there is no viscous normal stress term, and so we write

$$\Delta T_{rr} = - \sigma \left( \frac{1}{R_1} + \frac{1}{R_2} \right) = (-p)_{r = \alpha R}. \tag{2-3.22}$$

In place of Eq. 2-3.18 we obtain

$$\left\{ -\alpha R \frac{\partial}{\partial r} \frac{\gamma \rho (\alpha R)^2}{\mu} \psi - \frac{\sigma}{\gamma \alpha R \mu} \left[ 1 - k^2 (\alpha R)^2 \right] k^2 (\alpha R)^2 \psi \right\}_{r = \alpha R} = 0. \tag{2-3.23}$$

We now substitute $\psi$ into Eqs. 2-3.21 and 23 and obtain a pair of linear homogeneous algebraic equations in the constants $A$ and $B$. We set the determinant of the coefficients of $A$ and $B$ to zero and find a relationship between the "growth rate" parameter $\gamma$ and the "wavelength" of a surface disturbance, $\lambda$, in the dimensionless format

$$\mathsf{G}^2 \equiv \frac{\gamma^2 \rho (\alpha R)^3}{\sigma} = (\alpha\kappa)\left[1 - (\alpha\kappa)^2\right] \frac{I_1(\kappa)\, K_1(\alpha\kappa) - I_1(\alpha\kappa)\, K_1(\kappa)}{I_1(\kappa)\, K_0(\alpha\kappa) - I_0(\alpha\kappa)\, K_1(\kappa)}. \tag{2-3.24}$$

$\kappa$ is defined in terms of a wavelength $\lambda$ as

$$\kappa \equiv \frac{2\pi R}{\lambda} = kR. \tag{2-3.25}$$

Inspection of Eq. 2-3.24 shows that $\mathsf{G}$ is positive (hence $\gamma$ has a *real* and positive root) only for $\alpha\kappa < 1$.

We conclude that this physical system allows for the possibility of an unstable free surface, if it is subjected to any disturbances whose wavelengths exceed the (inner) circumference of the tube. This seems to be the first point at which we have introduced the notion that the surface is subject to *any* disturbance. In fact, we introduced that notion at the point that we wrote Eq. 2-3.5. The stream function for the static (stable) case is simply $\psi$ = constant, since it would follow from Eqs. 2-3.1 that there would be no motion. Equation 2-3.5 is equivalent to the idea that at time $t = 0$ there is a disturbance to the fluid corresponding to a stream function that is spatially periodic, with a wave number $k$:

$$\psi(0) = F(r)\, \exp(ikz). \tag{2-3.26}$$

We then assumed that the ensuing flow corresponded to a stream function of the form given by Eq. 2-3.5. Details aside, what followed was the analysis of how that stream function evolved in time: did it dampen (stability) or did it grow unbounded (instability)? The answer is that some disturbances—those for which $\alpha\kappa > 1$—will dampen, while others—those for which $\alpha\kappa < 1$—will grow unbounded.

We make one other assumption in this kind of stability analysis, and that is that the fluid is subjected to a "spectrum" of disturbances corresponding to a range of wave numbers $k$, and that if the system is unstable, it will exhibit dynamics controlled by the fastest-growing disturbance. Implicit here is the idea that for small enough disturbances, each "mode" of response (corresponding to each value of $k$) is independent of the others. That is why the restriction to small perturbations in $\psi$, $p$, and $h$ is made. *This is a stability theory for infinitesimal disturbances.* It tells us nothing about how the system will respond to large disturbances. In fact, by the time that the perturbation in the flow becomes large enough to distort the interface seriously, the analysis becomes suspect.

What we ask of this small-disturbance stability analysis is that it tell us *whether* a system is stable to small disturbances and, if not, what is the wavelength of the fastest-growing disturbance. Let's go back to the analysis, then, and examine the issue of the rate of growth of the fastest-growing disturbance. Returning to the case of the inviscid fluid, we find that Eq. 2-3.24 tells us that the growth rate G goes through a maximum for $\alpha\kappa$ on the range [0,1]. Figure 2-3.2 shows the function $G(\alpha\kappa)$ for one value of $\alpha$. This plot defines the wave number $\kappa^*$ of the fastest-growing disturbance (for that value of $\alpha$) and the growth rate $G^*$ corresponding to that disturbance. This leads to a plot of $G^*(\alpha)$, as shown in Fig. 2-3.3.

The stability theory is much more difficult to work out for the viscous liquid, and we will not present the details here. There is a simple analytical result available (Aul and Olbricht, 1990) for the limiting case of a thin viscous liquid layer ($\alpha \approx 1$) surrounding an inviscid core fluid (such as a gas). It may be written in the dimensionless format

$$G' \equiv \gamma^* \sqrt{\frac{\rho R^3}{\sigma}} = \sqrt{\frac{\rho R \sigma}{\mu^2}}\,\frac{(1-\alpha)^3}{12\alpha^4}$$

$$\text{for } 1-\alpha \ll 1 \text{ and high viscosity.} \tag{2-3.27}$$

The dimensionless group $\mu/\sqrt{(\rho R \sigma)}$ occurs often in viscous flow problems with

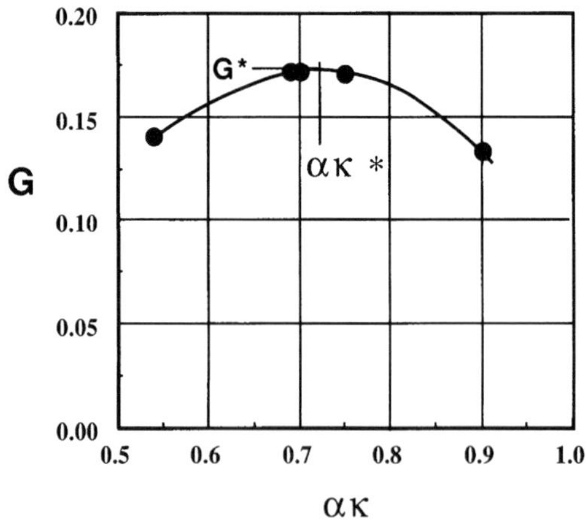


***FIGURE 2-3.2*** $G(\alpha\kappa)$ for $\alpha = 0.9$.

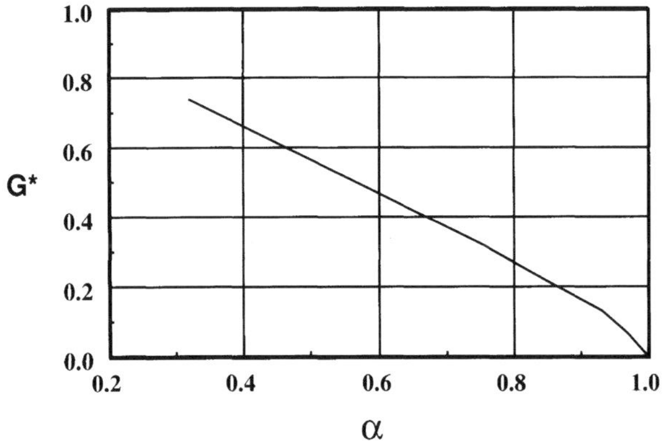


*FIGURE 2-3.3* The growth rate function G*(α).

surface tension, and it is called the Ohnesorge number, Oh:

$$\mathrm{Oh} \equiv \frac{\mu}{\sqrt{\rho R \sigma}} = \frac{\mathrm{We}^{1/2}}{\mathrm{Re}}. \tag{2-3.28}$$

As indicated above, the Ohnesorge number may be written in terms of the Reynolds and Weber numbers:

$$\mathrm{We} \equiv \frac{\rho U^2 R}{\sigma} \quad \text{and} \quad \mathrm{Re} \equiv \frac{\rho U R}{\mu}. \tag{2-3.29}$$

Note that the radius $R$ and not the diameter is used in these definitions.

Figure 2-3.4 shows (for one value of α) the behavior of Eq. 2-3.27 (the high-viscosity limit) along with the inviscid limit, recast in terms of $G'$. Somewhere in the neighborhood of $\mathrm{Oh} = 10^{-3}$ there is a transition from the viscous-dominated to the inertia-dominated (inviscid) theory.

This stability theory has been verified to various degrees through experiments by Goren (1962) and Aul and Olbricht (1990). The former coated the inside of a capillary with a uniform layer of viscous liquid and then observed the ensuing development of "bridges" across the tube, as suggested in Fig. 2-3.5. The "bridges" are created as the amplitudes of the annular "peaks" equal the tube radius, at which point coalescence occurs. The wavelength λ of the instability is obtained from observation of the distance separating the centers of successive bridges. If there is insufficient liquid in the annular film (α very close to unity) then it may not be possible for bridges to occur. The final state of the instability would be a periodic series of annular "collars," separated by "dry" tube surface.

Aul and Olbricht carried out an experiment in which the core fluid was a liquid flowing coaxially with the annular, more viscous layer. It can be shown that when the viscosity of the core liquid is much less than that of the annular liquid, and if the Reynolds number of the core flow is very small, the annular layer instability is in accordance with the models presented above, which ignore the dynamics of the core liquid. Although the core flow shears the annular layer and

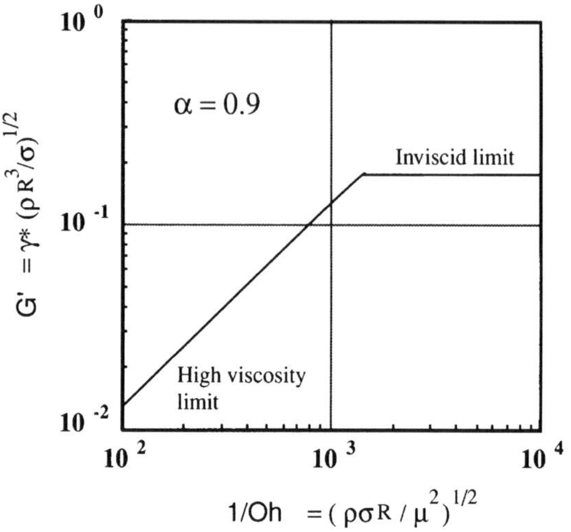


***FIGURE 2-3.4*** Dimensionless growth rate function G′, for $\alpha = 0.9$.

moves it down the tube (in accordance, to a first approximation, with the development of Section 2-2), some of the experiments described by Aul and Olbricht were carried out at such low flow rates that the annular layer could be regarded as stationary.

It is interesting to examine a set of data presented by Aul and Olbricht. A capillary of inside radius $R = 27$ μm was coated with a viscous liquid. The coating thickness was 1.8 μm initially ($\alpha = 0.933$). The annular liquid had the following physical properties: $\rho = 1$ g/cm$^3$ and $\mu = 1.73$ poise. The core liquid was water, and the interfacial tension was $\sigma = 3.5$ dyn/cm. These investigators found that the average wavelength of the bridges formed across the capillary was 225 μm, with a range from 200 to 280 μm. We may calculate the Ohnesorge number for this experiment and find its inverse to be

$$\frac{1}{\mathrm{Oh}} \equiv \sqrt{\frac{\rho R \sigma}{\mu^2}} = \sqrt{\frac{1 \times 0.0027 \times 3.5}{1.73^2}} = 0.056. \qquad (2\text{-}3.30)$$

Noting Fig. 2-3.4, we expect that the viscous-dominated theory (Eq. 2-3.27) would hold for the growth rate. Goren's theory leads to the prediction that the instability would appear at a wavelength corresponding to

$$\alpha \kappa^* = \frac{\sqrt{2}}{2} = 0.707 \quad \text{for } \alpha \to 1. \qquad (2\text{-}3.31)$$

(This is true regardless of the viscosity of the annular fluid.) This leads to the prediction that

$$\frac{\lambda}{2\pi R \alpha} = \frac{1}{0.707} \quad \text{or} \quad \lambda = \frac{2\pi \times 27 \times 0.933}{0.707} = 224 \ \mu\text{m}. \qquad (2\text{-}3.32)$$

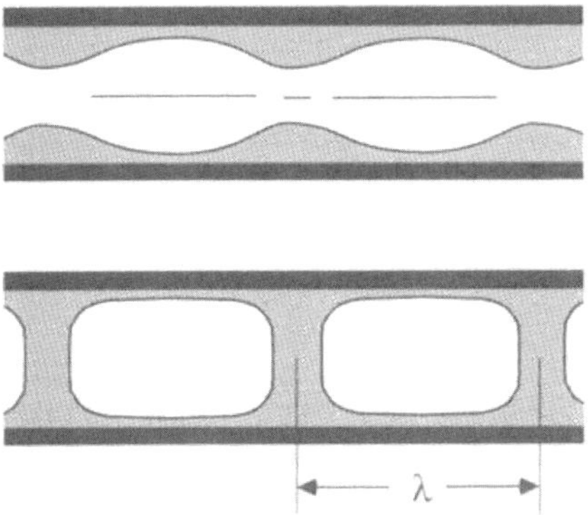


*FIGURE 2-3.5* Instability of the annular layer leads to bridges.

This is in excellent agreement with the data. No data were presented for the growth rate.

The analysis described here neglects any effect of gravity. Gravity plays a role in an especially important and interesting flow that arises when a cylindrical filament is continuously withdrawn from a pool of liquid with a horizontal free surface, as in wire coating. Then it is observed (Quere, 1990) that there is a critical entrained film thickness above which drops will form periodically along the axis of the entrained film. A detailed discussion of this phenomenon, and a numerical solution of the film evolution equation, is presented by Kalliadasis and Chang (1994b).

In these discussions of the stability of a thin annular film of liquid we have assumed the existence of the initial annular film. Now we examine the question of how one creates such a film in the interior of a capillary.

## 2-4 THE DEPOSITION OF A THIN ANNULAR FILM OF LIQUID INSIDE A CAPILLARY

Imagine that a long horizontal capillary is initially filled with some liquid. Subsequently, air is blown slowly and continuously through one end of the tube, displacing the liquid out of the other end. A classical set of data was obtained by Sir G. I. Taylor (1961) and is reproduced in Fig. 2-4.1.

The measured quantity was the fraction $m$ of liquid remaining after passage of the core fluid (air). From the geometry we find that $m$ is related to the parameter $\alpha$ of the previous section by

$$m = 1 - \alpha^2. \tag{2-4.1}$$

Over a range of viscosities, surface tensions, and capillary diameters the data appear to correlate in terms of the capillary number, defined by

$$\text{Ca} \equiv \frac{\mu U}{\sigma}, \tag{2-4.2}$$

where $U$ is the average velocity of the core fluid (the gas flow). The analysis of this fluid dynamics problem is very complex, and several approximate analytical analyses have been presented (Bretherton, 1961; Cox, 1962). We review the approach of Bretherton here in order to see the limitations of the method selected, as well as to examine the degree to which the model is able to mimic the data of Fig. 2-4.1. Figure 2-4.2 suggests the geometry for the analysis of the flow. The bubble leaves behind it a film of uniform thickness $h$. Observations suggest that the region of uniform thickness is a few radii behind the "nose," the moving front, of the bubble. The bubble surface is taken to have a uniform velocity $U$ in the axial direction, $z$.

Bretherton carried out an analysis based on planar lubrication theory. This begins with the assumption that the film thickness $h$ is so small compared to $R$ ($\alpha \approx 1$) that the film may be considered planar. The Navier–Stokes equations are a little simpler in Cartesian coordinates, and this simplifies the ensuing algebra. It is assumed further that the Reynolds number in the liquid (defined by $\text{Re} = \rho UR/\mu$, where $\mu$ is the viscosity of the fluid adjacent to the wall) is so small that the inertial terms, including the time derivatives, may be dropped from the Navier–Stokes equations. The essence of lubrication theory is in the assumption that the flow is nearly parallel to the solid surface, i.e., that $u_y << u_z$. Then the Navier–Stokes equations in the $z$ and $y$ directions simplify to

$$0 = -\frac{dp}{dz} + \mu \frac{d^2 u_z}{dy^2} \quad \text{and} \quad 0 = -\frac{dp}{dy}. \tag{2-4.3}$$

The right-hand half of this pair of equations tells us that $p$ is not a function of $y$, so $dp/dz$ is independent of $y$, and we may therefore integrate the left-hand half of this pair to find the velocity field in the form

$$u_z = \frac{1}{2\mu}\frac{dp}{dz}\left(y^2 - 2y_s y\right). \tag{2-4.4}$$

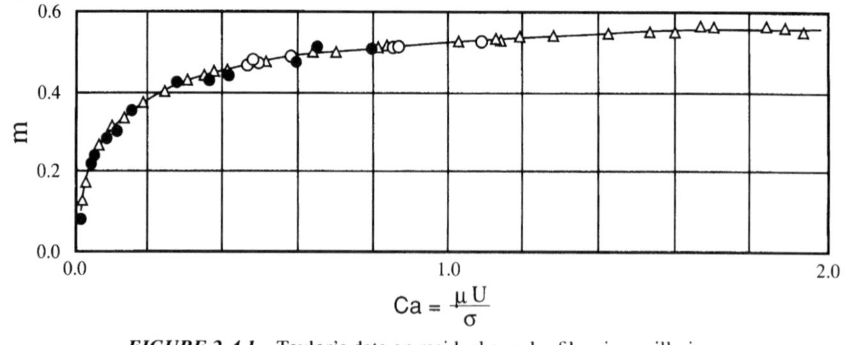


***FIGURE 2-4.1*** Taylor's data on residual annular films in capillaries.

Note that we use $y_s$ to refer to the position of the free surface in the $y$ direction, and $y_s = y_s(z)$. In obtaining Eq. 2-4.4 we have imposed two boundary conditions on the velocity field: no slip at the wall, and no shear stress at the free surface $y_s(z)$.

We must also impose a boundary condition on the normal stress at the free surface. This is the usual statement that the normal stress is continuous except for a jump associated with surface tension and the curvature of the surface. This would take the form, in Cartesian coordinates,

$$p = -\sigma \frac{d^2 y_s}{dz^2}. \tag{2-4.5}$$

In writing this simplified form of the Young–Laplace equation (see Eq. 2-3.12) we have used the planar approximation for the curvature in the $y$ direction (equivalent to setting $1/R_y = 0$), and we assumed that the slope of the free surface is small so that Eq. 2-3.14 simplifies to $1/R_z = -d^2y_s / dz^2$. This latter assumption will not be valid near the nose of the bubble.

We are neglecting the viscous normal stress contributions arising from terms $2\mu\, \partial^2 u_y / \partial y^2$ and $2\mu\, \partial^2 u_z / \partial z^2$. This implies that surface tension effects on the pressure dominate the effects of viscous stresses. This is equivalent to the assumption that the capillary number is small. Finally, the pressure $p$ is relative to the bubble pressure, which we arbitrarily set to zero.

If the bubble velocity is $U$, the motion of the bubble displaces the liquid at a volumetric rate (per unit width)

$$Q = U(y_s - h) \tag{2-4.6}$$

through the liquid layer of thickness $y_s$. But we may also find the same volumetric flow rate from the velocity field (Eq. 2-4.4) as

$$Q = \int_0^{y_s(z)} u_z\, dy = -\frac{y_s^3}{3\mu}\frac{dp}{dz}. \tag{2-4.7}$$

Now we differentiate Eq. 2-4.5 with respect to $z$, and use Eqs. 2-4.6 and 7 to find

$$\frac{d^3 y_s}{dz^3} = \frac{3\mu U}{\sigma}\frac{y_s(z) - h}{y_s^3} = 3\,\mathrm{Ca}\,\frac{y_s(z) - h}{y_s^3}. \tag{2-4.8}$$

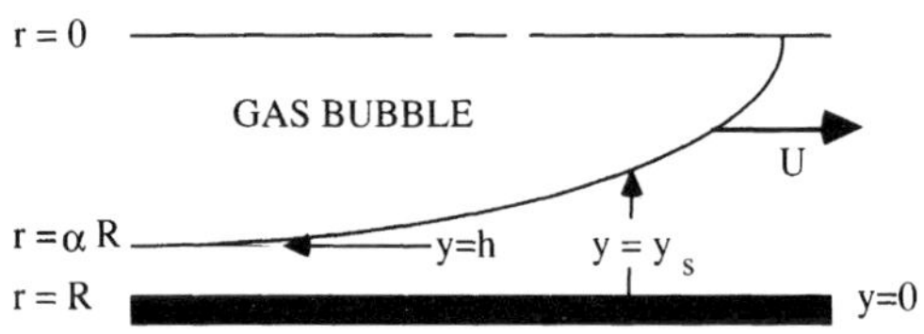


***FIGURE 2-4.2*** Geometry for analysis of film deposition.

We have now derived a differential equation for the free surface, which we can rewrite in a nondimensional format by defining

$$\eta \equiv \frac{y_s}{h} \quad \text{and} \quad \xi \equiv \frac{z}{h}(3\,\mathrm{Ca})^{1/3}, \tag{2-4.9}$$

after which we find

$$\frac{d^3\eta}{d\xi^3} = \frac{\eta - 1}{\eta^3}. \tag{2-4.10}$$

In the region approaching the nose of the bubble, where $y_s >> h$, we have $(\eta - 1)/\eta^3 << 1$, and so the general solution to Eq. 2-4.10 (setting the right-hand side exactly to zero) is

$$\eta = \tfrac{1}{2} a\,\xi^2 + b\xi + c. \tag{2-4.11}$$

The curvature of the free surface in the nose region then follows as

$$\frac{d^2 y_s}{dz^2} = \frac{a\,(3\,\mathrm{Ca})^{2/3}}{h}. \tag{2-4.12}$$

But we have assumed that the capillary number is small, which implies that viscous effects are secondary to surface tension effects. In other words, we are assuming that the viscosity does not have a significant effect on the dynamics of the flow, in the neighborhood of the nose of the bubble. This implies that the pressure field is uniform within the liquid film. The Young–Laplace equation, in turn, tells us that the only surface of revolution that corresponds to uniform pressure is spherical. (In the planar approximation the surface would be cylindrical.) Hence the curvature is simply $1/(R - y_s)$, or consistent with approximations already made, $1/R$. Thus we find

$$\frac{a\,(3\,\mathrm{Ca})^{2/3}}{h} \approx \frac{1}{R} \tag{2-4.13}$$

or

$$\frac{h}{R} = a\,(3\,\mathrm{Ca})^{2/3}. \tag{2-4.14}$$

The constant $a$ is unknown at this point, but we have a prediction that the deposited film thickness increases as the two-thirds power of the capillary number. This prediction must be tested in the range of very small capillary numbers, since the result is not expected to be valid outside that range. Taylor's data, for two liquids, and Bretherton's for a third liquid (all in the range of low capillary numbers) are replotted in Fig. 2-4.3 as a test of this prediction. The data confirm the expected slope, in the capillary number range of $10^{-4}$ to $10^{-2}$, and it is clear that by the time the capillary number is much beyond the order of 0.01 the data begin to fall below the predicted relationship. Deviations are seen at the other end as well. Bretherton's data, as well as additional data of Schwartz *et al.*

(1986), suggest that the film thickness reaches a lower limit at very small values of capillary number.

One shortcoming of the lubrication theory employed by Bretherton is that it is not possible to obtain the coefficient $a$ by analytical methods. Further, the range of validity of the key result (Eq. 2-4.14) is limited to very small values of Ca, since viscous effects are largely ignored in considering the behavior of the free surface. In contrast, Cox (1962) takes a different approach in which viscous effects are assumed dominant (large Ca) but still fails to obtain a complete analytical prediction of $h/R$ as a function of Ca. In the analysis of Cox, the shape of the interface near the region of uniform film thickness is assumed. Cox then derives a relationship between the (dimensionless) residual film thickness $h/R$ and the capillary number, but this relationship involves an unknown constant that arises in the assumption of a shape for the interface. When this constant is determined from experiments by matching photographs of the shape of the interface to the theory, very good correspondence of the theory to the data for $h/R$ is obtained, in the regime of *large* values of Ca. The extensive experimental observations of Taylor (see Fig. 2-4.1) show that for Ca > 2, $h/R$ becomes independent of Ca and approaches a value of 0.33. The theory of Cox, using the empirically determined coefficient as just described, predicts a constant value of $h/R = 0.37$.

At this stage we conclude that a relatively simple hydrodynamic theory leads to an equation (Eq. 2-4.14) that correlates the data of several investigators over several decades in the capillary number. Bretherton's theory fails at the high and low ends of the capillary number range that is experimentally accessible.

The challenge to understand the limitations to Bretherton's theory, and to extend it, has been taken up by several investigators. First, the approximations of Bretherton's theory which involve arguments regarding the relative importance of viscous and capillary forces have been examined and placed on a more rigorous foundation by Park and Homsy (1984). They used the method of

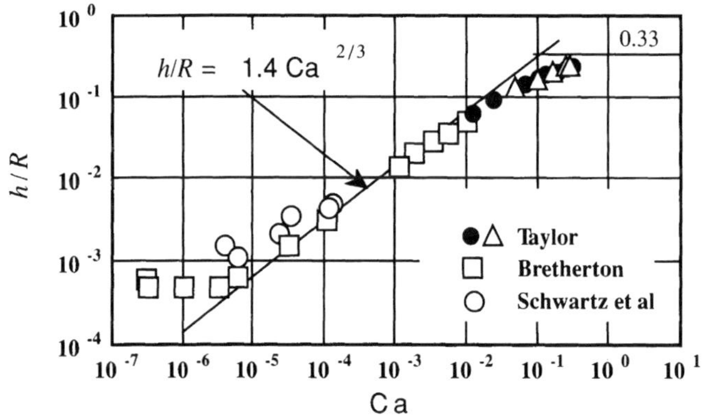


*FIGURE 2-4.3* Experimental test of Eq. 2-4.14.

matched asymptotic expansions, which they develop in some detail in the reference cited. [A somewhat simpler treatment by the same methods and applied to essentially the same problem is given by Wilson (1982).]

The most detailed treatment of the physics of this problem is given by Teletzke *et al.* (1988). They distinguish between *thick* thin films and *thin* thin films. Specifically, they note that as liquid film thicknesses approach molecular dimensions, intermolecular forces begin to become comparable to bulk viscous and capillary forces, and the fluid dynamics can be quantitatively and qualitatively affected by the action of these forces. The issue of whether the liquid wets the substrate is introduced, as well. Details aside, the final result is that one can explain the presence of a lower limiting film thickness as arising from the action of intermolecular forces.[1] The suggestion is also made that the presence of trace amounts of surface active impurities (which are difficult to avoid in all but the most meticulous of experimental conditions) gives rise to some of Bretherton's observations that the residual film thickness is greater than predicted by his theory.

Finally, we note here that an interesting and related problem has been studied and solved numerically by Tsai and Miksis (1994). They investigate the problem of a drop or bubble driven by a pressure gradient through a tube filled with liquid. The shape of the drop or bubble is calculated under creeping flow (low Reynolds number) conditions. Their work includes a numerical study of flow of the drop or bubble through a symmetric constriction in the tube radius. Under some conditions the drop or bubble will break into two or more pieces as it stretches through the constriction, and the numerical analysis gives conditions for the breakage of the drop or bubble.

## 2-5 A MORE CAREFUL DISCUSSION OF THE GEOMETRY AND DYNAMICS OF A FREE BOUNDARY PROBLEM

In the preceding discussion of film deposition we have used several (apparently rational) approximations to derive a differential equation (Eq. 2-4.10) for the film thickness as a function of position. It is not a trivial matter to arrive at an equation for the evolution of the free surface that is consistent with regard to the approximations used to simplify the equations of motion and the boundary conditions. Atherton and Homsy (1976) present a very detailed discussion of the derivation of equations of the class that concerns us here. At this point we want to go back a few steps and examine some of the details of the analysis of the film deposition problem. Our goal is to put some perspective on the nature of the approximations and the methods used to ensure consistent approximations. The algebra will be simpler, and the results essentially unchanged, if we consider a two-dimensional planar approximation to the axisymmetric case of interest to us. This will certainly be a good approximation in the region where the residual

[1]A treatment of this problem using matching asymptotic expansions, and including intermolecular forces, may be found in the work of Kalliadasis and Chang (1994a).

film is thin compared to the tube radius. In effect, we are repeating the discussion presented in Section 1-1 of Chapter 1, for the special case of Cartesian coordinates.

What comes out of the ensuing analysis is that the planar approximation is also accurate for the thick film region, so long as the capillary number is sufficiently small. With reference to Fig. 2-5.1, suppose that we consider the flow of a liquid bounded by a free surface that separates it from a gas phase. The position of the free surface is defined as (for a steady flow)

$$y_s = h(z). \tag{2-5.1}$$

The interface is characterized by a surface tension $\sigma$, and the liquid is Newtonian with a viscosity $\mu$. At some point on the free surface, where the height (in the axisymmetric case, the radius) is $h$, the surface has normal and tangential vectors **n** and **t**. In terms of the geometry the components of the normal vector, in the Cartesian coordinate system, are

$$\begin{aligned} n_z &= -h' \, (1 + h'^2)^{-1/2}, \\ n_y &= (1 + h'^2)^{-1/2}, \end{aligned} \tag{2-5.2}$$

where $h' = dh/dz$.

We can write the equations of conservation of mass and momentum for this flow, but our primary interest here is in the boundary conditions at a free surface with curvature. A purely kinematic condition is the assertion that fluid does not cross the free surface. We are considering only the steady-state case here, so the free surface is fixed in space. This implies that

$$\mathbf{u} \cdot \mathbf{n} = 0 = u_y n_y + u_z n_z \quad \text{along } h(z), \tag{2-5.3}$$

from which it follows that

$$u_y = h' u_z \quad \text{at } y = h(z). \tag{2-5.4}$$

The stress boundary conditions are more complicated. Physically, one boundary condition states that (for the case of a gas–liquid interface and low-speed relative motion) there is no shear stress at (tangential to) the interface. The other states that the stress normal to the interface must be balanced by a contribution

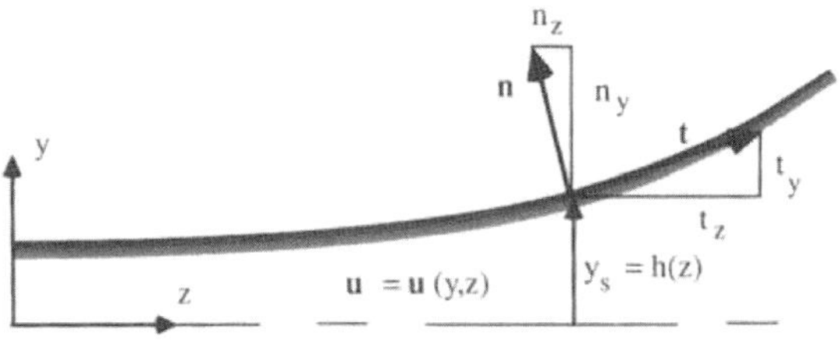


***FIGURE 2-5.1*** Geometry and dynamics near a free surface.

arising from surface tension, through the Young–Laplace equation. Let's set about writing these stress boundary conditions. We will suppose that the state of stress at some point on the free surface is determined by the stress tensor (within the liquid phase) **T**. We assume that the external fluid (taken to be a gas, here) is stress free. The stress vector normal to the free surface is given by **T·n**. The $z$- and $y$-directed components of **T·n** are

$$(\mathbf{T}\cdot\mathbf{n})_z = \mathsf{T}_{yz} n_y + \mathsf{T}_{zz} n_z$$

and

$$(\mathbf{T}\cdot\mathbf{n})_y = \mathsf{T}_{yy} n_y + \mathsf{T}_{yz}\, n_z . \tag{2-5.5}$$

The stress boundary conditions on the tangential and normal stresses are now written in the forms

$$(\mathbf{T}\cdot\mathbf{n})\cdot\mathbf{t} = 0$$

and

$$-(\mathbf{T}\cdot\mathbf{n})\cdot\mathbf{n} = \sigma\left(\frac{1}{R_1} + \frac{1}{R_2}\right). \tag{2-5.6}$$

From Fig. 2-5.1 we can note that

$$t_z = n_y \quad \text{and} \quad t_y = -\,n_z . \tag{2-5.7}$$

Now Eqs. 2-5.6 may be written as

$$\begin{aligned}(\mathbf{T}\cdot\mathbf{n})\cdot\mathbf{t} &= (\mathbf{T}\cdot\mathbf{n})_z\cdot t_z + (\mathbf{T}\cdot\mathbf{n})_y\cdot t_y = 0\\ &= \mathsf{T}_{yz} n_y^2 + \mathsf{T}_{zz} n_y n_z - \mathsf{T}_{yy} n_y n_z - \mathsf{T}_{yz} n_z^2\end{aligned}$$

and

$$\begin{aligned}(\mathbf{T}\cdot\mathbf{n})\cdot\mathbf{n} &= (\mathbf{T}\cdot\mathbf{n})_z\cdot n_z + (\mathbf{T}\cdot\mathbf{n})_y\cdot n_y = -\,\sigma\left(\frac{1}{R_1} + \frac{1}{R_2}\right)\\ &= \mathsf{T}_{yz} n_y n_z + \mathsf{T}_{zz} n_z^2 + \mathsf{T}_{yy} n_y^2 + \mathsf{T}_{yz} n_z n_y .\end{aligned} \tag{2-5.8}$$

We now introduce the Newtonian model for the stress components and after quite a bit of algebra we can write these equations in the following forms. For the boundary condition on the vanishing of the shear stress at the free surface we find

$$\mu\left(\frac{\partial u_z}{\partial y} + \frac{\partial u_y}{\partial z}\right)\left(1 - h'^2\right) - 2\mu\left(\frac{\partial u_z}{\partial z} - \frac{\partial u_y}{\partial y}\right) h' = 0. \tag{2-5.9}$$

For the condition on the normal stress (the second equation of Eqs. 2-5.8 above) the result is

$$\frac{2\mu}{\left(1 + h'^2\right)}\left[h'\left(\frac{\partial u_z}{\partial y} + \frac{\partial u_y}{\partial z}\right) - \frac{\partial u_y}{\partial y} - \frac{\partial u_z}{\partial z}\, h'^2\right] + p = \frac{-\,\sigma\eta''}{\left(1 + h'^2\right)^{3/2}} . \tag{2-5.10}$$

In the limit of the small-slope approximation, where all terms involving $h'$ may be neglected, we obtain the expected results:

$$\frac{\partial u_z}{\partial y} = 0 \quad \text{from Eq. 2-5.9} \tag{2-5.11}$$

and

$$-2\mu \frac{\partial u_y}{\partial y} + p = -\sigma h'' \quad \text{from Eq. 2-5.10.} \tag{2-5.12}$$

This latter equation is, in different notation, Eq. 2-4.5.

Now let's return to the formulation of the equations that describe the "Bretherton problem": the displacement of a viscous liquid by a gas bubble passing down the axis of a tube. We follow here the development given by Park and Homsy (1984), but not at their level of detail. They consider the problem of displacement of a viscous liquid by a gas bubble in a system confined between *parallel planes*. To the level of the approximations employed in the analysis, this turns out to yield the same equations that describe the bubble-in-a-tube problem. Figure 2-5.2 shows the geometry.

We use a moving coordinate system that immobilizes the bubble. The rigid boundary moves to the left at velocity $-U$. The flow field is divided into three regions. Far enough in the $-z$ direction (the "viscous" region) the film is of uniform thickness $H$ and the fluid moves in a lubrication-type flow with the velocity $-U$ at the rigid surface. The fluid does not move as a rigid body attached to the lower surface, however. The deformation field induced by the motion of the bubble creates a pressure gradient that pushes the fluid toward the left.

Far toward the right of the bubble the fluid is being pushed with respect to the rigid boundary. In a coordinate system fixed to that boundary the flow far to the right of the bubble would be Poiseuille flow. Near the bubble surface the fluid must accelerate into the thin-film viscous region through a region that we call

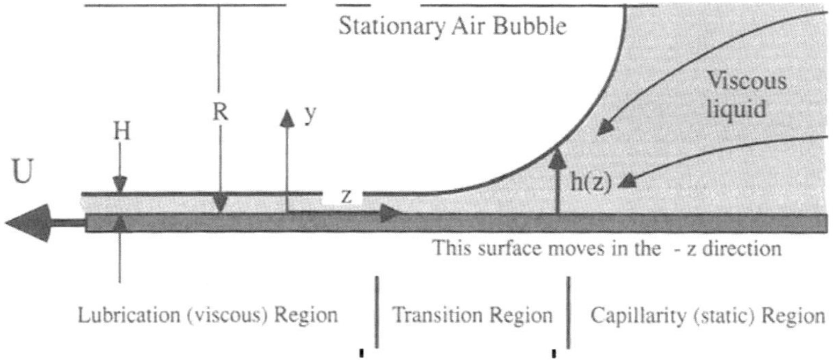


***FIGURE 2-5.2*** Flow regimes for analysis of deposition of a liquid film.

the "transition" region. It is in this region that viscous effects give rise to a pressure gradient that "pumps" fluid toward the lubrication region. Where the curvature of the bubble surface is significant, surface tension can alter the pressure field and affect the flow. If $H$ is small enough (< 1000 Å) van der Waals forces become significant and give rise to a "disjoining pressure" (see Chapter 9) that affects the flow. An analysis of this problem that includes disjoining pressure may be found in the paper of Kalliadasis and Chang (1994a).

### 2.5.1 The Transition Region

We begin with an analysis of the fluid dynamics in the transition region, where we expect that viscous and surface tension forces are comparable. We will write the Navier–Stokes equations and the continuity equation in nondimensional form. A natural choice of a length scale for nondimensionalizing the $y$ coordinate is the residual film thickness. Taking note of Eq. 2-4.14 we see that this leads to the definition

$$\frac{y}{R\mathrm{Ca}^{2/3}} \equiv \tilde{y}. \tag{2-5.13}$$

In the transition region where surface tension is important a natural choice of a pressure scale leads to

$$\frac{p}{\sigma/R} \equiv \tilde{p}. \tag{2-5.14}$$

The velocity $U$ seems appropriate for nondimensionalizing the $z$ component of velocity.

$$\frac{u_z}{U} \equiv \tilde{u}_z. \tag{2-5.15}$$

It is not so clear what we should choose as a length scale in the $z$ direction and a velocity scale in the $y$ direction. For the moment we will simply define the following:

$$\frac{u_y}{U_Y} \equiv \tilde{u}_y \quad \text{and} \quad \frac{z}{L_Z} \equiv \tilde{z}, \tag{2-5.16}$$

but we do not specify our choices of the scales $U_Y$ and $L_Z$. With these definitions the continuity equation becomes

$$\frac{\partial \tilde{u}_y}{\partial \tilde{y}} + \mathrm{N}_1 \frac{\partial \tilde{u}_z}{\partial \tilde{z}} = 0, \tag{2-5.17}$$

where $\mathrm{N}_1$ is defined as

$$\mathrm{N}_1 \equiv \frac{UR\,\mathrm{Ca}^{2/3}}{L_Z\,U_Y}. \tag{2-5.18}$$

Since the flow is two-dimensional in the transition region, both terms in Eq. 2-5.17 are nonzero and hence must be of the same order of magnitude. Thus $\mathsf{N}_1$ must be of order unity, and we arbitrarily set its value exactly to

$$\mathsf{N}_1 = 1 \tag{2-5.19}$$

This provides a constraint between the two unspecified scale factors: $L_Z$ and $U_Y$.

Now let's look at the $z$ component of the momentum equation. We assume that the Reynolds number of this flow is so small that the inertial terms may be ignored. Then we begin with

$$0 = -\frac{\partial p}{\partial z} + \mu\left(\frac{\partial^2 u_z}{\partial y^2} + \frac{\partial^2 u_z}{\partial z^2}\right), \tag{2-5.20}$$

which takes the nondimensional form

$$0 = -\frac{\mathsf{N}_2}{\mathsf{Ca}^{1/3}}\frac{\partial \tilde{p}}{\partial \tilde{z}} + \frac{\partial^2 \tilde{u}_z}{\partial \tilde{y}^2} + \mathsf{N}_2^2\frac{\partial^2 \tilde{u}_z}{\partial \tilde{z}^2}, \tag{2-5.21}$$

where

$$\mathsf{N}_2 \equiv \frac{\mathsf{Ca}^{2/3}}{L_Z / R}. \tag{2-5.22}$$

Since the flow is largely in the $z$ direction we expect that the dominant stress terms of this momentum balance are the pressure and the shear stress and that the third term of Eq. 2-5.21 is smaller than the other two. This suggests that an appropriate choice for $\mathsf{N}_2$ is one that makes the first two terms comparable in magnitude. The simplest choice is

$$\mathsf{N}_2 \equiv \mathsf{Ca}^{1/3}, \tag{2-5.23}$$

from which it follows that

$$\frac{L_Z}{R} \equiv \mathsf{Ca}^{1/3} \tag{2-5.24}$$

and, in view of Eq. 2-5.18 and 19,

$$\frac{U_Y}{U} \equiv \mathsf{Ca}^{1/3}. \tag{2-5.25}$$

Then Eq. 2-5.21 becomes

$$0 = -\frac{\partial \tilde{p}}{\partial \tilde{z}} + \frac{\partial^2 \tilde{u}_z}{\partial \tilde{y}^2} + \mathsf{Ca}^{2/3}\frac{\partial^2 \tilde{u}_z}{\partial \tilde{z}^2}. \tag{2-5.26}$$

Finally, this permits us to write the $y$ component of the momentum equation in the form

$$0 = -\frac{\partial \tilde{p}}{\partial \tilde{y}} + \mathsf{Ca}^{2/3}\frac{\partial^2 \tilde{u}_y}{\partial \tilde{y}^2} + \mathsf{Ca}^{4/3}\frac{\partial^2 \tilde{u}_y}{\partial \tilde{z}^2}. \tag{2-5.27}$$

Now we must examine the boundary conditions for this flow. On the solid (lower) surface we impose a no-slip condition:

$$\tilde{u}_y = 0 \quad \text{and} \quad \tilde{u}_z = -1 \quad \text{along } \tilde{y} = 0. \tag{2-5.28}$$

On the free surface we have the kinematic condition (Eq. 2-5.4) which, in nondimensional form, becomes simply

$$\tilde{u}_y = \tilde{h}' \, \tilde{u}_z \quad \text{at } \tilde{y} = \tilde{h}(\tilde{z}), \tag{2-5.29}$$

where now $'$ means $d/d\tilde{z}$ in all dimensionless equations and a dimensionless film thickness variable is defined as (note Eq. 2-5.13)

$$\tilde{h} \equiv \frac{h}{R\,\text{Ca}^{2/3}}. \tag{2-5.30}$$

The stress boundary conditions (Eqs. 2-5.9 and 10) become nondimensionalized to the forms

$$\left(\text{Ca}^{2/3}\frac{\partial \tilde{u}_z}{\partial \tilde{y}} + \frac{\partial \tilde{u}_y}{\partial \tilde{z}}\right)\left(1 - \tilde{h}'^2\right) - 2\left(\frac{\partial \tilde{u}_z}{\partial \tilde{z}} - \frac{\partial \tilde{u}_y}{\partial \tilde{y}}\right)\tilde{h}' = 0 \tag{2-5.31}$$

and

$$\frac{2\,\text{Ca}^{2/3}}{\left(1 + \text{Ca}^{2/3}\tilde{h}'^2\right)}\left[\tilde{h}'\left(\frac{\partial \tilde{u}_z}{\partial \tilde{y}} + \text{Ca}^{2/3}\frac{\partial \tilde{u}_y}{\partial \tilde{z}}\right) - \frac{\partial \tilde{u}_y}{\partial \tilde{y}} - \text{Ca}^{2/3}\frac{\partial \tilde{u}_z}{\partial \tilde{z}}\tilde{h}'^2\right]$$

$$+\tilde{p} = \frac{-\tilde{h}''}{\left(1 + \text{Ca}^{2/3}\tilde{h}'^2\right)^{3/2}}. \tag{2-5.32}$$

Having derived these equations, we are faced with the dilemma of solving them. Since there is no analytical solution we seek some approximation technique. We will use a *perturbation* technique, in which it is *assumed* that all variables can be expanded in power series in the single parameter that characterizes the system (the only parameter that appears in the equations above): the capillary number. Thus we write the following:

$$\tilde{h} = \sum_{n=0}^{\infty} \left(\text{Ca}^{1/3}\right)^n h_n(z),$$

$$\tilde{p} = \sum_{n=0}^{\infty} \left(\text{Ca}^{1/3}\right)^n p_n(y,z),$$

$$\tilde{\mathbf{u}} = \sum_{n=0}^{\infty} \left(\text{Ca}^{1/3}\right)^n \mathbf{u}_n(y,z). \tag{2-5.33}$$

The functions $h_n$, $p_n$, and $\mathbf{u}_n$ (= $[u_{yn}, u_{zn}]$ ) are the perturbation functions of order $n$, corresponding to the powers of capillary number in the series expansions above. They are dimensionless, as are their arguments, but we stop writing the ~ just to simplify the notation. A further notational simplification follows if we define

$$\delta \equiv \mathrm{Ca}^{1/3} \tag{2-5.34}$$

as the perturbation parameter.

The procedure is as follows. We substitute Eqs. 2-5.33 into the dynamic equations and boundary conditions given above, and we group terms with like powers of $\delta$ into separate sets of equations. We begin with the lowest-order equations, corresponding to the $n = 0$ terms. The results are as follows:

$$\frac{\partial u_{y0}}{\partial y} + \frac{\partial u_{z0}}{\partial z} = 0, \tag{2-5.35}$$

$$0 = -\frac{\partial p_0}{\partial z} + \frac{\partial^2 u_{z0}}{\partial y^2}, \tag{2-5.36}$$

$$0 = -\frac{\partial p_0}{\partial y}. \tag{2-5.37}$$

The boundary conditions are, to the lowest order,

$$u_{z0} = -1 \quad \text{and} \quad u_{y0} = 0 \quad \text{at } y = 0, \tag{2-5.38}$$

$$u_{z0}\, h_0' = u_{y0} \quad \text{and} \quad \frac{\partial u_{z0}}{\partial y} = 0 \quad \text{at } y = h_0(z), \tag{2-5.39}$$

$$-p_0 = h_0'' \quad \text{at } y = h_0(z). \tag{2-5.40}$$

We see that Eqs. 2-5.36 and 37 are identical to Eqs. 2-4.3, and Eq. 2-5.40 is the same as Eq. 2-4.5. Hence the lowest-order expansion (terms multiplying $\delta^0$) is exactly the problem set up by Bretherton. The solution for the velocity profile follows his analysis then, and in the dimensionless notation we end up with a differential equation for the film profile of the form:

$$\frac{d^3 h_0}{dz^3} = 3\,\frac{h_0 - t_0}{h_0^3}. \tag{2-5.41}$$

Note that this equation is of the same form as Eq. 2-4.10, although it corresponds to a slightly different nondimensionalizaton. We must keep in mind that $t_0$ is an unknown in this formulation. Because Eq. 2-5.41 is nonlinear we are not able to write down an analytical solution. It must be solved numerically. One boundary condition is that the solution for the film thickness in the transition region must match the thickness in the uniform region asymptotically as the lubrication region is approached. This implies

$$h_0 \to t_0 \quad \text{as} \quad z \to -\infty. \tag{2-5.42}$$

We could solve Eq. 2-5.41 numerically by a "marching" technique, in which we take $h_0 = t_0$ as an initial condition and assume initial values for the first and second derivatives of $h_0$. We then integrate in the $+z$ direction until we find, for some choice of $t_0$, a solution that satisfies boundary condition 2-5.40 at the point where the film thickness is the half-width of the channel, or (note Eq. 2-5.30)

$$h_0'' = -p_0 \quad \text{at } y = \frac{1}{\mathrm{Ca}^{1/3}}. \tag{2-5.43}$$

If we assume that the zeroth-order solution corresponds to the static meniscus, with a circular cross section, then the dimensionless pressure (Eq. 2-5.14, but keep the notation straight!) $p_0$ is just

$$p_0 = 1. \tag{2-5.44}$$

Hence the numerical task is simply to find, by trial and error, the value of $t_0$ that yields a solution to Eq. 2-5.41 that satisfies Eq. 2-5.43. A more detailed discussion of methods of solving this problem can be found in Park and Homsy (1984) and Teletzke *et al.* (1988). When this procedure is completed we obtain the desired result for the residual film thickness:

$$\frac{H_0}{R} = 1.337\ \mathrm{Ca}^{2/3}, \tag{2-5.45}$$

which compares quite well with the data shown in Fig. 2-4.3.

## 2-6 DEPLETION OF AN ANNULAR LIQUID FILM

We return finally to the problem introduced in Section 2-2, but relax the assumption that the annular liquid is replenished at the same rate as it is sheared out of the downstream end of the tube. We assume now that there is an initially uniform annular liquid film which is removed by the shearing action of the core flow. The annular film thickness is reduced in time, and the film thickness varies along the axis of the tube.

Figure 2-6.1 shows the situation of interest. A long tube of uniform circular cross section is coated with a thin, initially uniform film of an oil. By the qualifier "thin" we imply that the film thickness $H(z,t)$ always satisfies $H(z,t) << R$. The oil film is being constantly sheared by a turbulent immiscible flushant fluid, which in turn causes a shear flow in the film. Locally, the average velocity $U_f$ in the oil film, assuming a thin film exhibiting planar Couette flow, is

$$U_f = \frac{U_s}{2}, \tag{2-6.1}$$

where the velocity at the oil–flushant interface $U_s$ is given by the shear stress relation

$$\tau_0 = \mu_{oil}\, U_s/H. \tag{2-6.2}$$

We can find the shear stress $\tau_0$ from the definition of the friction factor for the flushing fluid:

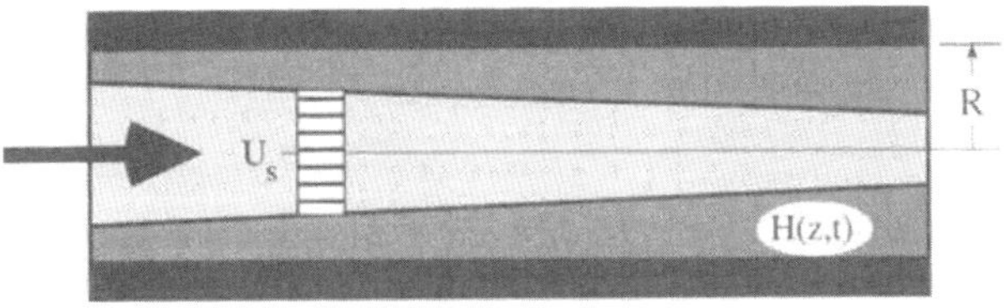


***FIGURE 2-6.1*** The annular liquid film is being sheared by the core flow. (The film thickness of the annular film is exaggerated in this sketch.)

$$f = \frac{\tau_0}{\frac{1}{2}\rho\left[Q/\pi R^2\right]^2}, \tag{2.6.3}$$

where $\rho$ is the density of the flushant, $Q$ is the flushant flow rate, and $R$ is the radius of the tube.

A local mass balance on the oil film shows that the transient film thickness is a function of axial position:

$$\frac{\partial H}{\partial t} + \frac{\partial q}{\partial z} = 0, \tag{2-6.4}$$

where $q$ is the volume flow rate (per unit width) within the annular film. If we assume that the flow in this film is locally planar Couette flow, then

$$q = \tfrac{1}{2} U_s H(z,t) = \frac{H^2(z,t)\,\tau_0}{2\mu_{oil}}. \tag{2-6.5}$$

Hence $H$ satisfies a partial differential equation of the form

$$\frac{\tau_0}{\mu_{oil}} H(z,t)\frac{\partial H}{\partial z} = -\frac{\partial H}{\partial t}. \tag{2-6.6}$$

While this differential equation admits an analytical solution ($H \quad \mu_{oil}\, z/\tau_0 t$), it is not a solution that satisfies physically meaningful boundary and initial conditions.

In view of the fact that we will measure the *average* $\overline{H}(t)$ in our experiments, we first average Eq. 2-6.4 over the tube length $L$ as follows:

$$\int_0^L \frac{\partial H}{\partial t}\,dz = -\int_{q_0}^{q_L} dq. \tag{2-6.7}$$

Then

$$\frac{d}{dt}\int_0^L H\,dz = q_0 - q_L$$

or

$$L\frac{d\overline{H}}{dt} = -\ q_L = \frac{-\ \tau_o H_L^2}{2\ \mu_{oil}}. \tag{2-6.8}$$

We have used the fact that no annular fluid enters the tube, so $q_0 = 0$. Equation 2-6.8 is a simple ordinary differential equation for the average film thickness, but its solution requires a value for the downstream film thickness, $H_L$. Since this parameter is itself unknown, we will solve Eq. 2-6.8 by first finding an approximation for $H_L$.

The approximation procedure begins with the assumption that the film thickness is everywhere uniform along the $z$ axis of the tube. (Of course, this approximation is inconsistent with Eq. 2-6.4.) Then an oil balance gives

$$U_f\, 2\pi RH = -2\pi RL\ dH/dt, \tag{2-6.9}$$

where $H$ is the (assumed uniform) film thickness at any time. The introduction of Eqs. 2-6.1 and 2 into Eq. 2-6.9 yields:

$$-\frac{dH}{dt} = \frac{H}{2L}U_s = \frac{H}{2L}\frac{\tau_0 H}{\mu_{oil}} \tag{2-6.10}$$

or

$$\frac{d(1/H)}{dt} = \frac{\tau_0}{2L\mu_{oil}}. \tag{2-6.11}$$

Now we rewrite Eq. 2-6.6 in the form

$$\frac{1}{H}\frac{\partial H}{\partial z} = \frac{\mu_{oil}}{\tau_o}\frac{\partial}{\partial t}\left(\frac{1}{H}\right) = \frac{1}{2L}. \tag{2-6.12}$$

This has the solution

$$\ln\left(\frac{H}{H_L}\right) = \frac{1}{2L}(z - L) \tag{2-6.13}$$

or

$$\frac{H}{H_L} = e^{-1/2}\, e^{z/2L}. \tag{2-6.14}$$

The average of $H$ then follows from

$$\frac{\overline{H}}{H_L} = e^{-1/2}\int_0^1 e^{z/2L}\, d\left(\frac{z}{L}\right) = 0.787 \tag{2-6.15}$$

or

$$H_L = 1.27\overline{H}. \tag{2-6.16}$$

Returning to Eq. 2-6.8 we may now write

$$\frac{d\overline{H}}{dt} = \frac{-0.807\tau_0 \overline{H}^2}{\mu_{oil}L}. \tag{2-6.17}$$

We find, finally,

$$\frac{H_0}{\overline{H}} - 1 = \frac{0.807\tau_0 H_0}{\mu_{oil}L} t,$$

or

$$\frac{1}{\varepsilon} - 1 = 0.807\ t^*, \tag{2-6.18}$$

where $H_0$ is the initial film thickness, $t^*$ is a dimensionless flushing time variable, and

$$\varepsilon \equiv \frac{\overline{H}}{H_0} \tag{2-6.19}$$

is the fractional reduction in the average film thickness. At this point we have a testable model. The numerical coefficient given in Eq. 2-6.18 (0.807) is approximate. Hence we will test the model by plotting data in the form $(1/\varepsilon - 1)$ vs. $t^*$.

Figure 2-6.2 shows data of Mickaily and Middleman (1993) which indicate that Eq. 2-6.18 fits the data quite well, with the coefficient 0.807 replaced by 0.59.

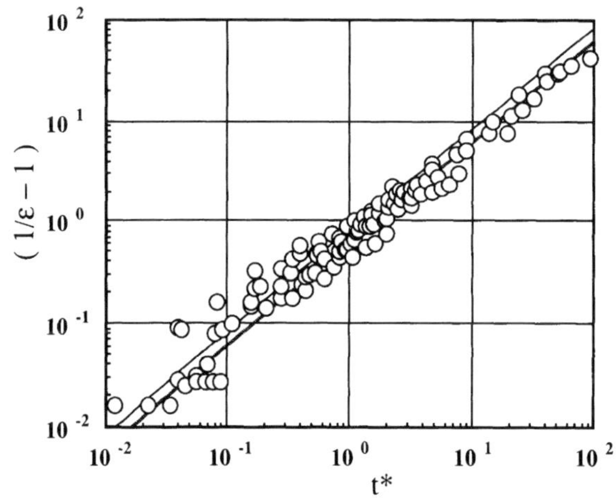


***FIGURE 2-6.2*** An experimental test of Eq. 2-6.18. The upper (thin) line is with the coefficient 0.807. The lower (heavy) line is with the coefficient 0.59. Data were taken in tubes with radii in the range 2 to 3 mm, with a length of 23 cm. Data from Mickaily and Middleman, 1993.

## Problems

P2-1 In deriving Eqs. 2-1.9 and 2-1.10 we treated the two flows as pressure-driven flows, with each fluid subjected to the same pressure gradient $\Delta P/L$. Carry out the following alternative approach. Consider the flow in the annular layer (fluid II) to be strictly a drag flow induced by the shear stress exerted on the interface between the two fluids. Does the final result for $Q^{II}$ differ from that given by Eq. 2-2.1? Should the flow of the annular layer be regarded as a combination of pressure and drag flow? Explain your answer.

P2-2 Give the asymptotic behavior of $\Psi$ (Eq. 2-1.14) in the limit of small values of $\eta$.

P2-3 Figure 2-1.2 shows that there is an optimum film thickness at which the ability of an annular layer to enhance the flow of the inner fluid is maximized. Find and plot the optimum $\eta$, and the value of $\Psi$ at the optimum, as a function of $M$.

P2-4 Recast Eq. 2-2.3 in the form of $\eta/\eta_0$ as a function of a dimensionless time $\tau$. Give the definition of $\tau$. What value of $\tau$ corresponds to a tenfold reduction in $\eta$ from its initial value $\eta_0$?

P2-5 Solve Eq. 2-2.1 for the case that $Q^{I}$ is somehow maintained at a constant value as $\eta$ changes. *Hint*: Use Eq. 2-1.13 to relate the changing $\Delta P$ to $\eta$, and then replace $\Delta P$ in Eq. 2-2.1.

P2-6 Develop a general expression for the radial velocity of the peak of a disturbance growing at the maximum growth rate. Give the specific result for the case of an inviscid annular layer. Suppose that a layer of water, 1 μm thick, is created on the inside of a 25-μm inside radius capillary. At what time would you expect to observe the liquid "bridge" across the radius? What important but unknown parameter of the instability process do you need to know to answer this question? How sensitive is your answer to the magnitude of this unknown parameter?

P2-7 In the experiment of Aul and Olbricht described in Section 2-3, the core fluid (water) had an average velocity through the capillary of 448 μm/s, and the peaks of the waves moved at a velocity of less than 1 μm/s.

(a) What is the Reynolds number of the core flow? Using the analysis of Section 2-2, predict the velocity of the interface of the annular layer, assuming that the layer is stable. Is the result consistent with the observation noted just above?

(b) Over what time scale would the annular layer thickness be reduced by 10% of its initial value?

(c) Calculate the maximum growth rate of the fastest-growing disturbance, and estimate the time required for a bridge to occur. How does that time compare to the time calculated in part (b) above?

P2-8 When is there enough liquid in a very thin annular film to bridge the radius? There is no simple answer to this question, but a simple geometrical analysis can aid the thinking on this issue. Suppose that the "bridge" is really a "disk" of liquid, of radius $R$ and width $W$. (This assumed geometry ignores the effects of the contact angle that the liquid makes with the inside surface of the capillary.) For values of $\alpha$ in the range $0.9 < \alpha < 0.99$, find the ratio $W/R$. What is the significance of small values of $W/R$ to the question of the formation of bridges?

P2-9 We want to coat the inside of a capillary with a thin layer of a silicone oil. The procedure will be as follows. We will fill the capillary completely with the oil and then, while the capillary is horizontal, we will displace the oil by maintaining a constant flow of air into one end of the tube. If we wish to have a film thickness of 1 μm, in a capillary of inside diameter 100 μm, what flow rate of air should be used? What is the Reynolds number of this air flow? If this air flow is maintained, how long a time is required for the film to show waves of amplitude equal to 10% of the film thickness? If this air flow is maintained, how long a time is required for the shear stress exerted by the air flow (neglected in the analysis of Bretherton) to drive 10% of the oil out of the tube? The silicone oil has a viscosity of 1 Pa-s, and an interfacial tension of 0.019 N/m. The density is 950 kg/m$^3$.

P2-10 Figure 2-4.3 suggests a limiting film thickness at very low values of the capillary number. What is the magnitude of the residual film thickness in these experiments? Does this suggest any mechanisms to explain the existence of a limiting value?

P2-11 Show that a circle of radius $R$ is a curve, $y(z)$, of constant curvature; that is, prove that

$$\frac{y''}{\left(1 + y'^2\right)^{3/2}} = \text{constant} = \frac{1}{R}. \qquad \text{(P2-11.1)}$$

P2-12 In discussing the static meniscus we use Eq. 2-5.43. What has become of the term $(1 + h'^2)^{3/2}$ that should be in the denominator?

P2-13 The numerical solution to Eq. 2-5.41 is obtained more efficiently if good estimates of the first two derivatives of $h_0$ are available in the lubrication region. Show that in the region where Eq. 2-5.42 is a good approximation, the solution to Eq. 2-5.41 that could satisfy Eq. 2-5.46 is

$$h_0 = t_0 + A \exp\left(3^{1/3} z\right).$$

What do different values of the constant $A$ correspond to, and why can we arbitrarily set it to $A = 1$?

P2-14 Note and discuss the differences between Eqs. 2-3.17 and 1-1.11.

P2-15 Read the paper by Arney *et al.* (1993). Does the model presented in Section 2-1 apply to their data? To answer, first recast the model of

Section 2-1 into a friction factor vs. Reynolds number relationship, and compare your result to the corresponding result of Arney *et al.*

## References


Arney, M. S., R. Bai, E. Guevera, D. D. Joseph, and K. Liu. (1993). Friction factor and holdup studies for lubricated pipelining—I. Experiments and correlations. *Int. J. Multiphase Flow.* **19**, 1061.

Atherton, R. W. and G. M. Homsy. (1976). On the derivation of evolution equations for interfacial waves. *Chem. Eng. Comm.* **2**, 57.

Aul, R. W., and W. L. Olbricht. (1990). Stability of a thin annular film in pressure-driven, low-reynolds number flow through a capillary. *J. Fluid Mech.* **215**, 585.

Bai, R., K. Chen, and D. D. Joseph. (1992). Lubricated pipelining. Stability of core-annular flow. Part 5. Experiments and comparison with theory. *J. Fluid Mech.* **240**, 97.

Bretherton, F. P. (1961). The motion of long bubbles in tubes. *J. Fluid Mech.* **10**, 166.

Cox, B. G. (1962). On driving a viscous fluid out of a tube. *J. Fluid Mech.* **14**, 81.

Everett, D. H., and J. M. Haynes. (1972). Model studies of capillary condensation I. Cylindrical pore model with zero contact angle. *J. Colloid Interface Sci.* **38**, 125.

Gauglitz, P. A., and C. J. Radke. (1988). An extended evolution equation for liquid film breakup in cylindrical capillaries. *Chem. Eng. Sci.* **43**, 1457.

Goren, S. L. (1962). The instability of an annular thread of fluid. *J. Fluid Mech.* **12**, 309.

Goren, S. L. (1964). The shape of a thread of liquid undergoing breakup. *J. Colloid Sci.* **19**, 81.

Hammond, P. S. (1983). Nonlinear adjustment of a thin annular film of viscous fluid surrounding a thread of another within a circular cylindrical pipe. *J. Fluid Mech.* **13**, 363.

Kalliadasis, S., and H.-C. Chang. (1994a). Apparent dynamic contact angle of an advancing gas-liquid meniscus. *Phys. Fluids* **6**, 12 .

Kalliadasis, S., and H.-C. Chang. (1994b). Drop formation during coating of vertical fibres. *J. Fluid Mech.* **261**, 135 .

Mickaily, E. S., and S. Middleman. (1993). Hydrodynamic cleaning of a viscous film from the inside of a long tube. *AIChE J.* **39**, 885.

Newhouse, L. A., and C. Pozrikidis. (1992). The capillary instability of annular layers and liquid threads. *J. Fluid Mech.* **242**, 193.

Park. C.-W., and G. M. Homsy. (1984). Two-phase displacement in Hele Shaw cells. *J. Fluid Mech.* **139**, 291.

Quere, D. (1990). Thin films flowing on vertical fibers. *Europhys. Lett.* **13**, 721.

Schwartz, L. W., H. M. Princen, and A. D. Kiss. (1986). On the motion of bubbles in capillary tubes. *J. Fluid Mech.* **172**, 259.

Taylor, Sir G. I. (1961). Deposition of a viscous fluid on the wall of a tube. *J. Fluid Mech.* **10**, 161.

Teletzke, G. F., H. T. Davis, and L. E. Scriven (1988). Wetting hydrodynamics. *Rev. Phys. Appl.* **23**, 989.

Tsai, T. M., and M. J. Miksis (1994). Dynamics of a drop in a constricted capillary tube. *J. Fluid Mech.* **274**, 197.

Wilson, S. D. R. (1982). The drag-out problem in film coating theory. *J. Eng. Math.* **16**, 209.

# 3 *Steady Laminar Axisymmetric Flow from a Tube: Exit Effects*

The *axisymmetric flow* of an isothermal incompressible Newtonian fluid issuing from a tube of uniform circular cross section represents another complex laminar flow for which an exact solution of the full Navier–Stokes equations is impossible. In this case the complexity is due not only to the presence of the inertial terms that reflect the dynamics of the exit region, but as well to the fact that the exit flow is a "free boundary" flow—one for which the interface of the free jet, the bounding surface of the flow, is itself unknown. Even in the limit of vanishing Reynolds number (creeping flow) this free boundary character makes an analytical approach impossible.

We consider two cases here. One is the jet issuing from a horizontal tube at a high enough speed (but still laminar) that its trajectory is approximately horizontal over the dynamic region of interest. The other is the vertical jet for which the key dynamic issue is the "draw-down" of the jet as gravity causes it to accelerate in the (axial) direction of flow.

## 3-1 THE HORIZONTAL JET

Figure 3-1.1 shows the geometry of the flow field. Gravity is neglected. If the surrounding fluid is air, the shear stress exerted by the surrounding air on the free surface of the liquid jet would be negligible. Hence at the boundary $z = 0$ the shear stress at the radial position $r = R_0$ changes discontinuously from a finite value to zero.

The simplest analysis we can display shows that we can expect the jet to have a downstream radius different from that of the tube from which the liquid issues. This follows from a simple and approximate momentum balance. We assume first that the liquid has a fully developed Poiseuille profile right at the tube exit.

$$\frac{u_z}{U_0} = 2\left[1 - \left(\frac{r}{R_0}\right)^2\right]. \tag{3-1.1}$$

Far downstream, if there is no shear stress exerted on the free surface of the jet, the velocity profile must flatten and the liquid then moves in plug flow:

$$u_z = U_\infty. \tag{3-1.2}$$

Now we may impose the principles of conservation of mass and axial momentum. For an incompressible fluid conservation of mass requires that

$$R_\infty^2 U_\infty = R_0^2 U_0, \tag{3-1.3}$$

while conservation of momentum is somewhat more complicated. If we assume that the velocity profile is the Poiseuille profile, right up to the tube exit, and if we ignore any other forces that might act on the boundaries of the jet, we simply equate the momentum fluxes at the tube exit and some position downstream where the profile has flattened, and find

$$\int_0^{R_0} 4\rho U_0^2 \left[1 - \left(\frac{r}{R_0}\right)^2\right]^2 r\,dr = \int_0^{R_\infty} \rho U_\infty^2 r\,dr. \tag{3-1.4}$$

Combining these last two equations we find

$$\chi \equiv \frac{R_\infty}{R_0} = \frac{\sqrt{3}}{2} = 0.866. \tag{3-1.5}$$

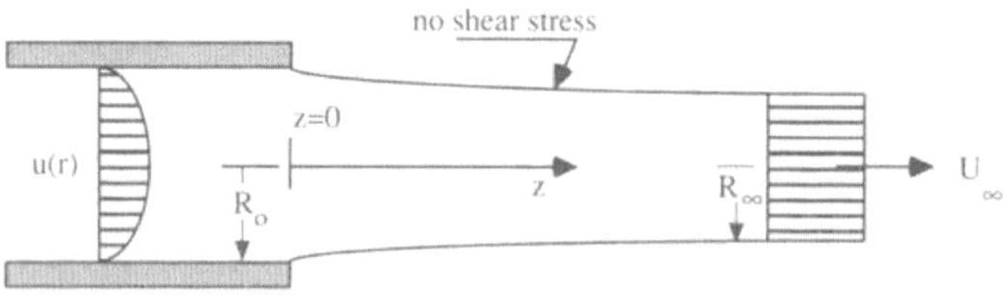


***FIGURE 3-1.1*** A jet of liquid leaving a tube.

This result was originally derived by Harmon (1955). Figure 3-1.2 shows experimental data (Middleman and Gavis, 1961) for several Newtonian liquids issuing in laminar flow from horizontal capillaries. [Goren and Wronski (1966) show additional data consistent with Fig. 3-1.2.] The predicted value of $\chi = 0.866$ is achieved only at large Reynolds numbers (though still well in the laminar flow regime).

Obviously, the simple momentum balance has failed to account for additional forces that are acting upon the jet. These forces are more significant as the Reynolds number gets smaller. The neglect of surface tension is one source of error in writing Eq. 3-1.4 in the simple form shown. However, this is a small error for most cases. A more significant error lies in the assumption that the velocity profile is given by Eq. 3-1.1 right up to the exit of the capillary. As soon as the jet is free of the confining wall of the capillary, the velocity profile is free to relax since there is no shear at the free boundary, and a no-slip condition of zero velocity on the boundary is no longer relevant. Hence the velocity profile along the jet axis, external to the tube, flattens toward the plug flow value of $u_z = U_\infty$. The rearrangement of the velocity profile can begin *before* the tube exit, because the vorticity associated with the axial deformation of the liquid at the tube exit can diffuse upstream and affect the velocity profile within the tube, just upstream from the exit. In order to account for these additional factors Eq. 3-1.4 must be rewritten as

$$\int_0^R \left[ \rho\, u_z^2(r,z=0) + p_0 - 2\mu \left(\frac{\partial u_z}{\partial z}\right)_{z=0} \right] r\, dr - \int_0^{R_\infty} \left(\rho U_\infty^2 + p_\infty\right) r\, dr = 2\pi R_0 \sigma - 2\pi R_\infty \sigma. \tag{3-1.6}$$

The terms on the right-hand side represent the axial components of surface forces that act on the boundaries of the jet. There is a force equal to $2\pi R_0 \sigma$ that

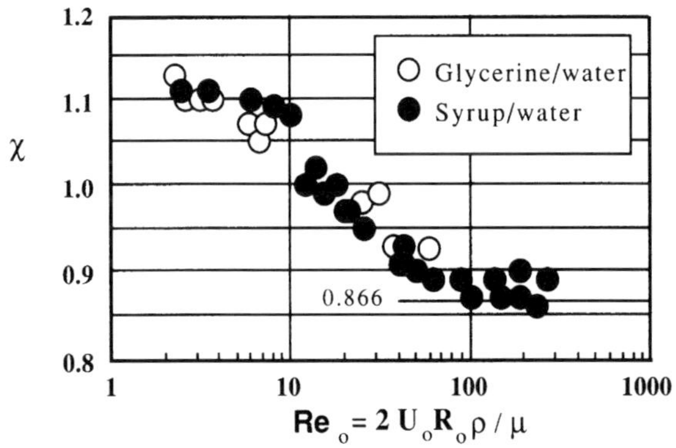


***FIGURE 3-1.2*** Expansion and contraction of capillary jets of newtonian liquids.

"pulls" the jet to the rim of the capillary, and an oppositely directed force $2\pi R_\infty \sigma$ that pulls on the perimeter of the jet at the downstream position where $R = R_\infty$.

According to the Young–Laplace equation the internal fluid pressures, relative to the atmospheric pressure external to the jet, are given by

$$p_0 = \frac{\sigma}{R_0} \quad \text{and} \quad p_\infty = \frac{\sigma}{R_\infty}. \tag{3-1.7}$$

In both of these expressions the surface curvature in a plane normal to the jet cross section, and aligned with the jet axis, is neglected as being small compared to the jet radius anywhere along the axis.

One cannot proceed further with Eq. 3-1.6 without a model for the exit velocity profile $u_z$ $(r, z = 0)$. Unfortunately, no analytical model is available to permit one to perform the integrations indicated in Eq. 3-1.6, and so no quantitative analytical model for comparison to Fig. 3-1.2 exists.

The goals of any theoretical analysis of this problem would include a prediction of the $\chi$ dependence on Re and We (see Problem P3-2), as well as predictions of the shape of the jet, and the downstream distance over which the change in jet radius is accomplished. Several analyses have dealt with one or more of these issues, and we review them here.

Goren and Wronski (1966) present several examples of measurements of the changing radius of a horizontal jet issuing from a capillary. Figure 3-1.3 shows two sets of data, selected to demonstrate both the expansion and contraction of a jet, depending upon the magnitude of the Reynolds number.

These authors also carry out an analytical approach to the prediction of the shape of a horizontal jet. They write the velocity and pressure distributions as perturbations about the final values in the downstream region. Hence they write

$$u_z = U_\infty + \hat{u} = U_\infty - \frac{1}{r}\frac{\partial \hat{\psi}}{\partial r}, \tag{3-1.8}$$

$$u_r = 0 + \hat{u}_r = \frac{1}{r}\frac{\partial \hat{\psi}}{\partial z}, \tag{3-1.9}$$

$$p = \frac{\sigma}{R_\infty} + \hat{p}, \tag{3-1.10}$$

$$R = R_\infty + \hat{R}. \tag{3-1.11}$$

In each case the perturbed quantity, denoted by the overmark ^, is taken to be small so that when these expressions are substituted into the Navier–Stokes equations the terms that are second order in these perturbed variables may be dropped from the equations. Note that the stream function is introduced, and as a result of the usual treatment of the Navier–Stokes equations that transforms them to stream function form we find

$$\nu D^4 \hat{\psi} - U_\infty \frac{\partial}{\partial z} D^2 \hat{\psi} = 0, \tag{3-1.12}$$

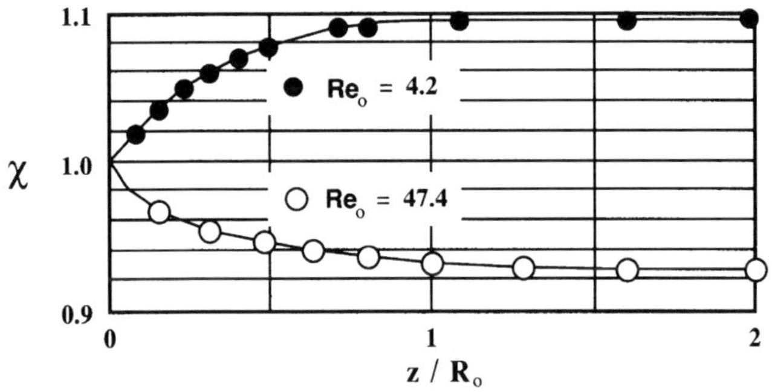


*FIGURE 3-1.3* Data of Goren and Wronski (1966) for laminar horizontal jets.

where, as in earlier chapters,

$$D^2\,\widehat{\psi} \;=\; \frac{\partial^2\widehat{\psi}}{\partial r^2} - \frac{1}{r}\frac{\partial\widehat{\psi}}{\partial r} + \frac{\partial^2\widehat{\psi}}{\partial z^2}. \tag{3-1.13}$$

Equation 3-1.12 is a Bessel equation, and the general solution, which involves the Bessel functions $J_1$ and $Y_1$, may be written as

$$\widehat{\psi} \;=\; \left\{\begin{array}{l} A\,\dfrac{\gamma\,r}{R_\infty}\,J_1\!\left(\dfrac{\gamma\,r}{R_\infty}\right) + B\,\dfrac{\gamma\,r}{R_\infty}\,Y_1\!\left(\dfrac{\gamma\,r}{R_\infty}\right) + \\ +\; C\,\dfrac{\gamma^*\,r}{R_\infty}\,J_1\!\left(\dfrac{\gamma^*\,r}{R_\infty}\right) + D\,\dfrac{\gamma^*\,r}{R_\infty}\,Y_1\!\left(\dfrac{\gamma^*\,r}{R_\infty}\right) \end{array}\right\} \exp\left(\frac{-\gamma\,z}{R_\infty}\right). \tag{3-1.14}$$

The parameter $\gamma$ is an eigenvalue for this boundary value problem, and

$$\gamma^{*\,2} \;=\; \gamma^2 + \tfrac{1}{2}\gamma\,\mathrm{Re}_\infty \quad \text{and} \quad \mathrm{Re}_\infty \equiv \frac{2R_\infty U_\infty}{\nu}. \tag{3-1.15}$$

Four boundary conditions are required in order to establish the coefficients $A$ through $D$ above. If the radial velocity is to vanish along the axis of the flow, and if the axial velocity is to remain finite along the axis, the coefficients $B$ and $D$ must be identically zero. The two remaining boundary conditions are on the stresses along the jet surface. One states that the shear stress vanishes at the free surface (the ambient medium is assumed to be incapable of exerting a shear stress) and the other condition states that there is a jump in normal stress across the free surface that arises because of surface tension and the curvature of the surface. In terms of the stream function these conditions may be shown to take the forms

$$\left\{D^2 - 2\,\frac{\partial^2}{\partial z^2}\right\}\widehat{\psi} \;=\; 0 \tag{3-1.16}$$

and

$$\int_{\infty}^{z} \frac{1}{r}\frac{\partial}{\partial r}\left\{D^2 - \frac{U_\infty}{\nu}\frac{\partial}{\partial z}\right\}\widehat{\psi}\, dz + 2\frac{\partial}{\partial r}\left(\frac{1}{r}\frac{\partial\widehat{\psi}}{\partial z}\right)$$

$$= \frac{\sigma}{\mu R_\infty^3 U_\infty}\left\{\widehat{\psi} + R_\infty^2 \frac{\partial^2\widehat{\psi}}{\partial z^2}\right\}. \tag{3-1.17}$$

The latter two conditions should be evaluated at the jet surface $R(z)$, but within the approximations employed here they are evaluated at $r = R_\infty$.

The jet radius may be written in terms of the stream function by using the kinematic condition that (for a steady-state flow)

$$u_r = U_\infty \frac{\partial\widehat{R}}{\partial z} = \left(\frac{1}{r}\frac{\partial\widehat{\psi}}{\partial z}\right)_{r = R_\infty} \tag{3-1.18}$$

or

$$\widehat{R} = \frac{1}{R_\infty U_\infty}\widehat{\psi}(R_\infty, z). \tag{3-1.19}$$

When the stream function from Eq. 3-1.14 (with $B = D = 0$) is substituted into Eqs. 3-1.16 and 17 a pair of linear homogeneous equations for $A$ and $C$ is obtained. By setting the determinant of the coefficients to zero an eigenvalue equation for the damping factor $\gamma$ is obtained. It may be written in the form

$$4\gamma^4\left[\frac{\gamma^* J_0(\gamma^*)}{J_1(\gamma^*)}\right] + 2\gamma^2\left(\gamma^{*2} - \gamma^2\right) - \left(\gamma^{*2} + \gamma^2\right)^2 \left[\frac{\gamma J_0(\gamma)}{J_1(\gamma)}\right] = \mathsf{S}_\infty\left(1 + \gamma^2\right)\gamma^2, \tag{3-1.20}$$

where $\mathsf{S}_\infty$ may be written (see Eq. 2-3.28)

$$\mathsf{S}_\infty = \mathsf{Oh}_\infty^{-2} = \frac{\rho R_\infty \sigma}{\mu^2}. \tag{3-1.21}$$

We see that $\mathsf{S}_\infty$ is just the square of the inverse of the Ohnesorge number (but based on the final radius).

There is an infinite number of roots to Eq. 3-1.20. Goren and Wronski assumed that the observed "decay" of the jet radius to its final value takes place according to the *smallest* value of $\gamma$ (the slowest decay rate). Thus Eq. 3-1.20 must be solved for the smallest eigenvalue, as a function of $\mathsf{Re}_\infty$ and $\mathsf{S}_\infty$. (They use $\mathsf{Re}_\infty$, the Reynolds number based on $R_\infty$ and $U_\infty$.) For very large values of $\mathsf{S}_\infty$, which from a practical point of view means for very low viscosity liquids, $\gamma$ is a single-valued function of $\mathsf{Re}_\infty$, as shown in Fig. 3-1.4. However, for smaller values of $\mathsf{S}_\infty$ that are more in the range of typical experimental conditions, the eigenvalues exhibit a peculiar behavior. This is illustrated by the limiting case $\mathsf{S}_\infty = 0$ in Fig. 3-1.5. For the Reynolds number range of about 16 to 20 the eigen-

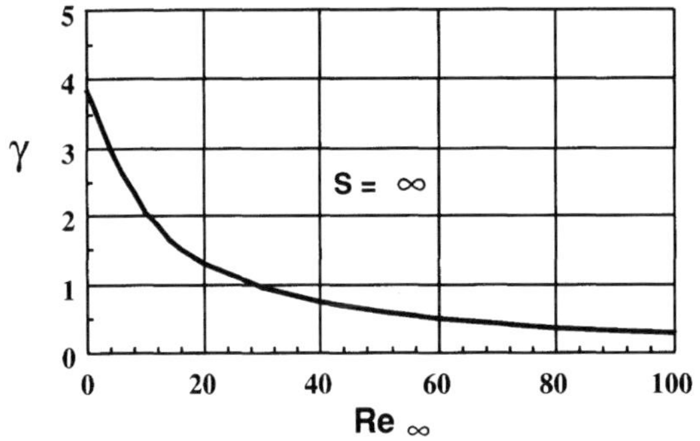


*FIGURE 3-1.4* The smallest eigenvalue $\gamma$ as a function of the Reynolds number, for $S = \infty$.

value $\gamma$ is multiple-valued. Further, there are bifurcation points from which complex values of $\gamma$ exist. The real parts of these complex roots are shown in Fig. 3-1.5.

The question to deal with here is whether this complicated behavior of $\gamma$ has any *physical* relevance. Goren and Wronski suggest that the mathematical behavior of the decay parameter may reflect real physical behavior by looking at measured values of $\gamma$ for a series of jets. These results are shown in Fig. 3-1.5 for liquids for which $S_\infty$ is in the range 0.2 to 0.3. What is observed is that there appear to be two distinct branches of the data for $\gamma$, and the Reynolds number region that separates these two branches is also the range over which the theory predicts unusual behavior for $\gamma$. Since this Reynolds number region is also the boundary between contracting and expanding jets, it appears that the rate at which the radius approaches its downstream value depends on whether the jet contracts or expands. Why this should be so is not clear.

The perturbation analysis of Goren and Wronski holds only for the final decay of the dynamics to the downstream plug flow state. Hence the model can-

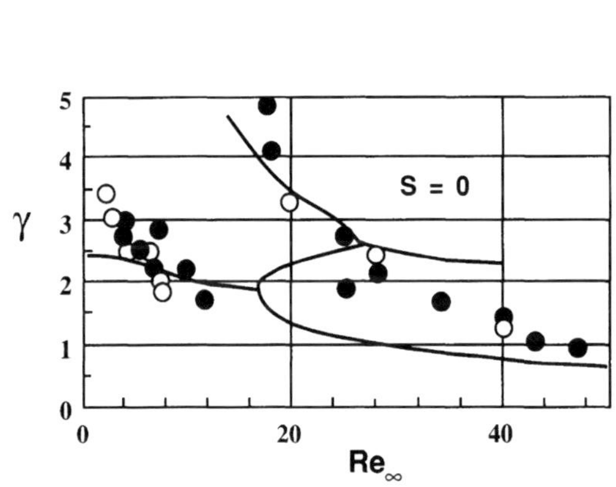


*FIGURE 3-1.5* The smallest eigenvalue $\gamma$ as a function of the Reynolds number, for $S_\infty = 0$, and data for $0.2 < S_\infty < 0.3$.

not be used to predict the jet shape over the whole region beyond the tube exit, and the data of Goren and Wronski show that the decay is not exponential near the tube exit. [Goren and Wronski also test a theory of Goren (1966) for the *initial* decay of the jet dynamics. This theory does not predict the rate of change of the jet radius to better than a factor of two.]

In summary, we have no quantitative analytical theory of the dynamics of a jet issuing from a horizontal tube. Both theory and observation agree that the jet reaches its final diameter within a downstream distance equivalent to a few tube diameters. Figure 3-1.6 shows experimental data (Middleman, 1964) for the jet "decay length" required to bring the diameter to within 1% of its final value, over a range of Reynolds numbers that includes both contracting and expanding jets.

Let's return to Eq. 3-1.6—the basic momentum balance for the dynamics of the jet. As noted earlier, the only way to predict the value of $\chi$, and the rate of approach of the jet radius to its final value, is to know the velocity profile right at the exit of the tube, including the axial deformation gradient $\partial u_z/\partial z$ at the exit. Such knowledge can only come from a full numerical solution to the dynamic equations, accounting for the diffusion (upstream into the tube) of the vorticity that accompanies the decay of the velocity profile of the jet in the region exterior to the tube. Such numerical solutions have been carried out by Nickell *et al.* (1974), and Omodei (1980). They successfully simulate the swelling of low Reynolds number jets, both with respect to the value of $\chi$ and with respect to the rate at which $\chi$ is achieved.

What comes out of the numerical analysis that is of special interest is an indication of the degree to which the exit conditions differ from the fully developed Poiseuille profile. For example, Fig. 3-1.7 shows the stress arising from the axi-

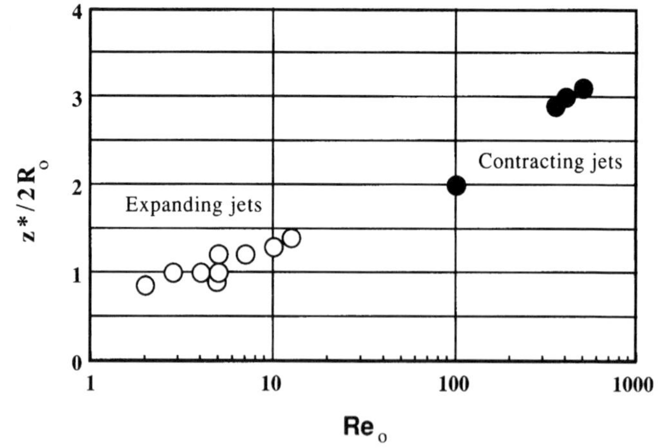


*FIGURE 3-1.6* Number of tube diameters downstream required to bring the jet diameter to within 1% of its final value.

dal deformation of the liquid at the exit plane, $z = 0$ (Nickell *et al.*, 1974). For the Poiseuille profile there is no axial deformation. What is plotted is a dimensionless axial stress N, defined as

$$\mathsf{N} \equiv \frac{2\mu\ (\partial u_z/\partial z)}{[\tau_{rz}]_{R0}}. \tag{3-1.22}$$

The axial stress has been normalized to the wall *shear* stress for the fully developed flow. What we see is that at the tube exit the axial stress is nonzero (hence there is significant departure from Poiseuille flow) and the magnitude of the axial stress is comparable to and even greater than the shear stress. If we return to Eq. 3-1.6 we see that what is important is the integral of this stress across the tube exit, or equivalently, the average of N over the cross section. Since there is both a positive and a negative region for N, we must see what the *integral* of N is. First, however, let's rewrite Eq. 3-1.6 in terms of this integral. We will ignore the effects of surface tension in order to simplify the discussion (see Problem P3-2).

We begin with the momentum equation in the form

$$\int_0^{R0} \left(\rho\, u_z^2(r,z=0) - \frac{4\mu U_0}{R_0}\mathsf{N}\right) r\, dr - \frac{\rho R_\infty^2 U_\infty^2}{2} = 0. \tag{3-1.23}$$

We will assume that the main source of the deviation from the predicted value of $\chi = 0.866$ arises from the axial stress term. Hence we will take the velocity profile at the exit to be the Poiseuille profile in calculating the first term on the left-hand side of Eq. 3-1.23. Then Eq. 3-1.23 becomes, after performing the indicated integrations,

$$\frac{1}{\chi^2} = \frac{4}{3} - \frac{8\,\overline{\mathsf{N}}}{\mathrm{Re}_0}. \tag{3-1.24}$$

If the average $\overline{\mathsf{N}}$ is positive, as seems reasonable from Fig. 3-1.7, then $\chi$ will exceed 0.866, as is observed in Fig. 3-1.2.

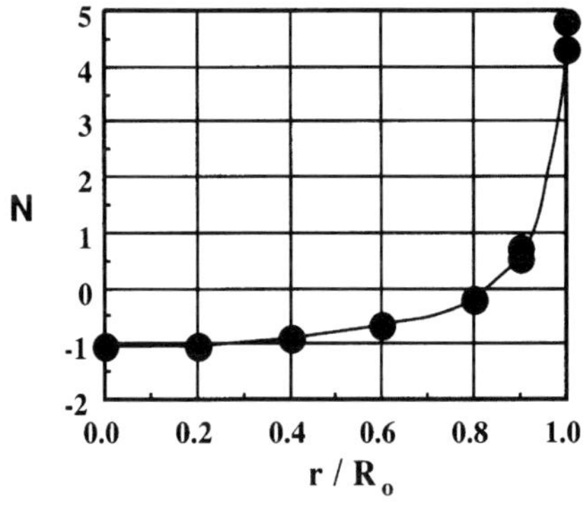


***FIGURE 3-1.7*** Axial stress at the tube exit.

Another feature of the dynamics that can be explored throught the numerical analyses is the role of surface tension. Omodei (1980) shows a large number of numerical simulations in which he explores the effect of surface tension through the parameter $S_0$ (see Eq. 3-1.21). In general, the effect of a finite value of $S_0$ is to reduce the expansion, or reduce the contraction, of horizontal jets. Figure 3-1.8 shows some results for a few cases corresponding to expanding jets. (Omodei finds that $Re_0 = 14.4$ is the boundary between expanding and contracting jets, in the absence of surface tension.) For larger Reynolds numbers, where the jet contracts, the effect of $S_0$ is small in the range of values that are experimentally attainable.

## 3-2 DRAW-DOWN OF A VERTICAL JET

The dynamics of the jet flow are altered significantly if the jet is vertical and pointing downward. The force of gravity accelerates or "draws" the jet, and the radius continually decreases along the jet axis (at least after some small distance downstream over which the jet *could* expand due to velocity profile rearrangement at low Reynolds numbers). The simplest analysis of jet draw-down under gravity follows from application of the Bernoulli equation (the energy equation) to a jet for which the effects of viscosity (profile rearrangement) and surface tension (pressure varying with jet radius) are neglected. For flat velocity profiles at the tube exit and along the jet axis the Bernoulli equation takes the form

$$\tfrac{1}{2}\rho U_0^2 = \tfrac{1}{2}\rho U_\infty^2 - \rho g z, \tag{3-2.1}$$

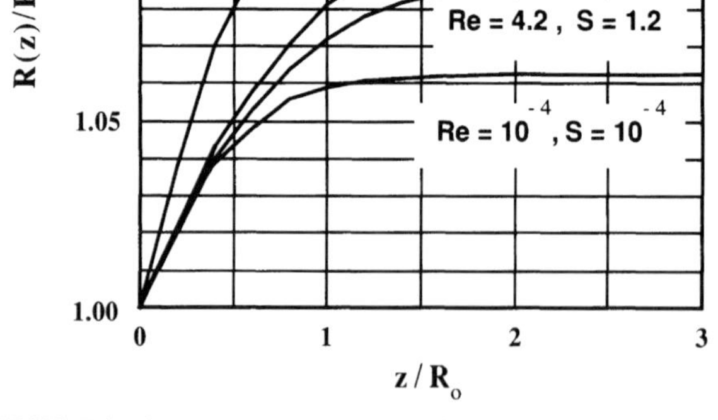


*FIGURE 3-1.8* Numerical simulations (Omodei) for horizontal jets.

where $z$ is the distance from the tube exit and is positive in the direction of flow. If we now combine this with the mass balance for an incompressible fluid,

$$U_0 R_0^2 = U_\infty R_\infty^2, \tag{3-2.2}$$

we find an equation for the jet radius along the axis as

$$\frac{R(z)}{R_0} = \left(1 + \frac{2\,g\,z}{U_0^2}\right)^{-1/4}. \tag{3-2.3}$$

This can be written in the form

$$\frac{R(z)}{R_0} = \left(1 + \frac{z/R_0}{\mathrm{Fr}_0}\right)^{-1/4}$$

$$= \left(1 + \frac{\mathrm{Re}_0}{\mathrm{Fr}_0}\,\zeta\right)^{-1/4} \quad \text{where} \quad \zeta \equiv \frac{z/R_0}{\mathrm{Re}_0}. \tag{3-2.4}$$

and the Froude number is

$$\mathrm{Fr}_0 \equiv \frac{U_0^2}{2gR_0}. \tag{3-2.5}$$

A set of experimental data obtained by Duda and Vrentas (1967) is shown in Fig. 3-2.1. Water issues vertically from a capillary with an inside diameter of $2R_0$ = 0.222 cm and a length of 25.5 cm, at a velocity of 38 cm/s. This corresponds to a Reynolds number of 834 and a Froude number of 6.5. The jet contracts rapidly, relative to the simple model given just above, because of the contraction associated with the velocity profile rearrangement just downstream of the tube exit.

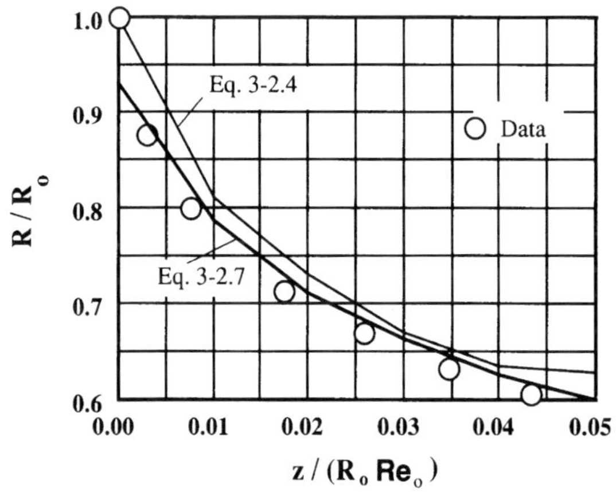


*FIGURE 3-2.1* Draw-down of a vertical laminar jet compared to Eqs. 3-2.4 and 3-2.7 ($\mathrm{Re}_0$ = 834, $\mathrm{Fr}_0$ = 6.5).

Equation 3-2.1 assumes that the exit velocity profile is flat. If the (more nearly) parabolic exit flow is accounted for in the Bernoulli equation, then the first term is changed since

$$\frac{\displaystyle\int_0^{R_0} \frac{1}{2}\, u_z^2(r)\, r\, dr}{\displaystyle\int_0^{R_0} r\, dr} = \frac{2}{3}\, U_0^2. \tag{3-2.6}$$

Equation 3-2.4 is then changed to the form

$$\frac{R(z)}{R_0} = \left(\frac{4}{3} + \frac{z/R_0}{\mathrm{Fr}_0}\right)^{-1/4} = \left(\frac{4}{3} + \frac{\mathrm{Re}_0}{\mathrm{Fr}_0}\,\zeta\right)^{-1/4}. \tag{3-2.7}$$

This result is also shown in Fig. 3-2.1 and provides a closer fit of the data, except at the tube exit, where the radius does not satisfy the correct initial condition.

An exact analytical solution of the jet draw-down problem is not possible, if the role of the velocity profile rearrangement is to be accounted for. A numerical solution to this problem is given in the work of Duda and Vrentas (1967). The velocity profile rearrangement is clearly seen in their simulations, an example of which is shown in Fig. 3-2.2. Even at $\zeta = 0.05$, which is about 20 diameters downstream, the velocity profile has not flattened to the extent expected of a horizontal jet of this Reynolds number (see Fig. 3-1.6).

To this point all of the discussion has been with regard to flows that are fully developed and parabolic just prior to the tube exit. Scriven and Pigford (1959)

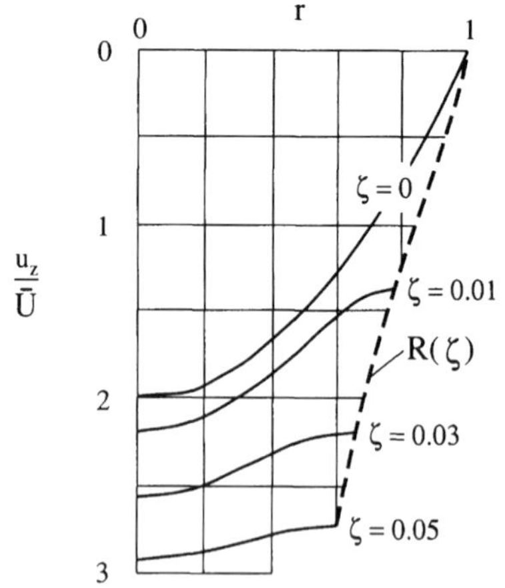


***FIGURE 3-2.2*** Axial velocity profiles downstream for a vertical jet ($\mathrm{Re}_0 = 834$, $\mathrm{Fr}_0 = 6.5$).

studied the draw-down of jets issuing from a nozzle designed to produce a *nearly flat* velocity profile at the exit. A set of their data is shown in Fig. 3-2.3, where we see that Eq. 3-2.4 , which is based on the assumption of a flat exit velocity profile, describes the draw-down rather well. The data are consistently about 2% below this simple theoretical model, which probably corresponds to a rapid and small degree of contraction associated with the departure of the exit profile from uniformity.

## 3-3 A NOTE ON DIMENSIONLESS GROUPS

In the analyses of this chapter we have been exposed to a veritable zoo of dimensionless "animals." But in accordance with the fundamental principle of dimensional analysis, the so-called Buckingham pi theorem, there is a limited number of independent dimensionless groups appropriate to describe the physics of the laminar jet with gravity acting upon it. This arises from the fact that there are six parameters that are relevant to the dynamics of the flow exterior to the tube: $\rho$, $\mu$, $\sigma$, $U_0$, $R_0$, and $g$. There are three fundamental dimensions required to define these groups: mass, length, and time. Hence the Buckingham pi theorem tells us that there will be three (six parameters minus three dimensions) *independent* dimensionless groups. We may take these to be

$$\mathrm{Re} \equiv \frac{2\rho UR}{\mu}, \qquad \mathrm{We} \equiv \frac{\rho U^2 R}{\sigma}, \qquad \mathrm{Fr} \equiv \frac{U^2}{2gR}$$

Among the literature cited in this chapter we can find a number of additional dimensionless groups used for presentation of theory and data. These include

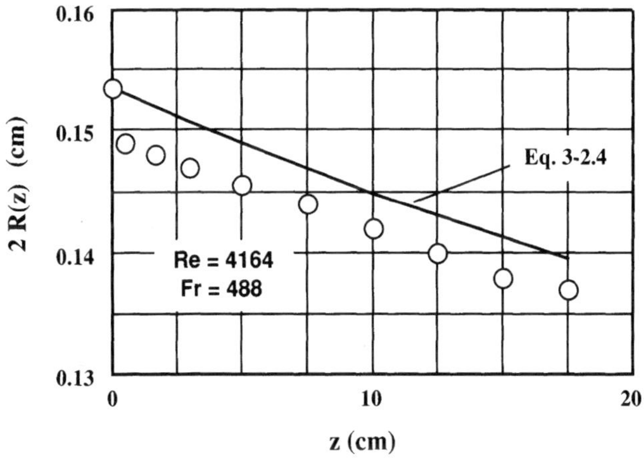


*FIGURE 3-2.3* Data of Scriven and Pigford (1959) for draw-down of a jet with a flat exit profile.

$$\mathrm{Ca} \equiv \frac{\mu U}{\sigma} = \frac{2\,\mathrm{We}}{\mathrm{Re}}, \qquad \mathrm{Bo} \equiv \frac{\rho g R^2}{\sigma} = \frac{\mathrm{We}}{2\mathrm{Fr}},$$

$$\frac{1}{\mathrm{Oh}} \equiv \frac{\sqrt{\rho\sigma R}}{\mu} = \mathrm{S}^{1/2} = \frac{\mathrm{Re}}{2\sqrt{\mathrm{We}}},$$

$$\mathrm{St} \equiv \frac{\mu U}{\rho g R^2} = \frac{\mathrm{Fr}}{\mathrm{Re}} = \frac{\mathrm{Ca}}{\mathrm{Bo}},$$

$$\mathrm{N_{pp}} \equiv \sigma\left(\frac{\rho}{\mu^4 g}\right)^{1/3} = \frac{\mathrm{Re}^{4/3}\,\mathrm{Fr}^{1/3}}{2\,\mathrm{We}}.$$

Groups we have not yet met are the Bond number Bo, the Stokes number St, and the physical property number $\mathrm{N_{pp}}$. Various authors will make choices among these groups in presenting their work. Note that some of these groups are *kinematic*—they depend on the speed of the jet—whereas others (Bo, Oh, and $\mathrm{N_{pp}}$) depend only on physical properties and, possibly, the radius of the tube.

## *Problems*

P3-1 In writing Eq. 3-1.4 we did not take into account the effect of an internal isotropic pressure in the liquid at the tube exit and far downstream. We ignored surface tension, and so we might assume that the absolute pressures at these two positions are identical. However, they act on different cross-sectional areas (since the radii are different) and so there *is* a net force in the axial direction. Consider the external pressure acting on the surface of the jet, in the axial direction, and show that this pressure exactly balances the force difference arising from the internal pressure, in the absence of surface tension.

P3-2 Suppose we add the contribution of surface tension to Eq. 3-1.4, but we continue to assume that the velocity profile is the Poiseuille profile right at the tube exit. Show that $\chi$ satisfies the following equation.

$$\frac{4}{3}\chi^2 - 1 + \frac{\chi^3 - \chi^2}{\mathrm{We}_0} = 0, \tag{P3-2.1}$$

where the Weber number is defined as

$$\mathrm{We}_0 \equiv \frac{\rho U_0^2 R_0}{\sigma}. \tag{P3-2.2}$$

P3-3 Derive the result given by Omodei as an alternative to Eq. P3-2.1 (but neglect surface tension).

$$\frac{4}{3}\chi^2 - 1 + \int_{-L_1/R_0}^{0} \frac{2}{R_0}\frac{\tau_{R0}(z)}{\rho U_0^2}\,dz + \frac{p(-L_1)}{\rho U_0^2} = 0, \tag{P3-3.1}$$

where $\tau_R$ is the shear stress along the inside surface of the tube and $-L_1$ is a distance (within the tube) far enough upstream along the axis that the flow is still fully developed Poiseuille flow. Of what value is this result? (What does Omodei use it for?)

P3-4 Derive Eq. 3-1.12.

P3-5 Derive Eqs. 3-1.16 and 3-1.17.

P3-6 For the data shown in Fig. 3-1.3, plot $\hat{R}/R_\infty$ vs. $z/R_\infty$. Do the data agree with the form of solution given in Eq. 3-1.14?

P3-7 A jet is formed under the following conditions.

$$\mu = 0.03 \text{ poise}, \qquad \rho = 1 \text{ g/cm}^3, \qquad \sigma = 30 \text{ dyn/cm},$$
$$R_0 = 0.05 \text{ cm}, \qquad U_0 = 200 \text{ cm/s}.$$

What are the values of the Reynolds number $\mathsf{Re}_0$ and the parameter $\mathsf{S}_0$ for this jet?

P3-8 Nickell *et al.* (1974) simulate the swelling of a Newtonian jet at a Reynolds number $\mathsf{Re}_0 < 0.001$ and $\mathsf{S}_0 < 10^{-5}$. A least-squares fit of their simulation for $R(z)/R_0$ is

$$\frac{R}{R_0} = 1 + 0.135\left[1 - \exp\left(-2.06\frac{z}{R_0}\right)\right]. \tag{P3-8.1}$$

To what degree is this result compatible with Fig. 3-1.5?

P3-9 Equation 3-2.7 does not satisfy the correct initial condition on the jet radius. Why? What is missing from the energy equation that leads to this error?

P3-10 The terms on the right-hand side of Eq. 3-1.6 are described in terms of the forces arising from surface tension "pulling" on the boundaries of the jet at the jet exit and far downstream. What we imply here is that surface tension is regarded as a force per unit of contact length (perimeter). Shouldn't the force term at the exit include the angle that the surface makes with the axis?

Suppose instead we considered the surface tension contribution through the Young–Laplace equation. Then we must calculate the net axial force arising from the pressure difference just beyond the exit and far downstream. Make that calculation, and compare the result to the terms on the right-hand side of Eq. 3-1.6.

## *References*


Adachi, K., and N. Yoshioka. (1984). Tube exit flows and laminar newtonian jets in the atmosphere. *Proc. IX Int. Congr. Rheology* **2**, 329.

Duda, J. L., and J. S. Vrentas. (1967). Fluid mechanics of laminar liquid jets. *Chem. Eng. Sci.* **22**, 855.

Goren, S. L. (1966). Development of the boundary layer at the free surface from a uniform shear flow. *J. Fluid Mech.* **25**, 87.

Goren, S. L., and S. Wronski. (1966). The shape of low-speed capillary jets of Newtonian liquids. *J. Fluid Mech.* **25**, 185.

Harmon, D. B. (1955). Drop sizes from low-speed jets. *J. Franklin Inst.* **259**, 519.

Middleman, S. (1964). Profile relaxation in newtonian jets. *IEC Fund.* **3**, 118.

Middleman, S., and J. Gavis. (1961). Expansion and contraction of capillary jets of Newtonian liquids. *Phys. Fluids* **4**, 355. Errata, *ibid.* **4**, 1450, 1961. See also Gavis, J. (1964). Contribution of surface tension to expansion and contraction of capillary jets. *Phys. Fluids* **7**, 1097.

Nickell, R. E., R. I. Tanner, and B. Caswell. (1974). The solution of viscous incompressible jet and free-surface flows using finite-element methods. *J. Fluid Mech.* **65**, 189.

Omodei, B. J. (1980). On the die-swell of an axisymmetric Newtonian jet. *Computers and Fluids* **8**, 275.

Scriven, L. E., and R. L. Pigford. (1959). Fluid dynamics and diffusion calculations for laminar liquid jets. *AIChE J.* **5**, 397.

Trang, C. T., and Y. L. Yeow. (1986). Extrudate swell of Newtonian and non-Newtonian fluids—the effect of gravitational body force. *J. Non-Newt. Fluid Mech.* **20**, 103.

Vrentas, J. S., C. M. Vrentas and A. F. Shirazi. (1985). Downstream boundary conditions for vertical jets. *AIChE J.* **31**, 1044.

# 4 The Transitions from Drops to a Coherent Jet and Back to Drops

*Axisymmetric flow of* an isothermal incompressible Newtonian fluid from a long tube of uniform circular cross section undergoes a series of transitions as the flow rate increases. At very slow efflux velocities the liquid leaves the tube as a series of periodic drops (Weis *et al.*, 1992). Eventually this dripping flow undergoes a transition, and a coherent laminar jet (the topic of the previous chapter) is achieved. In the absence of external forces on the surface of the jet, there is usually a point downstream along the jet axis where the coherent jet becomes unstable and breaks up into a stream of discrete drops. High speed flash photography (see Color Plates 1 and 2) reveals the complex structure of the jet in the region of the breakup. The goal of any analysis of these phenomena is the prediction of the critical speed at which each transition occurs, the dependence of the critical speed on tube diameter and liquid properties, and the size(s) of the drops formed from the jet. In the case of jet instability, the length of the coherent jet that precedes the first drop would be a desired output of an analytical model. All of these problems have received a great deal of attention, and each of them is of some considerable practical interest in a variety of areas of technology. We will discuss these problems in the order in which they occur as the jet velocity increases.

## 4-1 DRIPPING FROM A VERTICAL TUBE

Figure 4-1.1 suggests the dynamics of interest. At very low flow rates the liquid forms a drop attached to the tip of the tube. We want to predict the size of a drop that falls off of a tube under the action of gravity. We assume that the liquid is flowing through the tube or capillary at a very slow rate, and that as the drop grows its weight eventually overcomes the force of surface tension that holds the drop to the tip of the capillary, and the drop falls off. The capillary is vertical, pointed down, and the drop forms symmetrically on the tip. Figure 4-1.2 shows the detail of the contact region of the liquid with the capillary.

The contact line of the liquid with the capillary is a circle of diameter $D_0$. We assume that the drop is supported from the inner surface of the capillary, and so $D_0$ is the inside diameter. Along the three-phase (liquid, solid, gas) contact line, surface tension is defined as the force per unit length acting tangentially to the liquid–gas surface. Hence there is a vertical component of force given by

$$F_\sigma \cos \theta = \sigma \pi D_0 \cos \theta. \tag{4-1.1}$$

If the drop is in static equilibrium, then this force must just balance the weight of the drop, so we expect that the shape of the drop adjusts itself, for each drop volume $V$, so that

$$\sigma \pi D_0 \cos \theta = \rho g V. \tag{4-1.2}$$

If the drop volume is slowly increased, then the right-hand side of this equation increases, and there must be a corresponding change of the left-hand side. But the only parameter that can change is the angle $\theta$. The maximum value of $\cos \theta$ occurs when $\theta = 0$. The simplest hypothesis regarding the maximum stable drop size is that the drop will fall away from the capillary when $\theta = 0$, which is when

***FIGURE 4-1.1*** Drop hanging on the tip of a capillary.

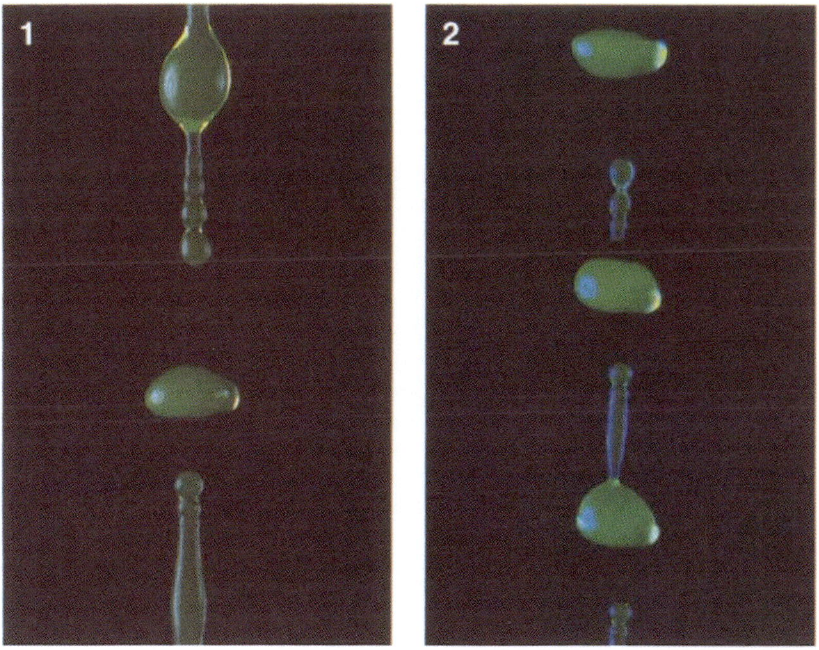


***COLOR PLATES 1 AND 2*** Stages in the breakup of a viscous jet. At some point large drops appear (Color Plate 1) and the connecting ligaments then (Color Plate 2) break up into satellite drops. See Chapter 4.

***FIGURE 5-6.5*** The variety of free surface shapes attained with a low Reynolds number impinging jet on a disk.

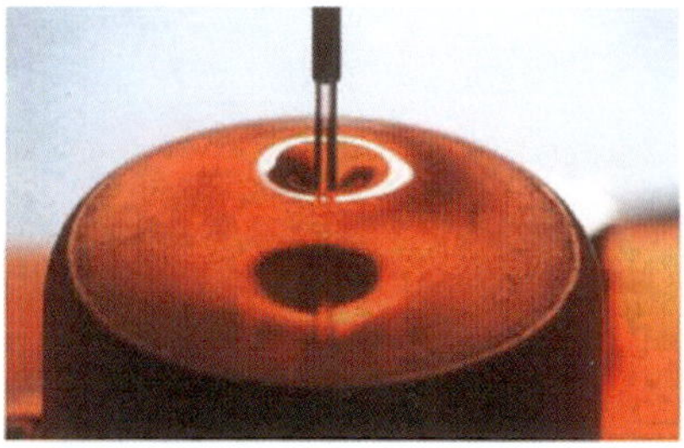

***FIGURE 5-6.6*** Detail of the flow field in the stagnation region of a laminar jet. The "critical dimple flow rate" has been exceeded.

***COLOR PLATE 3*** A water jet strikes a rigid planar surface. The radial outward flow from the stagnation point undergoes a circular hydraulic jump. Beyond the rim of the jump the film thickness has greatly increased and the flow is slowed. See Chapter 5.

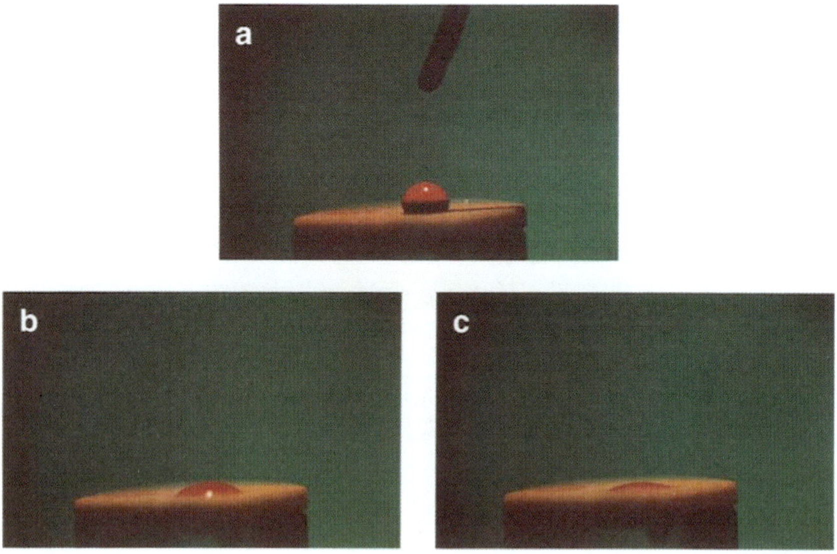


***FIGURE 8-5.2*** Wicking of a drop into a porous medium. To the side of the wicking drop is the "spot" remaining from a previously wicked drop.

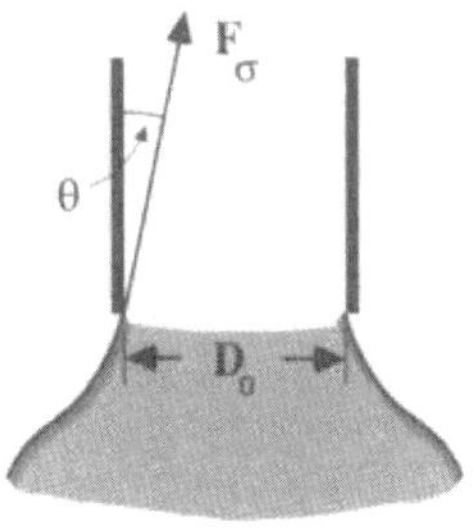


*FIGURE 4-1.2* Forces along the contact line.

the drop volume exceeds

$$V_{max} = \sigma\pi D_0/\rho g. \tag{4-1.3}$$

Figure 4-1.3 shows a photograph of a drop just prior to separation from the capillary.

If the released drop were a sphere of diameter $D_d$ then

$$V = \pi D_d^3/6. \tag{4-1.4}$$

This analysis yields a simple model of the size of the drop that falls from a slowly dripping capillary:

$$D_d = [6\sigma D_0/\rho g]^{1/3}. \tag{4-1.5}$$

We can nondimensionalize this expression to the form

$$D_d/D_0 = (6\sigma/D_0^2\rho g)^{1/3}. \tag{4-1.6}$$

An important dimensionless group that determines the nature of this process is the Bond number, defined for this problem as

$$\mathrm{Bo} = D_0^2\rho g/\sigma. \tag{4-1.7}$$

This simple model for drop size then takes the form, in terms of the Bond number,

$$D_d/D_0 = [6/\mathrm{Bo}]^{1/3}. \tag{4-1.8}$$

or

$$D_d/D_0 = 1.82\ \mathrm{Bo}^{-1/3}. \tag{4-1.9}$$

While this is a very simple model, its virtue lies in the fact that it is easily tested. Experiments have been conducted with a variety of liquids, over as wide a range of surface tensions as practical, in order to test this model. A constant flow rate of liquid is initiated into the capillary, and the falling drops are collected and weighed. From the measured density of the liquids it is then possible to calculate the volume and hence the drop diameter $D_d$. The data are presented in terms of the ratio $D_d/D_0$, and the model is tested by calculating the function

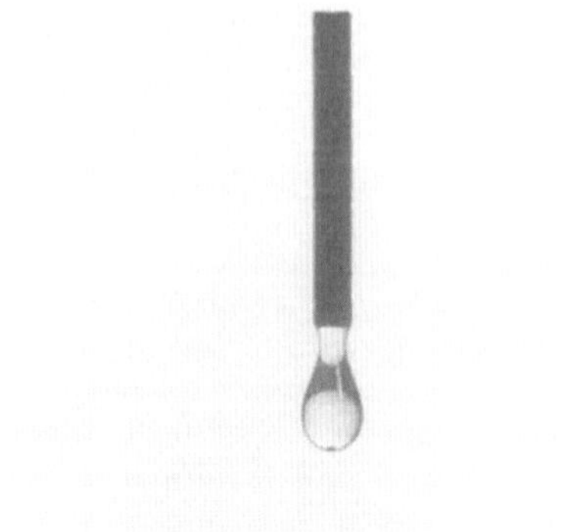

*FIGURE 4-1.3* A drop just before breakaway from the capillary tip.

$$\mathsf{N}_d = (D_d / D_0)\, \mathsf{Bo}^{1/3}. \tag{4-1.10}$$

If the model is correct we should find that $\mathsf{N}_d$ is independent of the liquid flow rate and that it has a value of 1.82 regardless of the liquid tested.

Figure 4-1.4 shows the data obtained in our own study (Mandrusov,1990). The results are in agreement with this model, except that the coefficient has a value of 1.6 instead of 1.82. This lower value arises because when the drop falls from the capillary tip it leaves some residual liquid attached to the tip (Wilson, 1988). Hence the drop volume is lower than predicted by this model.

The data in Fig. 4-1.4 are plotted as a function of the flow rate of the liquid, made dimensionless through the introduction of the Weber number, which is defined here as

$$\mathsf{We} = [\rho(4Q / \pi D_{ci}^2)^2\, D_{co}]/\sigma. \tag{4-1.11}$$

In this definition, $\rho$ is the liquid density, and the subscripts $i$ and $o$ on $D_c$ refer to the inner and outer diameters of the capillary, respectively. The reason for this distinction is that the drops are usually observed to fall from the capillary while in contact with the outer diameter.

Figure 4-1.4 demonstrates that the drop size is nearly independent of the drip rate until the Weber number becomes of order unity. Beyond that point the drop has enough inertia that it can separate from the capillary at a lower weight and hence at a lower drop size. When the Weber number exceeds a value of about $\mathsf{We} = 3$, dripping ceases and the liquid issues as a jet.

We conclude that the factors that affect the diameter of drops that fall from a capillary under the force of gravity can be understood in terms of a very simple model based upon a balance of forces at the tip of the capillary. The theory that results is not exact, but it does capture the essential physics of the process, failing only to yield the exact value of the coefficient in Eq. 4-1.9. Figure 4-1.4 may be used with confidence for the prediction of drop size from a knowledge of the

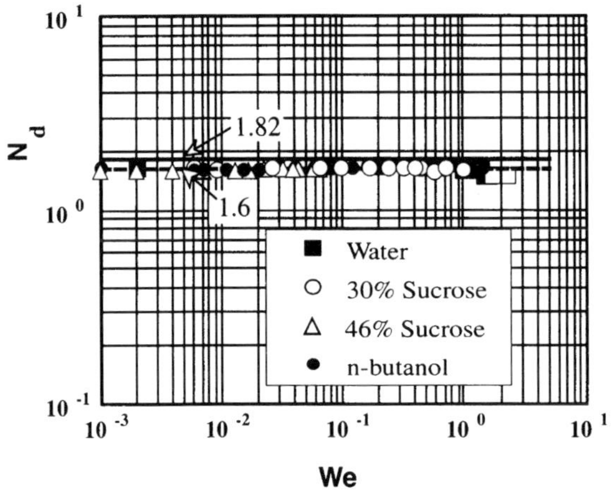


*FIGURE 4-1.4* Dimensionless drop size (see Eq.4-1.10) for several liquids, as a function of the Weber number.

liquid properties and the capillary inner and outer diameters. If the drop is observed to fall from the inner surface of the capillary, we simply replace $D_{co}$ with $D_{ci}$ in Eq. 4-1.11.

A much more detailed analysis of the critical conditions for drop breakoff is possible (Boucher and Evans, 1975), and we will return to this problem later, in Chapter 6.

## 4-2 DROPS ON DEMAND

In our analysis of the dripping problem we assumed that there is a constant supply of liquid to the capillary. Liquid accumulates in the forming drop, and eventually the drop separates from the tip of the capillary. There is another method of creating drops that warrants our consideration because of its commercial importance. This is the method used in one class of ink-jet printer, where drops are ejected from a tube *impulsively*, due to the imposition of a very short duration pressure rise on the liquid reservoir. The pressure can be generated by a piezoelectric source, or by rapidly raising the temperature of a surface in contact with the reservoir so as to cause formation of a vapor bubble, the growth of which ejects a drop. We refer to this as "drops on demand," a term used in several examples of published studies of this phenomenon (Fromm, 1984).

The dynamics of this flow are complicated not only by the free boundary nature but also by the fact that this is an unsteady flow. The liquid is quiescent and is then suddenly accelerated and ejected from the capillary. When the pressure is cut off, the liquid may continue to move forward because of its inertia.

For a given pressure pulse, the size of any ensuing drop is a complex function of the parameters of the system. Let's look first at a simpler problem—the transient start-up of the liquid in response to the onset of a pressure pulse. This will give us an idea of the time scale associated with transient flow of a liquid in a capillary.

We suppose that the liquid is confined to a reservoir to which is attached a capillary of some length $L$. A transient pressure rise is applied to the reservoir, and we will assume that the pressure history is that of a simple rectangular pulse; that is, the pressure satisfies the following functional relationship:

$$\begin{aligned} p &= 0 \quad \text{for } t < 0, \\ p &= P \quad \text{for } 0 \leq t \leq T_{\mathrm{p}}, \\ p &= 0 \quad \text{for } t > T_{\mathrm{p}}. \end{aligned} \tag{4-2.1}$$

We want to estimate the liquid volume ejected from a capillary when the reservoir is subjected to this pressure history. To do so we must solve for the transient start-up of this flow. We assume that the flow is laminar and that the nonlinear inertial terms are small compared to viscous effects, even though the nature of the flow is impulsive and over a very short time scale. (In an ink-jet printer, the pressure pulse is about $T_{\mathrm{p}} = 40\ \mu\mathrm{s}$.) We will evaluate this assumption later.

Thus we begin with the momentum balance in the form

$$\rho \frac{\partial u_z}{\partial t} = -\frac{\partial p}{\partial z} + \mu \frac{1}{r}\frac{\partial}{\partial r}\left(r \frac{\partial u_z}{\partial r}\right). \tag{4-2.2}$$

Notice that we are ignoring entrance effects and assuming that the flow is fully developed. During the time that the pressure is "on," the pressure gradient is $P/L$, so we must solve

$$\rho \frac{\partial u_z}{\partial t} = \frac{P}{L} + \mu \frac{1}{r}\frac{\partial}{\partial r}\left(r \frac{\partial u_z}{\partial r}\right). \tag{4-2.3}$$

An initial condition is

$$u_z = 0 \quad \text{at } t = 0 \quad \text{for} \quad r \text{ on } [0, R_0]. \tag{4-2.4}$$

The no-slip boundary condition is

$$u_z = 0 \quad \text{at} \quad r = R_0. \tag{4-2.5}$$

Symmetry along the axis requires

$$\frac{\partial u_z}{\partial r} = 0 \quad \text{at} \quad r = 0. \tag{4-2.6}$$

The solution for $u_z(r,t)$ may be written in the form (Bird *et al.*, 1960)

$$\Phi \equiv \frac{4\mu L u_z}{P R_0^2} = \left(1 - s^2\right) - 8 \sum_{n=1}^{\infty} \frac{J_0(\alpha_n s)}{\alpha_n^3 J_1(\alpha_n)} e^{-\alpha_n^2 \tau}, \tag{4-2.7}$$

where the eigenvalues $\alpha_n$ are the roots of

$$J_0(\alpha_n) = 0 \tag{4-2.8}$$

and dimensionless space and time variables are defined by

$$s \equiv \frac{r}{R_0} \quad \text{and} \quad \tau \equiv \frac{\mu t}{\rho R_0^2}. \tag{4-2.9}$$

In the ink-jet application that motivates this analysis, typical parameters would be

$$R_0 = 25\ \mu\text{m}, \qquad \rho = 1000\ \text{kg/m}^3, \qquad \mu = 0.01\ \text{Pa-s}.$$

From evaluation of the terms in the series of Eq. 4-2.7 we can see that steady flow is nearly achieved when

$$\alpha_1^2 \tau = 3$$

or

$$t = \frac{3\rho R_0^2}{(2.403)^2 \mu} = 32.5\ \mu\text{s}. \tag{4-2.10}$$

Since this is comparable to the duration of the pressure pulse, most of the flow occurs under transient conditions.

The volumetric flow rate at any time, relative to the steady value, is found to be

$$\frac{\int_0^1 \Phi s\, ds}{\int_0^1 (1 - s^2) s\, ds} = 1 - 32 \sum_{n=1}^{\infty} \frac{1}{\alpha_n^3 J_1(\alpha_n)} \varepsilon^{-\alpha_n^2 \tau} \int_0^1 s J_0(\alpha_n s)\, ds \tag{4-2.11}$$

or

$$\frac{\int_0^1 \Phi s\, ds}{\int_0^1 (1 - s^2) s\, ds} = 1 - 32 \sum_{n=1}^{\infty} \frac{1}{\alpha_n^4} e^{-\alpha_n^2 \tau}$$

$$= 1 - 0.96 e^{-5.77\tau}. \tag{4-2.12}$$

(Since this series converges rapidly, we use only the first term of the summation.)

By integrating the volume flow rate over the time of the pressure pulse, $T_p$, we find that the volume of liquid that crosses the exit of the tube in the time $T_p$ is (again, relative to that for steady flow)

$$\Psi \equiv \frac{V}{V_\infty} = 1 - \frac{0.166\rho R_o^2}{T_p \mu}\left[1 - \exp\left(-\frac{5.77 T_p \mu}{\rho R_o^2}\right)\right], \tag{4-2.13}$$

where

$$V_\infty \equiv \frac{\pi R_0^4 P T_p}{8\mu L}. \tag{4-2.14}$$

We can get some indication of the behavior of this system if we now select values for the length $L$ and the pressure pulse amplitude, $P$. We will use values of

$$L = 500\ \mu\text{m} \quad \text{and} \quad P = 10^5\ \text{N/m}^2\ (= 1\ \text{atm}) \tag{4-2.15}$$

and we find

$$V_\infty = 123 \times 10^{-15}\ \text{m}^3 = 123\ \text{picoliter}. \tag{4-2.16}$$

This corresponds to a spherical drop of diameter

$$D_{\text{drop}} = 62\ \mu\text{m}. \tag{4-2.17}$$

With the 40-μs pressure pulse specified earlier, we find that

$$\Psi = 0.75. \tag{4-2.18}$$

Hence the volume of liquid that passes the exit of the tube is 92 picoliter.

These simple calculations do not address the key issue for the dynamics of this flow:

> Does the liquid that issues from the tube actually form into a free droplet, or could it remain attached to the end of the tube?

This must depend on the inertia of the drop, relative to the restraining force of surface tension. Hence we would expect that the Weber number would play a central role in determining the fate of the ejected liquid. With the Weber number defined as

$$\text{We} = \frac{\rho U_\infty^2 R_0}{\sigma} \tag{4-2.19}$$

and using the steady mean velocity of the liquid

$$\frac{V_\infty}{\pi R_0^2 T_p} = U_\infty = \frac{P R_0^2}{8\mu L} \tag{4-2.20}$$

we find a value for the Weber number of the order of unity, for the conditions specified in this example. This means that the inertia of the liquid exiting the capillary is of the same order of magnitude as the force of surface tension holding it to the tip. Hence it is not clear, from this simple view, whether drops will in fact be formed under the conditions selected for this example. We should expect, however, that drops would detach from the capillary tip if there were a slight increase in flow rate over the value used here.

In order to go beyond these semiquantitative ideas it would be necessary to carry out numerical solutions of this impulsive free boundary flow. Fortunately, an example of such work exists (Fromm, 1984) and we review these results here. Details of the computational procedure are provided by Fromm, but will not be discussed here. Two sequences of Fromm's computations are shown in Fig. 4-2.1. They each correspond to the same value of the dimensionless parameter S (see end of Chapter 3) but represent ejection of liquid at two different pressures.

We should note here that Fromm's simulations do not correspond exactly to the pressure wave defined in Eq. 4-2.1. Instead, he imposes a brief interval of negative pressure, followed by a positive constant pressure for sufficient time that the volume of liquid ejected corresponds to a drop whose radius is that of the capillary. The pressure is then set negative again for a length of time such that the total "impulse" due to the pressure is zero; that is,

$$\int_0^T P(t)\, dt = 0. \tag{4-2.21}$$

In Fig. 4-2.1a, the first time shown is the point at which the positive pressure is turned off. If we ignore the initial brief negative pressure, this corresponds to

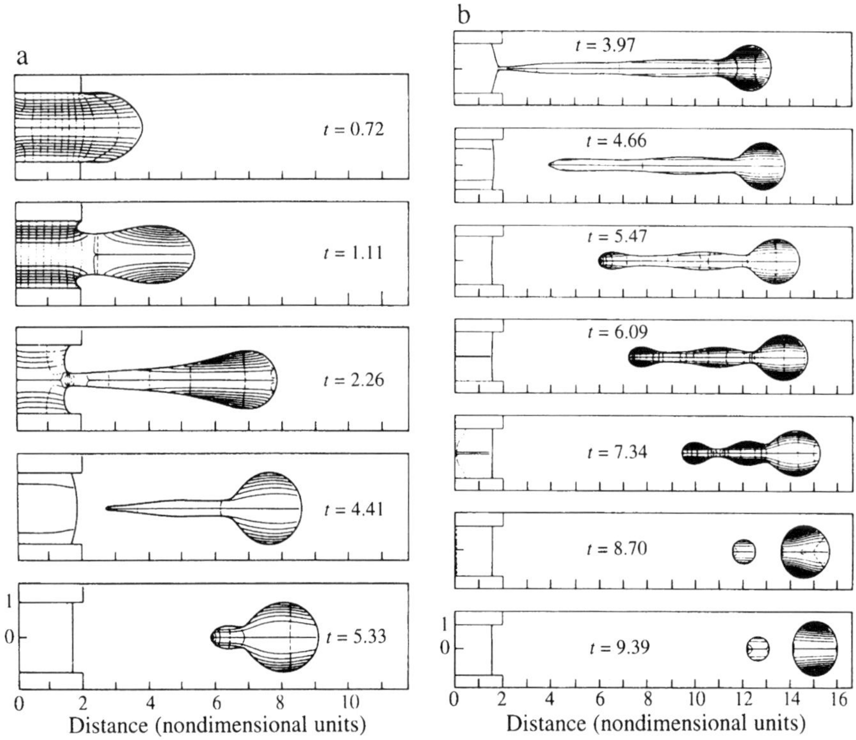


***FIGURE 4-2.1*** Time sequence of drop formation for S = 9. Copyright 1984 by International Business Machines Corporation; reprinted with permission.

(using information in Fromm)

$$T_p = 0.61\sqrt{\frac{\rho R_0^3}{\sigma}}. \tag{4-2.22}$$

The positive pressure has the magnitude (again, using Fromm's parameters)

$$P = 40\frac{\sigma}{R_0}. \tag{4-2.23}$$

Because of inertia the liquid moves forward to form a jet, and the jet detaches from the tip of the tube and a drop is formed. This drop has sufficient inertia on detachment that it continues to move forward, in this particular case, with a velocity of approximately

$$U_\infty = 0.6\sqrt{\frac{\sigma}{\rho R_0}}. \tag{4-2.24}$$

In Fig. 4-2.1b the views shown are late in the time sequence. This simulation corresponds to a higher pressure,

$$P = 100\frac{\sigma}{R_0}, \tag{4-2.25}$$

imposed for a time sufficient for the volume of liquid ejected to correspond to a drop whose radius is only three-fourths that of the capillary. What is interesting here is that by manipulating the operating variables in this way, a qualitatively different drop is formed. In this case we observe a so-called satellite drop that follows the larger, main, drop. In ink-jet printing this would be an undesirable feature, since the presence of the smaller drops would reduce the sharpness of the images formed from the main drops.

In Fig. 4-2.2 we see a sequence of high-speed video images of a jet of ink created in a thermal ink-jet printer (a Hewlett-Packard DeskJet 1200C color printer). The jet is initiated by an electrical pulse which rapidly heats ink in a reservoir and leads to nucleation of a vapor bubble. The growing bubble ejects the ink from an orifice with a diameter of 38 μm. The first image is taken 10 μs after the electrical pulse, and subsequent images are separated by 20 μs. Under the conditions of this particular event we observe a primary drop followed by a train of smaller drops. In subsequent sections we review theories and data relevant to the character of the drop breakup process.

## 4-3 STABILITY OF THE HORIZONTAL JET

In Chapter 3 we discussed several features of the horizontal jet. In particular, we saw that a laminar jet would exhibit a slight contraction or expansion of its diameter and that this change in diameter would take place over an axial length of the order of a few diameters (see Fig. 3-1.6). If we examine the laminar horizontal jet further along the direction of flow, we usually observe that the jet loses its coherence and becomes a stream of droplets periodically spaced along the axis. To the naked eye it may not be apparent that these drops exist, but a high-speed flash photograph (Fig. 4-3.1) reveals the details. An analysis of the dynamics of jet stability has as its primary goal the prediction of the axial distance downstream at which the first drops appear. A secondary goal is a determination of the conditions under which satellite drops appear and the relative sizes of the drops formed by the instability of the jet.

The primary source of this instability that leads to the transition from a coherent cylinder to a train of drops is the presence and action of surface tension. In fact, it turns out that under many conditions of practical interest, viscosity plays a secondary role and one can carry out a successful analysis on the assumption that the liquid is *inviscid*. The inviscid jet analysis is a classical problem, completed in the nineteenth century by Lord Rayleigh. We begin this section with a discussion of the stability of the inviscid jet.

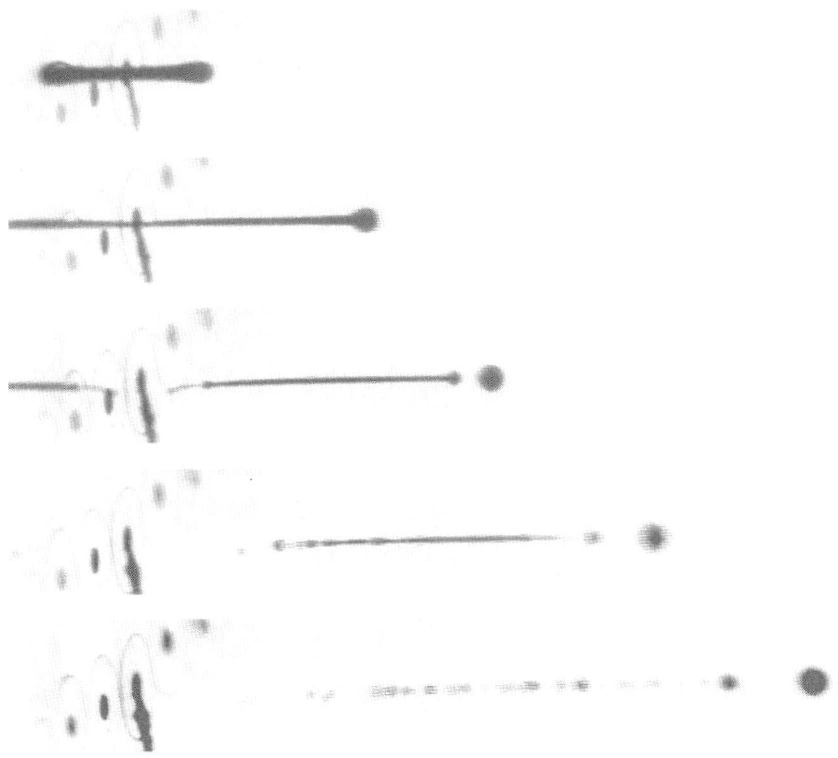

***FIGURE 4-2.2*** Ink jet ejected from an HP DeskJet 1200C printhead. See text for detailed discussion. The jet issues from left to right. The exit orifice is the dark oval shape from which the jet issues. The dark, nearly vertical streak is ink which has "drooled" onto the face of the printhead. To the left of the orifice is a reflection of the jet. Photographs kindly provided by Dr. Clayton Holstun of Hewlett–Packard Co., San Diego.

The liquid jet is assumed to be moving horizontally at a uniform velocity $U$. Gravity is ignored, as is any influence of the ambient medium. The latter assumption is reasonable for a low-speed jet, for which the shear stress exerted by the surrounding air is negligible. We adopt a cylindrical coordinate system fixed in the jet, that is, moving with the velocity $U$. In that coordinate system the undisturbed jet is stationary. Hence the velocity vector has zero axial and radial components, and the pressure is everywhere uniform and given by

$$p^0 = \frac{\sigma}{R}. \tag{4-3.1}$$

We first carry out a linearized stability analysis in which we assume that the velocity components, the pressure, and the jet radius are somehow perturbed only slightly from their stable values. The Navier–Stokes equations for the ensuing velocity, pressure, and radius that respond to that perturbation are to be solved.

We assume that the perturbed jet remains axisymmetric. Hence we may write the dynamic equations in the form

$$\rho \frac{\partial u_z}{\partial t} = -\frac{\partial p}{\partial z} + \mu \left[ \frac{1}{r} \frac{\partial}{\partial r} r \left( \frac{\partial u_z}{\partial r} \right) + \frac{\partial^2 u_z}{\partial z^2} \right] \tag{4-3.2}$$

and

$$\rho \frac{\partial u_r}{\partial t} = -\frac{\partial p}{\partial r} + \mu \left[ \frac{\partial}{\partial r} \left( \frac{1}{r} \frac{\partial}{\partial r} r u_r \right) + \frac{\partial^2 u_r}{\partial z^2} \right]. \tag{4-3.3}$$

Note that we have dropped the *nonlinear* terms from the left-hand side of the Navier–Stokes equations. We are assuming that inertial effects are not significant. This may be reasonable for viscous jets, but it is not clear that it is a good assumption for an inviscid jet. Note, as well, that we are keeping the acceleration terms, $\partial/\partial t$. Since a stability problem is by its very nature a transient problem, we cannot formulate it with the steady-state equations.

The continuity equation takes the form

$$\frac{1}{r} \frac{\partial}{\partial r} r u_r + \frac{\partial u_z}{\partial z} = 0. \tag{4-3.4}$$

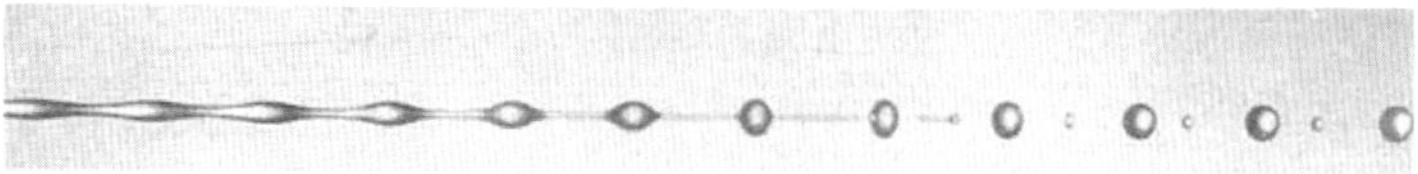

***FIGURE 4-3.1*** Photograph of an unstable laminar liquid jet (75% glycerine–water solution). From Goldin *et al.* (1969). Reprinted with the permission of Cambridge University Press.

### 4-3.1 Linear Stability of an Inviscid Jet

We begin with the inviscid analysis. The continuity equation is unaffected by any assumption about viscosity. If the viscosity is set to zero in the dynamic equations they take the forms

$$\frac{\partial u_r^0}{\partial t} = -\frac{1}{\rho}\frac{\partial p^0}{\partial r} \tag{4-3.5}$$

and

$$\frac{\partial u_z^0}{\partial t} = -\frac{1}{\rho}\frac{\partial p^0}{\partial z}. \tag{4-3.6}$$

We use the superscript $^0$ to denote the inviscid solution. We now define a so-called "velocity potential function" $\phi$ such that

$$u_r^0 = \frac{\partial \phi}{\partial r}, \quad u_z^0 = \frac{\partial \phi}{\partial z}. \tag{4-3.7}$$

Our task now is to find a function $\phi$ that yields velocities that satisfy the dynamic equations. When Eqs. 4-3.7 are substituted into both dynamic equations we obtain a pair of equations that are satisfied by a pressure function of the form

$$p^0 = -\rho\frac{\partial \phi}{\partial t} + \frac{\sigma}{R}. \tag{4-3.8}$$

The constant of integration has been selected as $\sigma/R$ so that the pressure is strictly due to surface tension in the stationary case ($\phi$=0).

When Eqs. 4-3.7 are substituted into the continuity equation we find that $\phi$ must be a solution to

$$\frac{1}{r}\frac{\partial}{\partial r}\left(r\frac{\partial \phi}{\partial r}\right) + \frac{\partial^2 \phi}{\partial z^2} = 0. \tag{4-3.9}$$

(This is why $\phi$ is called a potential function—it is a solution to the Laplace equation.) $\phi$ is expected to be a function of $r$, $z$, and $t$. Based on our observations we know that the disturbance to the radius is periodic along the $z$ axis and that it grows monotonically in time. We seek a solution that has the form

$$\phi = \Phi(r) \exp(ikz + \alpha t). \tag{4-3.10}$$

We choose to write the periodic $z$ dependence in the more compact format $\exp(ikz)$ instead of in terms of sine or cosine functions. The time dependence grows or dampens exponentially according to the magnitude and sign of $\alpha$, which is a growth rate parameter. We call $k$ the *wave number* and $\alpha$ the *growth rate* of the disturbance. We have hypothesized that $\phi$ has the form given in Eq. 4-3.10. In particular, we have assumed that the surface has been disturbed or perturbed in a spatially periodic manner. When we substitute this into the continuity equation (now in the form of Eq. 4-3.9) we find that the assumed forms of the $z$ and $t$ functions satisfy the equation and that $\Phi$ must be the solution to

$$\frac{1}{r}\frac{d}{dr}\left(r\frac{d\Phi}{dr}\right) - k^2\Phi = 0. \tag{4-3.11}$$

This is a Bessel equation whose solutions are the "hyperbolic" Bessel functions $I_0$ and $K_0$. The $K_0$ function is unbounded at $r=0$, so a solution involving $K_0$ cannot satisfy a boundary condition that requires that the velocity be bounded along $r=0$. Hence we write the solution in the form

$$\Phi = AI_0(kr), \tag{4-3.12}$$

where $A$ is a constant of integration. When this is introduced into Eq. 4-3.10 we may write the dynamic variables from Eqs. 4-3.6 and 8 as

$$p^0 = A\rho\alpha I_0(kr)\ \exp(ikz + \alpha t) + \frac{\sigma}{R}, \tag{4-3.13}$$

$$u_r^0 = AkI_1(kr)\ \exp(ikz + \alpha t), \tag{4-3.14}$$

$$u_z^0 = AikI_0(kr)\ \exp(ikz + \alpha t). \tag{4-3.15}$$

(We have used the fact that $dI_0/dr = I_1$, where $I_1$ is another Bessel function.) We will write the disturbance to the jet radius as

$$R^0 = R + \zeta^0(z,t), \tag{4-3.16}$$

where $\zeta^0$ is connected to the radial velocity of the surface by

$$\frac{\partial\zeta^0}{\partial t} = u_r^0 \quad \text{at} \quad r = R.^1 \tag{4-3.17}$$

This permits us to write the surface disturbance as

$$\zeta^0 = A\frac{k}{\alpha} I_1(kR)\ \exp(ikz + \alpha t). \tag{4-3.18}$$

The boundary condition on the pressure function follows from the Young–Laplace equation

$$p^0 = \sigma\left(\frac{1}{R_1} + \frac{1}{R_2}\right) \quad \text{at} \quad r = R. \tag{4-3.19}$$

The radii of curvature are given by the following expressions. Normal to the axis the cross section is circular, with radius $R^0$ given by Eq. 4-3.16. Hence

[1] This boundary condition, as well as the next one given by Eq. 4-3.19, should really be evaluated at $r = R + \zeta$. This complicates application of the boundary condition since $\zeta$ is unknown. We make the approximation that for $\zeta/R$ small enough, the simpler boundary condition will not give rise to a large error.

$$\frac{1}{R_1} = \frac{1}{R+\zeta^0} = \frac{1}{R\left(1+\zeta^0/R\right)} \approx \frac{1-\zeta^0/R}{R}. \tag{4-3.20}$$

In the plane of the axis the radius of curvature is given approximately by

$$\frac{1}{R_2} = -\frac{\partial^2\zeta^0}{\partial z^2}. \tag{4-3.21}$$

Thus the pressure is given by

$$p^0 = \frac{\sigma}{R} - \frac{\sigma}{R^2}\left(\zeta^0 + R^2\frac{\partial^2\zeta^0}{\partial z^2}\right). \tag{4-3.22}$$

This expression is valid only for small-amplitude disturbances such that the approximation that $\zeta^0/R << 1$ is valid. When this expression for $p^0$ is equated to that given in Eq. 4-3.13, and Eq. 4-3.18 is used for $\zeta^0$, we find the following:

$$\frac{\sigma}{R^2}\left(1-k^2R^2\right)\frac{k}{\alpha}I_1(kR) = \rho\alpha I_0(kR). \tag{4-3.23}$$

Note that the integration constant $A$ has disappeared. What we are left with is a relationship between the growth rate $\alpha$ and the wave number of a disturbance to the jet surface, $k$. Solving for $\alpha$, we find

$$\alpha^2 = \frac{\sigma k}{\rho R^2}\left(1-k^2R^2\right)\frac{I_1(kR)}{I_0(kR)}. \tag{4-3.24}$$

Recall (or look at Eq. 4-3.10) that if $\alpha$ is real and positive the disturbances grow exponentially. We find that the ratio of Bessel functions ($I_1/I_0$) is positive for all arguments. Hence $\alpha$ is real and positive so long as $kR < 1$. ( If $kR > 1$ then $\alpha^2$ is negative, $\alpha$ is imaginary, and hence the time dependence is periodic. ) We conclude that any periodic disturbance to the jet which has a wave number that satisfies the condition $0 < kR < 1$ will lead to instability of the jet. For a given value of $kR$ in that range the disturbance will grow at a rate given by Eq. 4-3.24.

Figure 4-3.2 shows the functional dependence of $\alpha$ on $kR$. Note that we have nondimensionalized $\alpha$ in this plot. The most significant feature of this plot is the occurrence of a maximum value of $\alpha$. The maximum occurs at a value of $kR$ = 0.69. Its value is

$$\alpha^{max} = 0.34\left(\frac{\sigma}{\rho R^3}\right)^{1/2}. \tag{4-3.25}$$

What is the physical meaning of this maximum?

When a jet of liquid issues from a capillary it is subject to environmental noise that arises from ever-present vibrations transmitted to the capillary through its support structure or through the air. These vibrations may correspond to a spectrum of frequencies at the exit of the capillary, and this spectrum in turn creates disturbances in the jet flow and radius that correspond to a spectrum of wavelengths. According to this mathematical model, some disturbances will damp out and others will grow. Among those that grow, there is one wavelength that grows most rapidly, at the rate corresponding to $\alpha^{max}$. In order to connect the theory to observations, we make the assumption that the jet breaks down from the most rapidly growing disturbance. Hence we expect to observe uniformly spaced drops, and the spacing should satisfy the following constraint:

$$\frac{L}{2\pi R} = \frac{1}{0.69} \tag{4-3.26}$$

or

$$L = 9.1R = 4.55D. \tag{4-3.27}$$

Since we assume that the observed disturbance corresponds to $\alpha^{max}$, then the radius of the disturbance is expected to grow as

$$\zeta^0 = \zeta_i \, e^{\alpha^{max} t}, \tag{4-3.28}$$

where $\zeta_i$ is the magnitude of the disturbance corresponding to $\alpha^{max}$. But we do not know the magnitude of this initial disturbance! It is too small to see on a photograph of a jet. It is possible to infer the magnitude of this intial disturbance in the following way. From Eq. 4-3.28 we can find the time for the disturbance to break up the jet, based on the assumption that this corresponds approximately to the time at which $\zeta^0 = R$. The result is

$$t^* = \frac{L^*}{U} = \frac{1}{\alpha^{max}} \ln \frac{R}{\zeta_i} = \frac{C}{\alpha^{max}}. \tag{4-3.29}$$

Note that if the jet is traveling at the constant speed $U$, then the distance from the capillary exit at which the first drops appear is $L^* = Ut^*$. The constant $C$ replaces the unknown ratio of the initial disturbance to the radius of the jet. If we now nondimensionalize the breakup length with the jet diameter we find

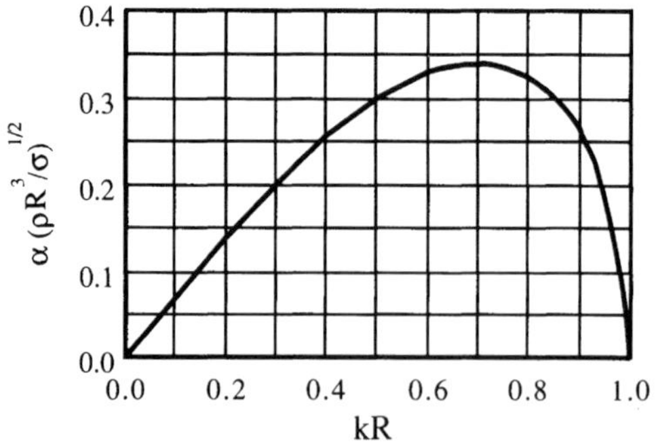


***FIGURE 4-3.2*** Dimensionless growth rate parameter as a function of the dimensionless wave number of a disturbance on an inviscid liquid jet.

$$\frac{L^*}{D} = 1.04C\left(\frac{\rho U^2 D}{\sigma}\right)^{1/2} = 1.04C\ \mathrm{We}^{1/2}. \tag{4-3.30}$$

(Note that we use the capillary *diameter D* in the Weber number, rather than the radius. We are, at this stage, ignoring the fact that the *jet* diameter and speed will be different from those in the capillary, because of the contraction of the jet just after it exits the capillary.)

At this stage we have a testable mathematical model. Figure 4-3.3 (Grant and Middleman, 1966) shows data for breakup length as a function of jet velocity, plotted according to the dimensionless groups suggested by Eq. 4-3.30. With the exception of one set of data the model correlates all of the data when a value of $C = 13$ is selected. The "exceptional" set of data is for a liquid with a viscosity of 0.16 Pa-s. The other data are for low-viscosity liquids, ranging from 0.001 to 0.026 Pa-s. It seems reasonable to assume that the effect of a high viscosity is to retard the rate of growth of the disturbance and thus lead to a larger breakup length. We conclude that the theory for an inviscid liquid does a very good job of predicting the observed parametric behavior of the breakup length, even up to viscosities an order of magnitude greater than that of water.

### *4-3.2 Linear Stability of the Viscous Jet*

Next we extend this stability theory to the viscous jet. One way to do this is to write the solutions for the velocity components as the sum of the inviscid

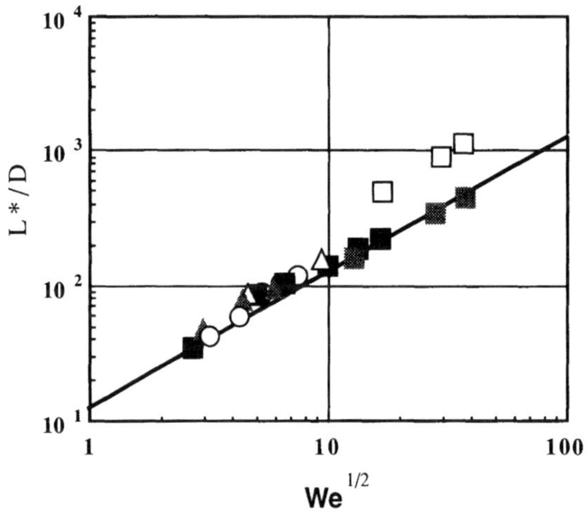


***FIGURE 4-3.3*** Breakup length data plotted as a test of Eq. 4-3.30. The line is for $C = 13$.

solutions plus a pair of additional functions:

$$u_r = u_r^0 + u_r^v,$$

$$u_z = u_z^0 + u_z^v. \tag{4-3.31}$$

The superscript $^v$ reminds us that this is the viscous contribution to the solution. When these equations are substituted into Eqs. 4-3.2 and 3, and the properties of the inviscid solutions are accounted for, the result is a pair of partial differential equations for the two viscous components of the velocity vector and the pressure. The third equation is supplied by the continuity equation. If a stream function is now introduced, defined as usual by the equations

$$u_r^v = \frac{1}{r}\frac{\partial\psi}{\partial z} \quad \text{and} \quad u_z^v = -\frac{1}{r}\frac{\partial\psi}{\partial r}, \tag{4-3.32}$$

we may eliminate pressure from the two momentum equations and obtain a single equation for the stream function. If we again assume that for the stability analysis we may write, in this case, $\psi$ in the form

$$\psi = \Psi(r)\ \exp(ikz + \alpha t), \tag{4-3.33}$$

we find, after quite a bit of algebra, that $\Psi$ must be the solution to

$$\frac{d^2\Psi}{dr^2} - \frac{1}{r}\frac{d\Psi}{dr} - \left(k^2 + \frac{\alpha}{\nu}\right)\Psi = 0. \tag{4-3.34}$$

The solution for the inviscid portion may now be written in the form (see Eq. 4-3.12)

$$\Phi = A_1\, I_0(kr) \tag{4-3.35}$$

and the solution to Eq. 4-3.34 may be found in the form

$$\Psi = A_2\, r\, I_1(k'r) \tag{4-3.36}$$

where $k'^2 \equiv k^2 + \alpha/\nu$.

As a result of all of this mathematical manipulation we may write the velocity and pressure fields in the forms

$$u_r = -ik\left[iA_1\, I_0'(kr) - A_2\, I_1(k'r)\right] \exp(ikz + \alpha t),$$

$$u_z = k\left\{iA_1\, I_0(kr) - A_2\left[\frac{I_1(k'r)}{kr} + \frac{k'}{k}\, I_1'(k'r)\right]\right\} \exp(ikz + \alpha t),$$

$$p = -\rho\alpha\, A_1\, I_0(kr)\ \exp(ikz + \alpha t). \tag{4-3.37}$$

(The prime marks $'$ on $I_0$ and $I_1$ denote derivatives with respect to their arguments.)

We now impose boundary conditions in order to find relationships among the integration constants $A_1$ and $A_2$, and the growth rate and wave number parameters $\alpha$ and $k$. The boundary conditions are for the *viscous* jet.

First, we set the shear stress at the jet surface equal to zero:

$$\tau_{rz} = \mu \left( \frac{\partial u_z}{\partial r} + \frac{\partial u_r}{\partial z} \right)_{r=R} = 0. \qquad (4\text{-}3.38)$$

Note that we apply the boundary condition to the undisturbed surface $r = R$ as an approximation. The result is a relationship between $A_1$ and $A_2$:

$$A_1 = -A_2 \frac{I_1(k'R)(k'^2 + k^2)}{2ik^2 I_1(kR)}. \qquad (4\text{-}3.39)$$

The second boundary condition is on the equality of the radial stress across the jet surface, with proper account taken of the surface tension contribution from the Young–Laplace equation:

$$[\tau_{rr} + p]_{r=R} = \sigma \left( \frac{1}{R_1} + \frac{1}{R_2} \right). \qquad (4\text{-}3.40)$$

The left-hand side is given by

$$\tau_{rr} + p = 2\mu k \left[ -kA_1 I_1'(kr) + ik'A_2 I_1'(k'r) + \frac{\alpha}{2\nu k} A_1 I_0(kr) \right] \exp(ikz + \alpha t) + \frac{\sigma}{R}. \qquad (4\text{-}3.41)$$

We include the $\sigma/R$ term so that Eq. 4-3.41 reduces to the correct result for the "static" case. For the right-hand side we take note of the facts that

$$\frac{1}{R_1} = \frac{1}{R + \zeta} = \frac{1}{R\,(1 + \zeta/R)} \approx \frac{1}{R} - \frac{\zeta}{R^2} \qquad (4\text{-}3.42)$$

for a circular cross section, where $\zeta$ is the deviation of the radius from the value $R$, and

$$\frac{1}{R_2} \approx -\frac{\partial^2 \zeta}{\partial z^2}. \qquad (4\text{-}3.43)$$

for small slopes $\partial\zeta / \partial z$. Hence we need an expression for the displacement of the radius, $\zeta$. This follows from the kinematics as

$$\frac{\partial \zeta}{\partial t} = (u_r)_{r=R} \qquad (4\text{-}3.44)$$

or

$$\zeta = \frac{ik}{\alpha} \left[ iA_1 I_1(kR) + A_2 I_1(k'R) \right] \exp(ikz + \alpha t). \qquad (4\text{-}3.45)$$

This result, with Eqs. 4-3.42 and 43, lets us write Eq. 4-3.40 as a transcendental equation for the growth rate parameter $\alpha$:

$$\alpha^2 + \frac{2\nu k^2}{I_0(kR)}\left[ I_1'(kR) - \frac{2kk'}{k^2+k'^2}\frac{I_1(kR)}{I_1(k'R)} I_1'(k'R)\right]\alpha$$

$$= \frac{\sigma k}{\rho R^2}\left[1-(kR)^2\right]\frac{I_1(kR)}{I_0(kR)}\frac{k'^2-k^2}{k'^2+k^2}. \tag{4-3.46}$$

Although this is written in the form of a quadratic equation for $\alpha$ it is much more complex, since $k'$ is a function of $\alpha$, and so $\alpha$ appears in the arguments of several Bessel functions. If $kR$ is less than unity a reasonable approximation for the Bessel functions is possible, with the result that Eq. 4-3.46 takes a simpler form:

$$\alpha^2 + \frac{3\mu(kR)^2}{\rho R^2}\alpha = \frac{\sigma}{2\rho R^3}\left[1-(kR)^2\right](kR)^2. \tag{4-3.47}$$

As before, $\alpha$ has a maximum value for $kR$ on the range [0,1], and it is easily found to be

$$\alpha^{\max} = \left[\left(\frac{8\rho R^3}{\sigma}\right)^{1/2} + \frac{6\mu R}{\sigma}\right]^{-1}. \tag{4-3.48}$$

(For $\mu = 0$, Eq. 4-3.48 differs slightly from Eq. 4-3.25, which does not involve an approximation for the Bessel function term.) If the path from Eq. 4-3.29 to Eq. 4-3.30 is followed for the case of the viscous jet we find the breakup length to be of the form

$$\frac{L^*}{D} = C\left(\mathrm{We}^{1/2} + \frac{3\,\mathrm{We}}{\mathrm{Re}}\right) = C\,\mathrm{We}^{1/2}\left(1 + 3\,\mathrm{Oh}\right), \tag{4-3.49}$$

where the Ohnesorge number is based on the capillary *diameter*, rather than on the radius as in Chapter 3.

Experimental data for viscous Newtonian liquids are shown in Fig. 4-3.4 (Kroesser and Middleman, 1969). The data are in very good agreement with Eq. 4-3.49 when the *jet* diameter and velocity are used, rather than the capillary diameter and average velocity in the capillary. (In other words, account is taken, in plotting these data, of the contraction of the jet that occurs over a distance that is short in comparison to the breakup length. The difference in the coefficient $C$ in this plot, compared to that of Fig. 4-3.3, arises from this change.) The data cover a range of Ohnesorge numbers of $0.28 < \mathrm{Oh} < 1.03$.

### *4-3.3 A One-Dimensional Approximation to the Jet Equations*

An alternative approach to solving the Navier–Stokes equations with the continuity equation is possible, which utilizes a one-dimensional approximation to these equations that is based on the idea that the radial velocity is very small compared to the axial velocity. The result is a partial differential equation with only one space variable, a reduction in dimensionality that makes it easier to obtain analytical solutions useful for computational purposes.

The case of the inviscid jet is easier to start with. We begin with the continuity and $z$ directed momentum equations, but in the absence of viscosity and gravity. We maintain the presence of the nonlinear inertial terms for the moment. Thus we begin with

$$\rho\left(\frac{\partial u_z}{\partial t} + u_r \frac{\partial u_z}{\partial r} + u_z \frac{\partial u_z}{\partial z}\right) = -\frac{\partial p}{\partial z} \tag{4-3.50}$$

and

$$\frac{\partial u_z}{\partial z} + \frac{1}{r}\frac{\partial (r\, u_r)}{\partial r} = 0. \tag{4-3.51}$$

An approximation to these equations can be developed along several lines. As they stand, we have two equations (plus the radial momentum equation, which we have not written here) in the three unknown functions, $u_r$, $u_z$, and $p$. One possible direction for simplification is to reduce the "dimensionality" of the problem by, somehow, getting rid of the radial component of velocity. For example, we may integrate the continuity equation across the cross section of the jet:

$$\int_0^{R(z,t)} r\left[\frac{\partial u_z}{\partial z} + \frac{1}{r}\frac{\partial (r\, u_r)}{\partial r}\right] dr = \frac{\partial}{\partial z}\int_0^R u_z r\, dr + \int_0^R \frac{\partial (r\, u_r)}{\partial r}\, dr$$
$$= \frac{\partial}{\partial z}\left(\frac{1}{2}R^2 u\right) + R\,(u_r)_R = \frac{\partial}{\partial z}\left(\frac{1}{2}R^2 u\right) + R\frac{\partial R}{\partial t}$$
$$= \frac{\partial}{\partial z}\left(\frac{1}{2}R^2 u\right) + \frac{1}{2}\frac{\partial R^2}{\partial t}. \tag{4-3.52}$$

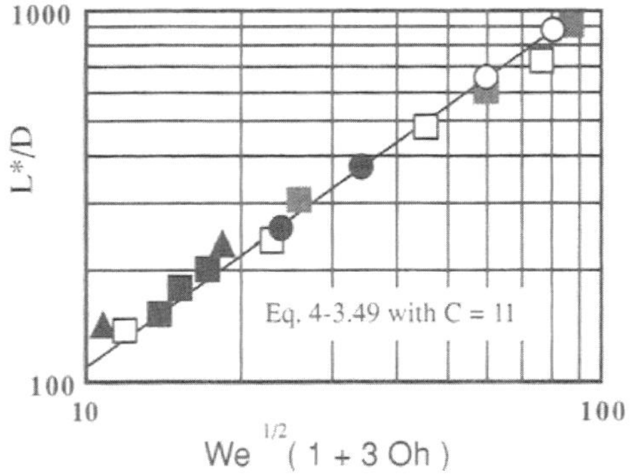


***FIGURE 4-3.4*** Breakup length for viscous Newtonian jets, testing Eq. 4-3.49.

In this equation we introduce the average axial velocity $u$ across the radius:

$$u \equiv \frac{2}{R^2}\int_0^{R(z,t)} u_z(r,z,t)r\,dr. \tag{4-3.53}$$

We may then write the continuity equation in the form

$$\frac{\partial R^2}{\partial t} + \frac{\partial}{\partial z}\left(R^2 u\right) = 0. \tag{4-3.54}$$

Manipulation of Eq. 4-3.50 is a little more difficult. We could proceed along the following line. We could assume that the radial velocity term in the inertial terms on the left-hand side of Eq. 4-3.50 is negligible in comparison to the other inertial terms. We might try to support that argument with the idea that if the change of radius $\partial R/\partial z$ were small in some sense, then the radial velocity would vanish exactly. Making that assumption, and using Eq. 4-3.53, we convert Eq. 4-3.50 to the form

$$\rho\left(\frac{\partial u}{\partial t} + u\frac{\partial u}{\partial z}\right) = -\frac{\partial p}{\partial z}. \tag{4-3.55}$$

The problem with this approach is that it is not at all clear how much of an error we make with this approximation. For example, we see from the continuity equation in the form of Eq. 4-3.51 that if $u_r$ were everywhere zero, then $u_z$ would not vary in the $z$ direction, and $\partial R/\partial z$ would vanish. Obviously, this is inconsistent with the physics of interest to us. Attempts at formalizing this kind of approach, and making consistent approximations, can be found in the works of Matovich and Pearson (1969), Lee (1974), Pimbley (1976), Pimbley and Lee (1977), and in the greatest detail Green (1976) and Bogy (1978). We will simply take Eqs. 4-3.54 and 55 as a starting point.

The dependent variables are now the averaged axial velocity $u$ and the jet radius $R$. The dimensionality of the problem has been reduced because these variables depend only on one space variable, $z$, and on time $t$. We must add an equation for the pressure $p(z,t)$ in order to close this set of equations. For the inviscid jet the pressure term is written in terms of the surface tension contribution only:

$$p = \sigma\left(\frac{1}{R_{\mathrm{N}}} + \frac{1}{R_{\mathrm{T}}}\right), \tag{4-3.56}$$

where the principal radii of curvature of the jet surface are defined on Fig. 4-3.5.

From this geometry we can find

$$R_{\mathrm{N}} = R\left[1 + \left(\frac{\partial R}{\partial z}\right)^2\right]^{1/2}$$

and

$$R_T = \frac{\left[1 + (\partial R/\partial z)^2\right]^{3/2}}{-\partial^2 R/\partial z^2} \tag{4-3.57}$$

Now we suppose that just downstream of the capillary exit the jet has the uniform velocity $U_0$ and radius $R_0$. We will write the velocity of the jet, in the region where the jet is perturbed, in the form

$$v = u(z,t) - U_0 \tag{4-3.58}$$

and the jet radius will be written as

$$R = R_0\,(1 + \delta). \tag{4-3.59}$$

We change to a moving coordinate system of velocity $U_0$, so we have a new axial coordinate defined by

$$\eta = z - U_0\, t. \tag{4-3.60}$$

Now the system of equations becomes transformed to

$$\frac{\partial v}{\partial t} + v\frac{\partial v}{\partial \eta} = -\frac{1}{\rho}\frac{\partial p}{\partial \eta} \tag{4-3.61}$$

and

$$\frac{\partial R}{\partial t} + v\frac{\partial R}{\partial \eta} + \frac{R}{2}\frac{\partial v}{\partial \eta} = 0. \tag{4-3.62}$$

**The Linear Analysis** Lee (1974) carried out a linearization of these equations on the assumption that the variables $u/U_0$ and $\delta$ were small compared to unity. When these assumptions are introduced into Eqs. 4-3.61, 62, 56, and 57

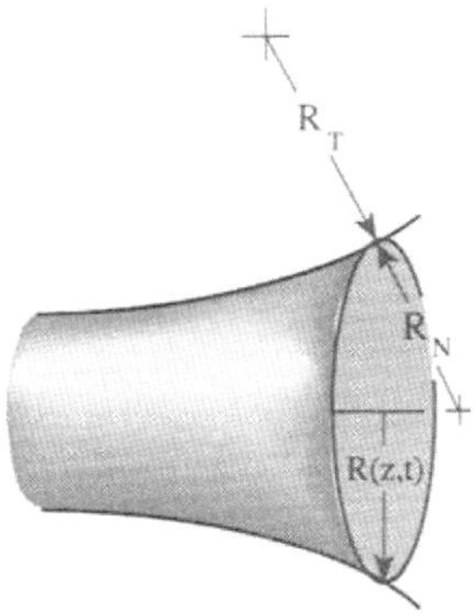


*FIGURE 4-3.5* Geometry of the jet surface.

and quadratic terms in these small quantities are dropped, the result is

$$\frac{\partial v}{\partial t} = \frac{\sigma}{\rho R_0}\left(\frac{\partial \delta}{\partial \eta} + R_0^2 \frac{\partial^3 \delta}{\partial \eta^3}\right) \tag{4-3.63}$$

and

$$\frac{\partial v}{\partial \eta} = -2\frac{\partial \delta}{\partial t}. \tag{4-3.64}$$

We may now eliminate the velocity between these two equations and find a single linear equation for the perturbed radius $\delta$:

$$\frac{\partial^2 \delta}{\partial t^2} + \frac{\sigma}{2\rho R_0}\frac{\partial^2 \delta}{\partial \eta^2} + \frac{R_0 \sigma}{2\rho}\frac{\partial^4 \delta}{\partial \eta^4} = 0. \tag{4-3.65}$$

If we now write the perturbation to the jet radius in the form

$$\delta(\eta,t) = \delta_0 \exp(\alpha t + ik\eta) \tag{4-3.66}$$

and substitute this into Eq. 4-3.65 we find (see Eq. 4-3.24)

$$\frac{2\rho R_0^3 \alpha^2}{\sigma} = (R_0 k)^2\left[1 - (R_0 k)^2\right]. \tag{4-3.67}$$

The growth rate parameter $\alpha$ obeys an equation very similar in form to that given by the classical analysis of Rayleigh, and we find that numerically the result for $\alpha^{max}$ is little different from that given in Eq. 4-3.25. Hence the breakup length is only a little different from that given in Eq. 4-3.30.

In the analysis that leads to the result above it is assumed that the initial disturbance to the jet is to its *radius* and that it is of the form of Eq. 4-3.66 for $t = 0$. Another form of perturbation is possible, and this is to the *velocity* of the jet as it emerges from the capillary. For example, we could perturb the jet sinusoidally according to

$$v(\eta,0) = \Delta u \sin(2\pi\eta/\lambda). \tag{4-3.68}$$

Here $\Delta u$ is the magnitude of the disturbance velocity. The disturbed radius of the jet will then be sought in the form

$$\delta(\eta,t) = A(t) \cos(2\pi\eta/\lambda). \tag{4-3.69}$$

When this is substituted into Eq. 4-3.65 we find a differential equation for the amplitude of the disturbance, $A(t)$:

$$\frac{d^2A}{dt^2} - \beta^2\frac{4\pi^2}{\lambda^2}\left(1 - \frac{4\pi^2}{\lambda^2}R_0^2\right)A = 0, \tag{4-3.70}$$

where

$$\beta^2 \equiv \frac{\sigma}{2\rho R_0}. \tag{4-3.71}$$

The solution for $A(t)$ is

$$A(t) = C_1 \cosh \alpha t + C_2 \sinh \alpha t \tag{4-3.72}$$

with the growth rate parameter given by

$$\alpha^2 = \frac{\sigma}{2\rho R_0} \frac{4\pi^2}{\lambda^2} \left(1 - \frac{4\pi^2}{\lambda^2} R_0^2\right). \tag{4-3.73}$$

(This follows from Eq. 4-3.67, with wave number $k$ replaced by $2\pi/\lambda$.)

From Eqs. 4-3.69 and 72 we may write the radius of the jet in the form

$$\delta(\eta,t) = \left(C_1 \cosh \alpha t + C_2 \sinh \alpha t\right) \cos(2\pi\eta/\lambda). \tag{4-3.74}$$

We must determine the constants $C_1$ and $C_2$ from initial conditions on $\delta$ and $\partial\delta/\partial t$ at time zero: $\delta(\eta,0) = 0$ at $t = 0$. (It is the velocity, and *not the radius*, that is initially disturbed.) So it follows from Eq. 4-3.74 that

$$C_1 = 0. \tag{4-3.75}$$

For $\partial\delta/\partial t$ we return to Eq. 4-3.64 and note that

$$\frac{\partial\delta(\eta,0)}{\partial t} = -\frac{1}{2} \frac{\partial v(\eta,0)}{\partial\eta} = -\Delta u \frac{\pi}{\lambda} \cos\left(\frac{2\pi\eta}{\lambda}\right) \tag{4-3.76}$$

or, using Eq. 4-3.74,

$$\alpha C_2 = -\Delta u \frac{\pi}{\lambda}. \tag{4-3.77}$$

Therefore the disturbance to the jet radius grows as

$$\delta(\eta,t) = \Delta u \frac{\pi}{\alpha\lambda} \cos\frac{(2\pi\eta)}{\lambda} \sinh \alpha t. \tag{4-3.78}$$

The breakup time is defined as the time at which the minimum in $\delta(\eta,t)$ (which occurs at $\lambda/\eta = 2$) is equal to the initial jet radius. Then the breakup length follows as

$$L^* = \frac{U_0}{\alpha} \sinh^{-1}\left(\frac{\alpha\lambda}{\pi\Delta u}\right). \tag{4-3.79}$$

We should pause for a moment to recall the physics of the case that we are examining here. In previous analyses of the stability of the jet it was assumed that the *radius* was subject to a *spectrum* of disturbances and that the jet broke down under the influence of the most rapidly growing disturbance (i.e., at $\alpha^{max}$.) In the case being illustrated here, we assume that the jet is subjected to a single perturbation to the *velocity*, of a known and controlled amplitude and wavelength, and that the jet breaks down according to the growth rate at this wavelength. That is why $\lambda$ and $\alpha$ both appear in Eq. 4-3.79. We could use Eq. 4-3.73 to find the wavelength of the most rapidly growing disturbance, and then $\alpha^{max}$ would follow.

To summarize briefly, we find that the one-dimensional approximation to the jet equations yields a stability result that is practically identical to that of Rayleigh, for the inviscid jet. We can use this approximation method to explore the stability of the jet to other forms of disturbance than that to the radius. From a practical point of view it is more important to study the case of *velocity* perturbation because that is most easily achieved and controlled in the laboratory realizations of these jets.

If we now wish to examine the effect of viscosity we must obtain the corresponding one-dimensional equations for the *viscous* jet. Equation 4-3.51 is of course still valid, so we still have Eq. 4-3.54. We must add to Eq. 4-3.62 the appropriate contributions that arise from viscous effects within the jet. When this is done we find (see Problem P4-5)

$$\frac{\partial^2\delta}{\partial t^2} + \frac{\sigma}{2\rho R_0}\frac{\partial^2\delta}{\partial\eta^2} + \frac{R_0\sigma}{2\rho}\frac{\partial^4\delta}{\partial\eta^4} - \frac{3\mu}{\rho}\frac{\partial^3\delta}{\partial\eta^2\partial t} = 0. \tag{4-3.80}$$

Equations similar to this are to be found in several related studies of liquid jet dynamics. Bogy (1978) derives an equation similar to this using what is called Cosserat theory. His equation has some additional terms, corresponding to viscous and inertial effects—ignored here—associated with the *radial* flow in the jet. Matovich and Pearson (1969) give what amounts to the steady-state version of Eq. 4-3.80. If we go back to a surface disturbance of the form of Eq. 4-3.66 and follow the subsequent steps of the stability analysis, but include the viscous term, we find that Eq. 4-3.80 leads to exactly Eq. 4-3.47.

We conclude that the one-dimensional linear approximation to the jet equations yields essentially identical results to those of the two-dimensional theories for both the inviscid and viscous jets.

### *4-3.4 The Nonlinear Analysis of Jet Stability*

An important feature of jet breakup is the existence of satellite drops, the smaller drops that appear in Fig. 4-3.1. The *linear* stability analyses for inviscid or viscous jets fail to predict satellite drops—they are a nonlinear phenomenon. We would like to be able to predict the existence of drops of different sizes on a single jet, their relative sizes, and possibly define conditions under which their appearance could be avoided. Hence we must turn to nonlinear analyses of the jet dynamics.

One way to do a nonlinear analysis is to use a full numerical solution to the Navier–Stokes equations, along the lines of the analysis of Section 4-2 for the impulsive drop on demand flow (Lafrance, 1975). An alternative that utilizes a one-dimensional approximation to the jet equations is also possible, and we illustrate that first.

Because of the linearity of Eq. 4-3.65, a sinusoidal initial disturbance remains sinusoidal, and the jet ultimately breaks down into an infinite train of uniformly sized drops. Of course, by the time the disturbance is near the breakup point, the linear assumptions no longer hold. Hence a nonlinear analysis is necessary if we hope to describe the observation of satellite drops. Lee (1974) has done this for the inviscid jet by carrying out a numerical solution of Eqs. 4-3.61 and 62 (with 4-3.56 and 57 for the pressure term). The viscous case has been done, using the one-dimensional jet equations, by Bousfield and Denn (1987) and Bousfield *et al.* (1990). Other approaches will be cited as appropriate.

Let's look first at some of the observations, particularly with respect to the key phenomenon—satellite drop formation. Experiments have been carried out with low-viscosity liquids by Lafrance (1975) and by Rutland and Jameson (1970). For small initial disturbances *to the jet radius* measurements were made of the radii of the main and satellite drops. The results are shown in Fig. 4-3.6 for jets perturbed at different wave numbers. We see that the relative size of the primary and satellite drops depends strongly on the wave number of the disturbance. As the dimensionless wave number approaches a value of approximately $kR_0 = 0.7$, the satellite drops become very small compared to the primary drops. It is these observations that we would hope to predict with a nonlinear stability analysis.

In the one-dimensional approximation we must solve the integrated forms of the continuity equation (Eq. 4-3.54)

$$\frac{\partial R^2}{\partial t} + \frac{\partial}{\partial z}\left(R^2 u\right) = 0. \tag{4-3.81}$$

and the momentum equation

$$\rho\left(\frac{\partial u}{\partial t} + u\frac{\partial u}{\partial z}\right) = -\frac{\partial p}{\partial z} + 3\mu\frac{\partial^2 u}{\partial z^2}. \tag{4-3.82}$$

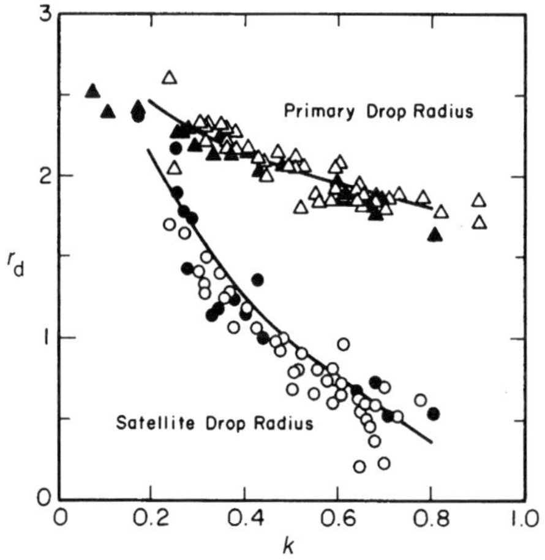


***FIGURE 4-3.6*** Dependence of drop radius on wave number of disturbance. Reprinted with permission from Bousfield *et al.* 1986. In their notation, $k$ is a dimensionless wave number: $k = 2\pi R_0/\lambda$, and $r_d$ is $R_{drop}/R_0$.

In the nonlinear case the pressure arising from surface tension must be written as in Eq. 4-3.56, with Eqs. 4-3.57 for the curvature terms. It is convenient to nondimensionalize these equations in the following manner. From the parameters that characterize the jet ($\rho$, $\sigma$, $R_0$, $U_0$) we can form the following dimensionless variables:

$$\tilde{t} \equiv \frac{t}{\sqrt{\rho D_0^3 / \sigma}}, \quad \tilde{z} \equiv \frac{z}{D_0}, \quad \tilde{u} \equiv \frac{u}{\sqrt{\sigma/\rho D_0}},$$

$$\tilde{p} \equiv \frac{p}{\sigma/D_0}, \quad \tilde{R} \equiv \frac{R}{R_0}. \tag{4-3.83}$$

(The use of the capillary diameter rather than the radius is an arbitrary choice.) Then Eqs. 4-3.81 and 82 become

$$\frac{\partial \tilde{R}^2}{\partial \tilde{t}} + \frac{\partial}{\partial \tilde{z}}\left(\tilde{R}^2 \tilde{u}\right) = 0, \tag{4-3.84}$$

$$\frac{\partial \tilde{u}}{\partial \tilde{t}} + \tilde{u}\frac{\partial \tilde{u}}{\partial \tilde{z}} = -\frac{\partial \tilde{p}}{\partial \tilde{z}} + 3\,\mathrm{Oh}\,\frac{\partial^2 \tilde{u}}{\partial \tilde{z}^2}, \tag{4-3.85}$$

where the Ohnesorge number is defined with the jet diameter as the length scale:

$$\mathrm{Oh} \equiv \frac{\mu}{\sqrt{\sigma D_0 \rho}}. \tag{4-3.86}$$

It may seem odd that this nondimensionalization does not introduce a dynamic group such as the Weber number. (Note that the Ohnesorge number has no velocity in it.) However, we must consider the initial conditions for the stability analysis, for example (Eq. 4-3.68)

$$R(z,0) = R_0,$$
$$u(z,0) = U_0 + \Delta u \sin(2\pi z/\lambda), \tag{4-3.87}$$

where $\Delta u$ is the magnitude of the disturbance velocity. When these are nondimensionalized we find

$$\tilde{R} = 1$$

$$\tilde{u}(\tilde{z},0) = \frac{U_0}{\sqrt{\sigma/\rho D_0}} + \frac{\Delta u}{U_0}\sin\frac{(2\pi z)}{\lambda}$$
$$= \mathrm{We}^{1/2} + \frac{\Delta u}{U_0}\sin\left(2kR_0\tilde{z}\right). \tag{4-3.88}$$

Now, and as expected, a Weber number appears (based on diameter, not radius) and a dimensionless velocity perturbation $\Delta u/U_0$ appears, as well.

Numerical solutions of Eqs. 4-3.84 and 85 were carried out by Bousfield, using initial perturbations to either the velocity (as in Eqs 4-3.87) or the radius. The details and results are presented in Bousfield *et al.*(1986) and Bousfield *et al.* (1990). Of particular interest to us are the predictions of the one-dimensional theory for the size and occurrence of satellite drops. These predictions are shown on Fig. 4-3.6, for the case of radius perturbations, and the agreement with the corresponding data is excellent. Bousfield *et al.* (1986) also show that full numerical solutions of the dynamic equations, using a finite-element method, agree very closely with the results based on the one-dimensional model.

As noted earlier, an important issue is the possibility of the elimination of satellite drops. Based on the dimensional analysis just above, we would expect that the presence or absence of satellite drops would depend on the set of parameters Oh, We, $kR_0$, and (for velocity perturbations) $\Delta u/U_0$. Figure 4-3.7 (from Bousfield *et al.*, 1990) shows examples of one-dimensional simulations of jet radius at the point in time where $R(z)$ has vanished somewhere along a wavelength. This is taken as the breakup time. In Fig. 4-3.7a breakup occurs leaving a finite amount of liquid between the primary drops. This will form a satellite drop. At the higher Ohnesorge number shown in Fig. 4-3.7b the drops break, leaving a very thin "thread" connecting them. Surface tension forces (pressure) drive this small amount of liquid into the primary drops, and no satellite is seen. Bousfield indicates that satellite drops are eliminated at Ohnesorge numbers of the order of unity.

The control of satellite drops is demonstrated and discussed in an excellent series of papers by Chaudhary *et al.* (1980a–c). They impose a periodic pressure pulsation on the fluid just before the orifice and examine the effect of frequency on satellite drop formation. The experimental work complements a theoretical analysis based on solutions of the nonlinear equations for an inviscid axisymmetric jet.

An interesting and related problem arises when a liquid is ejected *upward* from a vertical capillary in a gravitational field. This is a so-called capillary fountain, and numerical solutions and some very nice photographs can be found in a paper by Schulkes (1994).

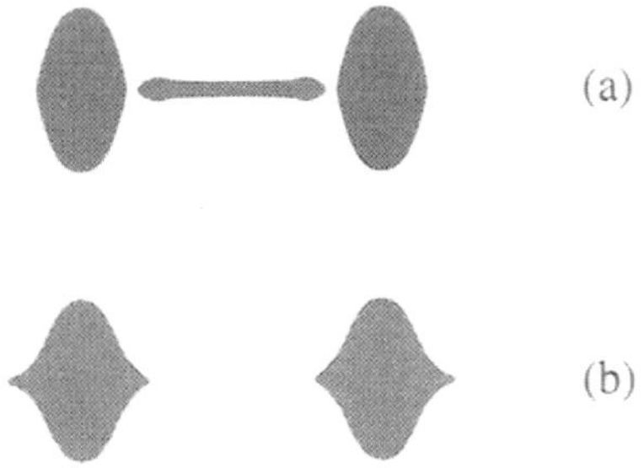


***FIGURE 4-3.7*** Numerical simulations of drop breakup. Formation of satellite drops depends on the Ohnesorge number, $\text{Oh} = \mu/(\rho\sigma d)^{1/2}$, where $d$ is the initial diameter. (a) Oh = 0.4; (b) Oh = 2.8. From Bousfield *et al.* (1990). Reprinted with the permission of Cambridge University Press.

## *Problems*

P4-1 For the transient problem illustrated in Section 4-2, what is the Reynolds number of this flow at equilibrium?

P4-2 Wilson (1988) presents a mathematical model and some data on the dripping of liquids from a capillary. How do his data compare to the observations in Fig. 4-1.3? What is the Weber number range for his data? Put his model in a dimensionless format.

P4-3 Manfre (1966) presents data on drop formation and writes the drop volume in the form

$$V = K(\sigma\pi D_o/\rho g), \tag{P4-3.1}$$

where $D_o$ is the outside diameter of the capillary and $K$ is an empirical coefficient. For several different liquids, Manfre finds that $K$ is in the range $0.6 < K < 0.85$. What $K$ value corresponds to the data of Fig. 4-1.4?

P4-4 Find $\alpha^{max}$ from Eq. 4-3.67 and compare it to Rayleigh's result, Eq. 4-3.25.

P4-5 Derive the viscous term in Eq. 4-3.80. Go back to Eq. 4-3.50. If viscosity is included, it takes the form

$$\rho\left(\frac{\partial u_z}{\partial t} + u_r\frac{\partial u_z}{\partial r} + u_z\frac{\partial u_z}{\partial z}\right) = \frac{\partial T_{zz}}{\partial z}, \tag{P4-5.1}$$

where

$$T_{zz} = -p + 2\mu\frac{\partial u_z}{\partial z} \tag{P4-5.2}$$

is the total axial stress arising from the deformation of the liquid. (It does not include the contribution of surface tension.) In a viscous Newtonian liquid the pressure is defined as the mean normal stress:

$$p = -\tfrac{1}{3}\left(T_{zz} + T_{rr} + T_{\theta\theta}\right). \tag{P4-5.3}$$

For axisymmetric flow there is no $\theta$ component of stress. If we neglect the radial flow we can also set $T_{rr} = 0$. Then we find

$$p = -\frac{T_{zz}}{3} \tag{P4-5.4}$$

and

$$T_{zz} = 3\mu\frac{\partial u_z}{\partial z}. \tag{P4-5.5}$$

We make use of Eq. 4-3.64, in this linear analysis. Equation 4-3.80 then follows, with the appropriate change in notation.

P4-6 In the nondimensionalization shown in Eqs. 4-3.83 we did not use the viscosity. Suppose the variables are made dimensionless using the following:

$$\tilde{t} \equiv \frac{t}{\mu R_0 / \sigma}, \quad \tilde{z} \equiv \frac{z}{R_0}, \quad \tilde{u} \equiv \frac{u}{U_0},$$

$$\tilde{p} \equiv \frac{p}{\sigma / R_0}, \quad \tilde{R} \equiv \frac{R}{R_0}. \tag{P4-6.1}$$

How are Eqs. 4-3.84 and 85 changed? Use Eq. 4-3.87 as the initial condition.

## *References*


Bird, R. B., W. E. Stewart and E. N. Lightfoot. (1960). *Transport Phenomena*. John Wiley, N.Y., p. 129.

Bogy, D. B. (1978). Use of one-dimensional Cosserat theory to study instability in a viscous jet. *Phys. Fluids* **21**, 190.

Boucher, E. A., and M. J. B. Evans. (1975). Pendent drop profiles and related capillary phenomena. *Proc. R. Soc. Lond. A*. **346**, 349.

Bousfield, D. W., R. Keunings, G. Marucci, and M. M. Denn. (1986). Nonlinear analysis of the surface tension driven breakup of viscoelastic filaments. *J. Non-Newt. Fluid Mech.* **21**, 79.

Bousfield, D. W. and M. M. Denn. (1987). Jet breakup enhanced by an initial pulse. *Chem. Eng. Comm.* **53**, 61.

Bousfield, D. W., I. H. Stockel, and C. K. Nanivadekar. (1990). The breakup of viscous jets with large velocity modulations. *J. Fluid Mech.* **218**, 601.

Chaudhary, K. C., and L. G. Redekopp. (1980a). The nonlinear capillary instability of a liquid jet. Part 1. Theory. *J. Fluid Mech.* **96**, 257.

Chaudhary, K. C., and T. Maxworthy. (1980b). The nonlinear capillary instability of a liquid jet. Part 2. Experiments on jet behavior before droplet formation. *J. Fluid Mech.* **96**, 275.

Chaudhary, K. C., and T. Maxworthy. (1980c). The nonlinear capillary instability of a liquid jet. Part 3. Experiments on satellite formation and control. *J. Fluid Mech.* **96**, 287.

Fromm, J. E. (1984). Numerical calculation of the fluid dynamics of drop-on demand jets. *IBM J. Res. Dev.* **28**, 322.

Goedde, E. F., and M. C. Yuen. (1970). Experiments on liquid jet instability. *J. Fluid Mech.* **40**, 495.

Goldin, M., J. Yerushalmi, R. Pfeffer and R. Shinnar, "Breakup of a Laminar Capillary Jet of a Viscoelastic Fluid," *J. Fluid Mech.* **38**, 712(1969).

Grant, R. P., and S. Middleman. (1966). Newtonian jet stability. *AIChE J.* **12,** 669.

Green, A. E. (1976). On the non-linear behaviour of fluid jets. *Int. J. Eng. Sci.* **14**, 49.

Kroesser, F. W., and S. Middleman. (1969). Viscoelastic jet stability. *AIChE J.* **15**, 383.

Lafrance, P. (1975). Nonlinear breakup of a laminar liquid jet. *Phys. Fluids* **18**, 428.

Lee, H. C. (1974). Drop formation in a liquid jet. *IBM J. Res. Dev.* **18**, 364.

Mandrusov, E. (1990). Unpublished data, Univ. of California, San Diego.

Manfre, G. (1966). Rheological aspects of drop formation. *J. Appl. Phys.* **37**, 1955.

Matovich, M. A., and J. R. A. Pearson. (1969). Spinning a molten threadline. *Ind. Eng. Chem. Fund.* **8**, 512.

McCarthy, M. J., and N. A. Molloy. (1974). Review of stability of liquid jets and the influence of nozzle design. *Chem. Eng. J.* **7**, 1.

Pimbley, W. T. (1976). Drop formation from a liquid jet. A linear one-dimensional analysis considered as a boundary value problem. *IBM J. Res. Dev.* **20**, 148.

Pimbley, W. T., and H. C. Lee. (1977). Satellite droplet formation in a liquid jet. *IBM J. Res. Dev.* **21**, 21.

Rayleigh, Lord (J. W. Strutt). (1879). On the instability of jets. *Proc. Lond. Math. Soc.* **10**, 4. See also On the instabiilty of a cylinder of viscous liquid under capillary force. *Phil. Mag.* **34**, 145. In addition, see *The Theory of Sound,* vol. II. Dover, New York, 1945.

Rutland, D. F., and G. J. Jameson. (1970). Theoretical prediction of the sizes of drops formed in the breakup of capillary jets. *Chem. Eng. Sci.* **25**,1689.

Schulkes, R. M. S. M. (1994). The evolution of capillary fountains. *J. Fluid Mech.* **261**, 223.

Sterling, A. M., and C. A. Sleicher. (1975). The instability of capillary jets. *J. Fluid Mech.* **68**, 477.

Weis, D. M., P. F. Dunn, and M. Sen. (1992). The periodicity of droplets emanating from interconnected orifices. *Exp. Fluids* **13**, 257.

Wilson, S. D. R. (1988). The slow dripping of a viscous fluid. *J. Fluid Mech.* **190**, 561.

# 5 Impingement of a Coherent Jet Normal to a Rigid Surface

## *5-1 INTRODUCTION*

We now consider the behavior of a coherent vertical jet that impinges continuously on a rigid planar surface that is normal to the axis of the jet. Figure 5-1.1a is a schematic that shows the key features of the dynamics of such a system. Figure 5-1.1b is a high-speed flash photograph of such a jet. (Another view of an impinging jet is presented on Color Plate 3.) The primary issue here is with respect to the thickness of the liquid film that flows radially across the planar surface. Several observations are possible, depending upon the conditions of the flow.

We assume that the planar surface is a smooth disk concentric with the axis of the jet, or else a plane of infinite extent. Then the ensuing flow should be axisymmetric. If the jet flow is laminar and steady (but not "creeping") the typical observation is that the radial flow across the disk consists of two regions. Near the jet axis the flow is rapid and the liquid film is quite thin. If the radius of the disk is large enough there will be a narrow annular region over which the film thickness rapidly increases. Beyond that "point," called the radius of the "hydraulic jump," the thickened liquid film moves much slower in order to preserve mass flow. Of particular interest is the dependence of the jump radius $r_h$ on

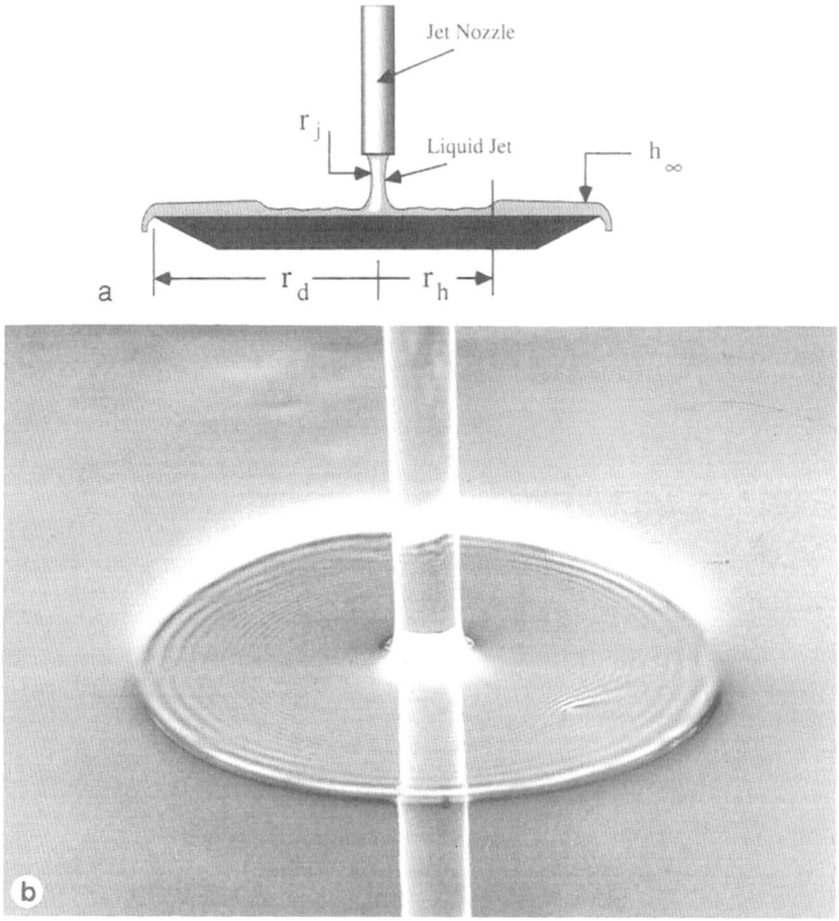


***FIGURE 5-1.1(a, b)*** A liquid jet impinges on a circular concentric disk.

the flow rate $Q$ and radius of the jet $r_j$, the radius of the disk $r_d$, and of course on the physical properties of the liquid. We will see that this is a very complex problem which in certain respects has largely defied analysis. Because the physics of this flow is quite complicated, we begin by taking the problem apart, in a sense, and examining some isolated features of this flow field.

We can begin with a mechanical energy balance—the Bernoulli Equation—for this flow. Figure 5-1.2 shows the control volume over which the balance is made.

Kinetic energy flows into the control volume across the plane labeled A and leaves across the plane B. We will assume that the pressure within the jet and the film is atmospheric, and therefore ignore any pressure contribution to the energy balance. We ignore the contribution of surface tension to the energy bal-

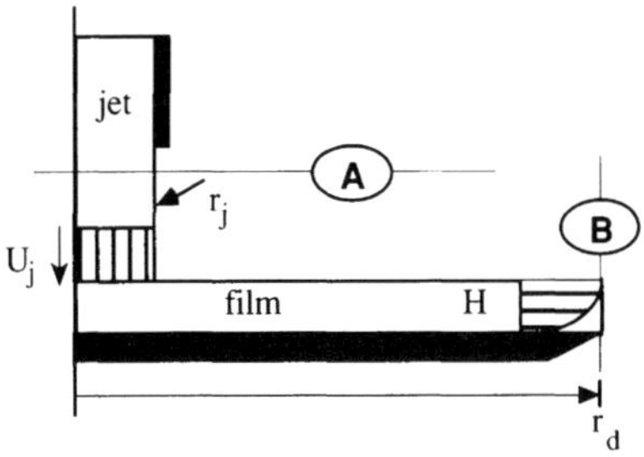


*FIGURE 5-1.2* Energy balance, liquid jet on a disk.

ance. We will take the plane A close to the point of impact. Then $U_j$ of the jet is the velocity just before impact, which may be greater than that of the jet leaving the capillary because of the acceleration due to gravity. We ignore any further gravitational effects, since the vertical length scale (the film thickness) is typically of the order of millimeters. Finally, there will be some contribution to the energy balance from the dissipation of kinetic energy through viscous drag at the disk surface and from the complex flow in the region where the liquid "turns" from an axial flow to a radial flow. We write the steady-state Bernoulli equation as

$$\Delta \frac{1}{2} \frac{\langle u^3 \rangle}{\langle u \rangle} + \widehat{E}_v = 0, \tag{5-1.1}$$

where $\hat{E}_v$ is the irreversible loss of energy due to viscous friction. The average velocities are defined as

$$\langle u^n \rangle \equiv \frac{1}{H} \int_0^H u_r^n \, dz. \tag{5-1.2}$$

In the entrance region (across the plane A) we will assume that the velocity profile across the radius of the jet is flat. Then

$$\frac{\langle u^3 \rangle}{\langle u \rangle}_A = U_j^2. \tag{5-1.3}$$

Because of viscous effects there will be a velocity profile across the film, at $r = r_d$, so we write

$$\frac{\langle u^3 \rangle}{\langle u \rangle}_B = K_{31} U_{f,d}^2, \tag{5-1.4}$$

where $K_{31}$ is a coefficient that depends on the detail of the velocity profile, and $U_{f,d}$ is the average velocity of the liquid at the edge of the disk. (See Problem P5-1.)

Conservation of mass leads to the simple relationship

$$\pi r_j^2 U_j = 2\pi r_d H U_{f,d}. \tag{5-1.5}$$

This permits us to write Eq. 5-1.1 as

$$U_j^2\left(1 - \frac{K_{31}}{4}\frac{r_j^4}{r_d^2 H^2}\right) = 2\widehat{E}_v. \tag{5-1.6}$$

For an inviscid liquid ($\widehat{E}_v = 0$, $K_{31} = 1$) this leads to

$$H = \frac{1}{2}\frac{r_j^2}{r_d}, \tag{5-1.7}$$

since $K_{31} = 1$. This result predicts that the film thickness decreases with increasing disk radius for an inviscid jet of fixed radius. We will find that this is contrary to observations. The model fails as a consequence of our neglect of the viscous effects of the disk (whieh is a no-slip boundary) in retarding the radial flow. Hence we have to account somehow for the viscous flow across the disk.

The viscous dissipation term may be written in the form

$$\widehat{E}_v = \frac{\mu}{w}\int \Phi_v \, dV, \tag{5-1.8}$$

where $w$ is the mass flow rate:

$$w = \pi r_j^2 \rho U_j = 2\pi r_d H \rho U_{f,d}. \tag{5-1.9}$$

The dissipation function $\Phi_v$ for a Newtonian fluid is given by (see, e.g., Bird *et al.*, 1960)

$$\Phi_v = 2\left[\left(\frac{\partial u_r}{\partial r}\right)^2 + \left(\frac{u_r}{r}\right)^2 + \left(\frac{\partial u_z}{\partial z}\right)^2\right] + \left(\frac{\partial u_r}{\partial z} + \frac{\partial u_z}{\partial r}\right)^2 \tag{5-1.10}$$

for an axisymmetric flow with no $\theta$ component of velocity. Since the flow field is unknown we cannot perform the indicated integration in Eq. 5-1.8.

It is useful at this stage to nondimensionalize these equations. We define

$$\tilde{\mathbf{u}} = \frac{\mathbf{u}}{U_j}, \quad (\tilde{r}, \tilde{z}) = \left(\frac{r}{r_d}, \frac{z}{r_d}\right). \tag{5-1.11}$$

We may then write the dissipation function in the form

$$\widehat{E}_v = \frac{2\mu U_j}{\rho r_j}\int_0^{H(\tilde{z})/r_d}\int_0^1 \tilde{\Phi}_v \, \tilde{r} \, d\tilde{r} \, d\tilde{z} = \frac{4U_j^2}{\mathrm{Re}_j}\Psi\left(\frac{r_d}{r_j}, \frac{H}{r_j}\right), \tag{5-1.12}$$

where $\Psi$ is the unknown integral of the dissipation function $\Phi_v$. (The Reynolds number is defined, as usual, as $\mathrm{Re} = 2\rho U_j r_j / \mu$, using the jet *diameter* as the length scale.) This lets us write Eq. 5-1.6 in the form

$$1 - \frac{K_{31}}{4}\frac{r_j^4}{r_d^2 H^2} = \frac{8}{\mathrm{Re}_j}\Psi\left(\frac{r_d}{r_j}, \frac{H}{r_j}\right) \tag{5-1.13}$$

or

$$\frac{H^2}{r_j^2} = \left\{\frac{4}{K_{31}}\frac{r_d^2}{r_j^2}\left[1 - \frac{8}{\mathrm{Re_j}}\Psi\left(\frac{r_d}{r_j}, \frac{H}{r_j}\right)\right]\right\}^{-1}. \tag{5-1.14}$$

Because $H$ is in the argument of the function $\Psi$, we do not have an explicit solution for $H$. However, it appears that Eq. 5-1.14 has the property that as Reynolds number gets smaller, that is, as viscous effects become more important, the film thickness $H$ increases.

## *5-2 A BOUNDARY LAYER MODEL* (Watson, 1964)

We illustrate here an approximate approach to determining some features of this flow. We follow the analysis of Watson (1964), which is based on a boundary layer approximation to the flow field. Implicit in Watson's analysis is the idea that the jet flow is at a high enough Reynolds number that the stagnation region is inviscid and that viscous effects occur near the lower boundary as the radial flow is retarded. Figure 5-2.1 indicates the geometry of the flow field. There are two regions into which the flow may be separated, based on the development of the boundary layer height $\delta(r)$. Near the stagnation region where $\delta << H$ a sizable portion of the flow is inviscid and unaffected by the presence of the lower surface. At some radial position $r_0$ downstream the boundary layer covers the entire film thickness.

We begin by assuming that the usual boundary layer approximations hold, so that the velocity field is defined by the solutions to

$$\frac{\partial r u_r}{\partial r} + \frac{\partial r u_z}{\partial z} = 0 \tag{5-2.1}$$

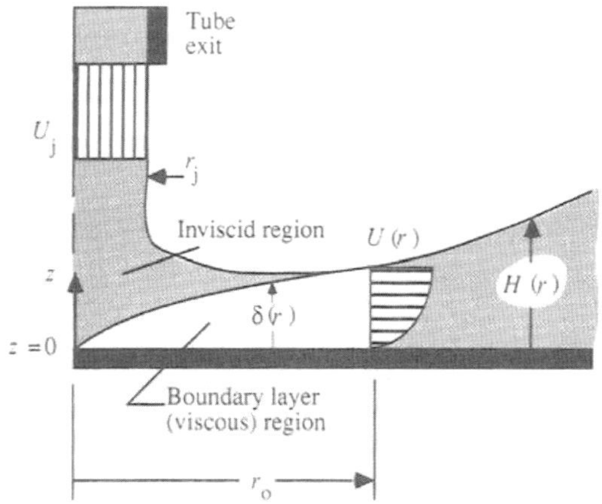


***FIGURE 5-2.1*** Radial flow across a disk, inviscid and boundary layer regions.

and

$$u_r \frac{\partial u_r}{\partial r} + u_z \frac{\partial u_r}{\partial z} = \nu \frac{\partial^2 u_r}{\partial z^2}. \tag{5-2.2}$$

Boundary conditions on velocity include the no-slip condition along the lower rigid boundary:

$$u_r = u_z = 0 \quad \text{at} \quad z = 0. \tag{5-2.3}$$

We assume that there is no shear stress along the free surface $H(r)$. Hence another boundary condition is that

$$\frac{\partial u_r}{\partial z} = 0 \quad \text{at} \quad z = H(r). \tag{5-2.4}$$

The volume flow rate from the jet satisfies

$$2\pi r \int_0^{H(r)} u_r \, dz = Q. \tag{5-2.5}$$

Watson derived a similarity solution for the flow by assuming

$$u_r = U(r) f(\eta), \tag{5-2.6}$$

where $U(r)$ is the speed of the free surface and $\eta = z/H(r)$. The boundary conditions (Eqs. 5-2.3 and 4) require that $f(\eta)$ satisfy

$$f(0) = 0, \quad f(1) = 1, \quad f'(1) = 0, \tag{5-2.7}$$

where $f'$ is $df/d\eta$. From Eq. 5-2.5 we find

$$2\pi r U H \int_0^1 f(\eta) \, d\eta = Q. \tag{5-2.8}$$

Now Eq. 5-2.1 will lead to the result

$$u_z = U H' \eta f(\eta), \tag{5-2.9}$$

where $H' = dH/dr$. With these results we may write the equation of motion in the form

$$\nu f'' = H^2 \frac{dU}{dr} f^2(\eta). \tag{5-2.10}$$

If a similarity solution exists, then, it is necessary for $H^2\, dU/dr$ to be a constant, independent of $r$, so that only the independent variable $\eta$ appears in the defining equation for $f$. We write this requirement as

$$H^2 \frac{dU}{dr} = -\frac{3}{2} C^2 \nu, \tag{5-2.11}$$

where $C$ is some constant. This permits us to write Eq. 5-2.10 as

$$2f'' + 3C^2 f^2 = 0. \tag{5-2.12}$$

This has the solution

$$f'^2 = C^2(1 - f^3) \tag{5-2.13}$$

or

$$C\eta = \int_0^f (1 - x^3)^{-1/2}\, dx. \tag{5-2.14}$$

In this format we have $x$ as a dummy variable of integration, and $\eta$ is obtained numerically by picking various $f$ values for the upper limit of the integral. Thus we find $\eta(f)$ rather than $f(\eta)$. The constant $C$ follows from the boundary condition $f(1) = 1$ as

$$C = \int_0^1 (1 - x^3)^{-1/2}\, dx = 1.402. \tag{5-2.15}$$

Furthermore we may show that

$$\int_0^1 f(\eta)\, d\eta = \frac{1}{C}\int_0^1 f(1 - f^3)^{-1/2}\, df = \frac{2\pi}{3\sqrt{3}\, C^2}. \tag{5-2.16}$$

Then Eq. 5-2.8 becomes

$$rU(r)H(r) = \frac{1}{C}\int_0^1 f\, d\eta = \frac{3\sqrt{3}\, C^2}{4\pi^2} Q. \tag{5-2.17}$$

This lets us integrate Eq. 5-2.11 to find

$$U(r) = \frac{27\, C^2}{8\pi^4} \frac{Q^2}{\nu\left(r^3 + b^3\right)} \tag{5-2.18}$$

and

$$H(r) = \frac{2\, \pi^2}{3\sqrt{3}} \frac{\nu\left(r^3 + b^3\right)}{rQ}. \tag{5-2.19}$$

In these equations, $b$ is a constant of integration with units of length. Since $b$ is not known, the theory is incomplete at this point. However, we note that Eq. 5-2.19 has the interesting property that $H(r)$ goes through a minimum at the radial position

$$r_{\min} = 2^{-1/3}\, b, \tag{5-2.20}$$

where $H$ has the value

$$H_{\min} = \frac{2^{1/3}\pi^2}{\sqrt{3}} \frac{\nu b^2}{Q}. \tag{5-2.21}$$

Clearly, an important order of business is finding a value for $b$. This can be done by carrying out an approximate solution to the boundary layer equations by the integral method. If we denote the boundary layer thickness as $\delta(r)$ and consider the region $r < r_0$ of Fig. 5-2.1 we may integrate the radial component of the momentum equation (Eq. 5-2.2) to find (Problem P5-3)

$$\left(\frac{d}{dr} + \frac{1}{r}\right) \int_0^{\delta(r)} \left(U_j u_r - u_r^2\right) dz = \nu \left(\frac{\partial u_r}{\partial z}\right)_{z=0}. \tag{5-2.22}$$

We write the velocity field in the boundary layer form (cf. Eq. 5-2.6)

$$u_r = U_j f(z/\delta) \tag{5-2.23}$$

and take the surface velocity $U_j$ to be independent of $r$ and simply given by the jet velocity. After performing the indicated operations we find a differential equation for the boundary layer thickness $\delta$:

$$\frac{2\left(\pi - C/\sqrt{3}\right)}{3\sqrt{3}\, C^2} U_j^2 \left(\frac{d\delta}{dr} + \frac{\delta}{r}\right) = \frac{\nu U_j C}{\delta}. \tag{5-2.24}$$

The solution for $\delta$, satisfying the boundary condition that $\delta$ be finite at $r = 0$, is

$$\delta^2 = \frac{\sqrt{3}\, C^3}{\pi - C/\sqrt{3}} \frac{\nu r}{U_j}. \tag{5-2.25}$$

When the radial position $r_0$ is reached, all of the flow is in the boundary layer. Hence it follows from Eq. 5-2.17 that

$$r_0 U_j \delta(r_0) = \frac{3\sqrt{3}\, C^3}{4\pi^2} Q. \tag{5-2.26}$$

Using Eq. 5-2.25 we find

$$r_0^3 = \frac{9\sqrt{3}\, C\left(\pi - C\sqrt{3}\right)}{16\pi^3} \frac{Q r_j^2}{\nu}. \tag{5-2.27}$$

For $r < r_0$ the volume flow rate is the sum of the flows within and without the boundary layer:

$$Q = 2\pi r \left[ U_j \delta \int_0^1 f(\eta)\, d\eta + U_j (H - \delta) \right]. \tag{5-2.28}$$

This permits us to write the liquid film thickness in the form

$$H(r) = \frac{r_j^2}{2r} + \left(1 - \frac{2\pi}{3\sqrt{3}C^2}\right)\delta(r) \tag{5-2.29}$$

and $\delta$ is given explicitly by Eq. 5-2.25.

Finally, we may obtain the parameter we have been seeking—the constant of integration $b$ that appeared in Eq. 5-2.18 (and 19). We go back to Eq. 5-2.18 and state that the free surface velocity $U(r)$ in that equation must just equal the jet velocity $U_j$ at the radial position $r = r_0$. The result is

$$b^3 = \frac{9\sqrt{3}\,C(3\sqrt{3}C - \pi)}{16\pi^3}\frac{Qr_j^2}{\nu} \tag{5-2.30}$$

and, using the known value for $C$,

$$\frac{b}{r_j} = 0.567\left(\frac{Q}{r_j\nu}\right)^{1/3}. \tag{5-2.31}$$

The Reynolds number for this flow may be written conveniently in terms of $Q$ as

$$\mathrm{Re}_j = \frac{2Q}{\pi r_j \nu}. \tag{5-2.32}$$

With these results in hand, we may write expressions for the film thickness profile, using Eq. 5-2.19 for the region $r > r_0$ and Eq. 5-2.29 for $r < r_0$. In the course of doing this we take note of the fact that these equations may be rearranged in terms of two dimensionless variables:

$$\Gamma \equiv \frac{H}{r_j}\mathrm{Re}_j^{1/3} \tag{5-2.33}$$

and

$$\zeta \equiv \frac{r}{r_j}\mathrm{Re}_j^{-1/3}. \tag{5-2.34}$$

Figure 5-2.2 shows the film thickness profile corresponding to Watson's theory.

It is important to recall some of the assumptions that have led to this model of the radial flow across a surface. The primary restriction is that the flow in the boundary layer must be laminar. However, if the Reynolds number for the jet is too small, it is unlikely that the idea of an inviscid region where the jet flow "turns" into the radial flow makes any sense physically. Hence we can expect that there would be some range of Reynolds numbers, unknown at this point, over which this model could be expected to describe observations.

A second restriction arises from the fact that the theory presented here neglects any effect of gravity in Eq. 5-2.2. As the film thickness grows large (Fig. 5-2.2) we should expect that hydrostatic pressure plays an increasingly important role in the dynamics of the radial flow. Thus we should expect some limitation based on a local Froude number defined with $H$ as the length scale (and hence unknown *a priori*):

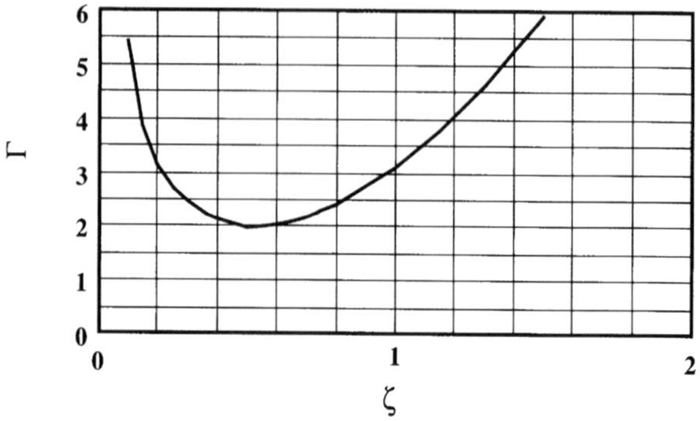


***FIGURE 5-2.2*** Film profile for a laminar jet—Watson's theory.

$$\mathrm{Fr_j} \equiv \frac{U_j^2}{gH}. \tag{5-2.35}$$

We will come back to this point shortly, in order to discuss the "hydraulic jump."

A "thought experiment" suggests another potential problem with the theory. Suppose we have a lower surface that is unbounded in the radial direction. As the liquid moves radially across that surface it slows down rapidly. Note that Eq. 5-2.18 shows us that $U(r)$ decays as $r^{-3}$. There seems to be nothing in the mathematical model that suggests that the film profile depends on the size of the disk across which the flow is occurring. Since a larger disk would support a larger film thickness far downstream, we might expect that the downstream film thickness would somehow affect the upstream flow field, perhaps through hydrostatic effects. Intuition is an individual thought process, but the intuition of the author is that the size of the disk would somehow affect the flow. An experimental study is certainly in order on this point.

## *5-3 THE HYDRAULIC JUMP*

We have mentioned the neglected role of gravity in the flow model. We have also ignored the presence of the hydraulic jump, and the film thickness model gives no hint of a region of sudden increase in $H(r)$. This seems like the appropriate point to introduce gravity into the dynamics. The goal is to obtain an estimate of the position at which the jump occurs. To do this we must introduce some physical statement regarding the cause of the jump. Figure 5-3.1 will help in this regard.

The simplest approach, initially, is to consider a momentum balance on an *inviscid* flow. For a flat (inviscid) velocity profile we already have the result given earlier (Eq. 5-1.7) which we now write in the form for $H(r)$ just upstream

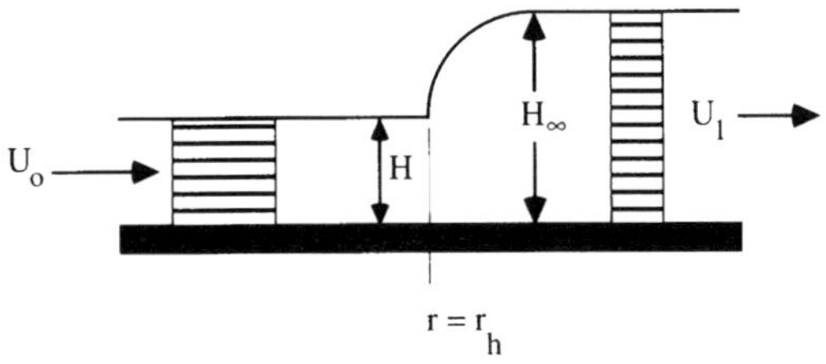


*FIGURE 5-3.1* Sketch for consideration of the hydraulic jump at $r_h$.

of the jump:

$$H = \frac{1}{2}\frac{r_j^2}{r_h}. \tag{5-3.1}$$

If we write the statement of conservation of momentum across the jump, we equate the rate of change of momentum, due to the slowing down of the flow, to the force associated with the difference in pressure that arises from the elevation change:

$$\rho\left(U_0^2 H - U_1^2 H_\infty\right) = \int_0^{H_\infty} P_> \, dz - \int_0^{H} P_< \, dz. \tag{5-3.2}$$

The pressure distribution on either side of $r = r_j$ is found from

$$\begin{aligned} P_< &= \rho g H - \rho g z \quad \text{for} \quad r < r_h, \\ P_> &= \rho g H_\infty - \rho g z \quad \text{for} \quad r > r_h. \end{aligned} \tag{5-3.3}$$

The result of performing the indicated integrations is

$$\rho\left(U_0^2 H - U_1^2 H_\infty\right) = \tfrac{1}{2}\rho g\left(H_\infty^2 - H^2\right). \tag{5-3.4}$$

If we introduce the flow rate

$$Q = \pi r_j^2 U_j = 2\pi r_h H U_0, \tag{5-3.5}$$

impose mass conservation in the form

$$2\pi r_h H U_0 = 2\pi r_h H_\infty U_1, \tag{5-3.6}$$

and note that for an inviscid fluid the Bernoulli equation requires that $U_0 = U_1$, we find

$$\left(\frac{Q}{2\pi r_h}\right)^2\left(\frac{1}{H} - \frac{1}{H_\infty}\right) = \frac{1}{2} g\left(H_\infty^2 - H^2\right). \tag{5-3.7}$$

This is a relationship between the radius of the jump, $r_h$, and the depth downstream, $H_\infty$. Noting that $H$ depends on $r_h$ through Eq. 5-3.1, and making the approximation $H^2 << H_\infty^2$, we may write an expression for the jump radius in the form

$$\frac{r_h H_\infty^2 g r_j^2}{Q^2} + \frac{r_j^2}{2\pi^2 r_h H_\infty} = \frac{1}{\pi^2}. \tag{5-3.8}$$

This may be regrouped into the following format:

$$\left(\frac{r_h}{r_j}\right)\left(\frac{H_\infty}{r_j}\right)\frac{1}{\mathrm{Fr}_{j,\infty}} + \frac{1}{2}\left(\frac{r_j}{r_h}\right)\left(\frac{r_j}{H_\infty}\right) = 1, \tag{5-3.9}$$

where a Froude number based on the height $H_\infty$ is defined (cf. Eq. 5-2.35):

$$\mathrm{Fr}_{j,\infty} \equiv \frac{U_j^2}{gH_\infty}. \tag{5-3.10}$$

Note two features of this model: it is not explicit in $r_h$, and it requires knowledge of $H_\infty$. Nevertheless, it is a testable model if data are available for both $r_h$ and $H_\infty$.

Figure 5-3.2 shows an example of experimental data obtained using low-speed jets of water. The Y axis is the left-hand side of Eq. 5-3.8. It is clear that the data do not correspond to $\mathrm{Y} = 1/\pi^2$.

As the Reynolds number decreases, the data depart from the inviscid model to an increasing degree. This suggests that the neglect of viscous effects may be inappropriate. We may return to Eq. 5-3.4 but introduce, as in the approach of Watson, the following changes. We neglect $H^2$ with respect to $H_\infty^2$, on the basis of typical observations. We continue to assume that the velocity profile beyond the jump is flat, with $u_r = U_1$. This ignores the viscous effects of the flow beyond the jump. (See Problem P5-6.) The momentum flow approaching the jump will be written using the velocity profile established in Section 5-2. Hence we write

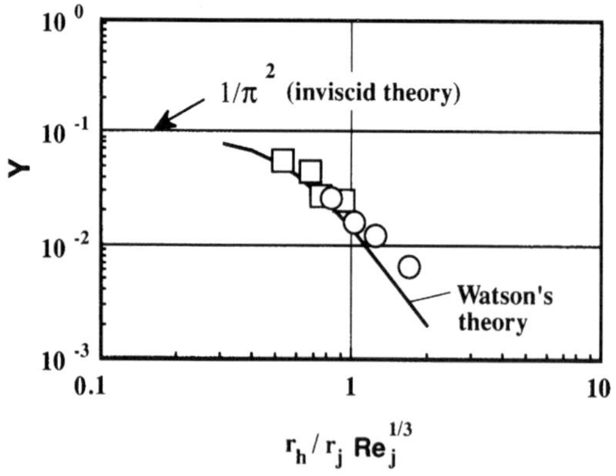


*FIGURE 5-3.2* Test of Eq. 5-3.8 and Eq. 5-3.12a. (Unpublished data of Lesniewski and Middleman.)

$$\tfrac{1}{2}\rho g H_\infty^2 = \rho \int_0^H u_r^2\, dz - \rho U_1^2 H_\infty. \tag{5-3.11}$$

The velocity profile depends on whether the jump is before or after the position $r_0$ given by Eq. 5-2.27. In either case, we perform the indicated integration. The details are presented in Watson (1964), and the results are the following:

$$\left(\frac{r_h}{r_j}\right)\left(\frac{H_\infty}{r_j}\right)\frac{1}{\mathrm{Fr}_{j,\infty}} + \frac{1}{2}\left(\frac{r_j}{r_h}\right)\left(\frac{r_j}{H_\infty}\right) = \left[0.260\left(\frac{r_h}{r_j}\right)^3 \mathrm{Re}_j^{-1} + 0.287\right]^{-1}$$

$$\text{for} \quad \left(\frac{r_h}{r_j}\right)\mathrm{Re}_j^{-1/3} \geq 0.272 \tag{5-3.12a}$$

and

$$\left(\frac{r_h}{r_j}\right)\left(\frac{H_\infty}{r_j}\right)\frac{1}{\mathrm{Fr}_{j,\infty}} + \frac{1}{2}\left(\frac{r_j}{r_h}\right)\left(\frac{r_j}{H_\infty}\right) = 1 - 1.02\left(\frac{r_h}{r_j}\right)^{3/2}\mathrm{Re}_j^{-1/2}$$

$$\text{for} \quad \left(\frac{r_h}{r_j}\right)\mathrm{Re}_j^{-1/3} \leq 0.272. \tag{5-3.12b}$$

Whichever the regime of the two choices above, the prediction is that the hydraulic radius satisfies a complex functional relationship in the form

$$\frac{r_h H_\infty^2 g r_j^2}{Q^2} + \frac{r_j^2}{2\pi^2 r_h H_\infty} = F\left[\left(\frac{r_h}{r_j}\right)\mathrm{Re}_j^{-1/3}\right]. \tag{5-3.13}$$

(That is why the $x$ axis of Fig. 5-3.2 was selected in the form shown.) On Fig. 5-3.2 we show the line corresponding to Eq. 5-3.12a. The data agree reasonably well with the predicted behavior. Watson's theory seems to account for the primary physical phenomena that control the dynamics of this flow. More detailed studies indicate some failures of the model and outline approaches to improvement of the model (Bowles and Smith, 1992).

In developing Eqs. 5-3.12, Watson assumes that the flow in the radially spreading film is laminar. He offers an aproximate criterion for laminar flow, which takes the form

$$\mathrm{Re}_j < 1.64 \times 10^4. \tag{5-3.14}$$

Then the use of Eq. 5-3.12a is subject to the criterion

$$\frac{r_h}{r_j} \geq 6.9, \tag{5-3.15}$$

while Eq. 5-3.12b is used when $r_h/r_j < 6.9$.

Watson's theory takes $H_\infty$ as a parameter, but in reality $H_\infty$ is determined by the flow field and the geometry of the surface on which the jet impacts. There is no extant theory for *a priori* prediction of $H_\infty$ (or equivalently, the Froude number) from knowledge of the Reynolds number and the geometry of the flow.

Equations 5-3.12a and b can be rearranged algebraically to the forms

$$\frac{\mathsf{RH}}{\mathsf{Fr}_{j,\infty}} + \frac{1}{2\,\mathsf{RH}} = \frac{0.26}{\mathsf{R}^3 + 0.287} \quad \text{for } \mathsf{R} > 0.27 \tag{5-3.16a}$$

and

$$\frac{\mathsf{RH}}{\mathsf{Fr}_{j,\infty}} + \frac{1}{2\,\mathsf{RH}} = 1 - 1.02\mathsf{R}^{3/2} \quad \text{for } \mathsf{R} < 0.27 \tag{5-3.16b}$$

by defining

$$\mathsf{R} \equiv \left(\frac{r_h}{r_j}\right) \mathsf{Re}_j^{-1/3} \tag{5-3.17}$$

and

$$\mathsf{H} \equiv \left(\frac{H_\infty}{r_j}\right) \mathsf{Re}_j^{1/3}. \tag{5-3.18}$$

In what follows we will consider only jets for which the Reynolds number satisfies the restriction on Eq. 5-3.16a.

Figure 5-3.3 shows the behavior of the hydraulic jump radius, in the form of a plot of R vs. $\mathsf{Fr}_{j,\infty}/\mathsf{H}$, with H as a parameter. We see that for sufficiently large values of H, the curves become independent of H. This observation suggests a simpler form of solution to Eq. 5-3.16a, for the dependence of the jump radius on parameters. When both H and R are large, a good approximation to Eq. 5-

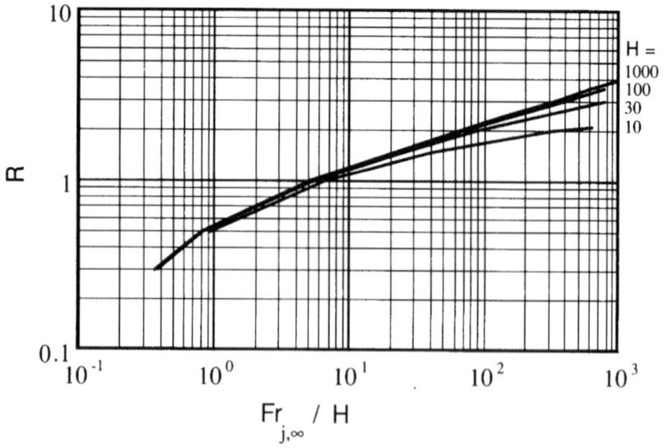


***FIGURE 5-3.3*** Solutions to Eq. 5-3.16a.

3.16a leads to a result that is explicit in the jump radius:

$$\frac{r_h}{r_j} = 0.714\mathsf{G}^{1/4}, \qquad (5\text{-}3.19)$$

where a new dimensionless group G is defined as

$$\mathsf{G} \equiv \mathsf{Re}_j \mathsf{Fr}_{j,\infty} \frac{r_j}{H_\infty}. \qquad (5\text{-}3.20)$$

This result is shown in Fig. 5-3.4, along with some of our data.

We noted earlier that the jet radius $r_j$ just above the point of impact appears in Watson's theory. This is different from the tube radius, $r_{tube}$, because the gravitational acceleration of the jet reduces the jet radius. For sufficiently large "standoff" distances $L$ of the tube exit from the impact plane, the jet radius can be considerably smaller than that of the tube. A simple application of the Bernoulli (energy) equation for an inviscid fluid leads to the result (see Eq. 3-2.3)

$$\frac{r_j}{r_{tube}} = \left(1 + \mathsf{Fr}_L^{-1}\right)^{-1/4}, \qquad (5\text{-}3.21)$$

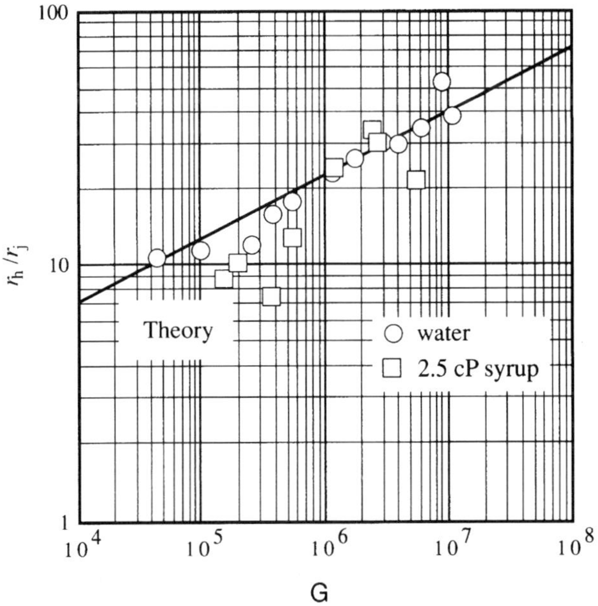


*FIGURE 5-3.4* The approximation corresponding to Eq. 5-3.19, and unpublished data of Ovalle and Middleman.

where a Froude number based on the standoff distance $L$ is defined as

$$\mathrm{Fr}_L = \frac{Q^2}{2\pi^2 g L r_{\mathrm{tube}}^4}. \tag{5-3.22}$$

For a viscous jet issuing from a laminar tube flow an additional correction is required that accounts for the rearrangement of the velocity profile of the jet subsequent to leaving the tube (Middleman and Gavis, 1961). In the range of Reynolds numbers of interest to us here, this would require that we multiply the tube radius $r_{\mathrm{tube}}$ in Eq. 5-3.21 by the factor 0.866.

It should be clear from the discussion of this section that a knowledge of the film thickness $H_\infty$ is essential to the development of a predictive model for the jump radius. Numerical modeling of this feature of the flow exists (Rahman *et al.*, 1991) but it is essentially untested. [Bowles and Smith (1992) predict the film thickness profile in the neighborhood of the jump, but use the position of the jump, based on the experiments of Craik *et al.* (1981), as input data to their model.]

It is possible to control the downstream film thickness $H_\infty$ by placing a rim or lip along the perimeter of the impingement disk. This has been done in several studies (e.g., Errico, 1986; Liu and Lienhard V, 1993). These two studies are especially interesting for their investigation of the details of the flow in the region of the jump.

Finally, we note that numerical solutions of the equations that describe an impinging jet have been presented. Chaudhry (1994) presents the results of a solution of the Navier–Stokes equations for an axisymmetric jet. Turbulence is accounted for using a $k$–$\varepsilon$ model for the Reynolds stresses. The results compare quite well to observations. The downstream depth is specified as a boundary condition, and the solution provides the position of the jump. Higuera (1994) solves the boundary layer equations that describe the hydraulic jump from a laminar two-dimensional *planar* jet flowing over a wide horizontal finite plate. He predicts the position of the jump *and* the thickness of the film at the downstream edge of the flow. Hence his prediction of the position of the jump is free of any assumption about the downstream thickness.

## *5-4 DISPLACEMENT OF A THIN VISCOUS FILM BY AN IMPINGING TURBULENT JET*

Liquid jets are often used for surface cleaning. Figure 5-4.1 depicts the situation where a turbulent jet is used to erode or shear an adherent liquid film off of a planar surface. Our goal is to develop a model from which the rate of thinning of the lower layer may be predicted in terms of jet parameters and the adherent liquid layer properties.

The underlying viscous film is removed by the action of the turbulent jet through two mechanisms. The jet creates a radial pressure distribution in the region of the stagnation point, and the jet exerts a shear stress on the interface of the two liquids. It is necessary to have a model for these characteristics of the jet before a model for flow in the viscous layer can be developed. The required information about the jet is reviewed in Yeckel and Middleman (1987). In particular, the radial pressure distribution is given by (see Section 5-5 )

$$p(r) = \frac{1}{2}\rho U_j^2\left[1 - \left(\frac{a^*r}{2r_j}\right)^2\right], \tag{5-4.1}$$

where $a^*$ is a dimensionless empirical parameter, and the shear stress is given by

$$\tau(r) = 0.656a^{*3/2}\rho U_j^2 \, \mathrm{Re}_j^{-1/2}\frac{r}{r_j}. \tag{5-4.2}$$

(See Problems P5-7 and 8.) The Reynolds number is defined as in Eq. 5-2.32.

In the analysis that follows we assume that the underlying film is so viscous that, with respect to the turbulent jet, it behaves like an immobile surface. Hence Eqs. 5-4.1 and 2, which apply to a jet impinging on a planar rigid surface, will be taken to apply to the case where the jet impinges on a slowly moving thin viscous film. Then these equations serve as boundary conditions on the upper surface of the viscous film. The lower boundary of the viscous film is simply the no-slip rigid surface.

If the film is thin and viscous we may neglect inertial effects, including time derivatives, in the Navier–Stokes equations for the film, and begin our analysis with the following momentum and mass balances:

$$\frac{\partial p}{\partial r} = \mu_f \frac{\partial^2 u_r}{\partial z^2}, \tag{5-4.3}$$

$$\frac{\partial p}{\partial z} = 0, \tag{5-4.4}$$

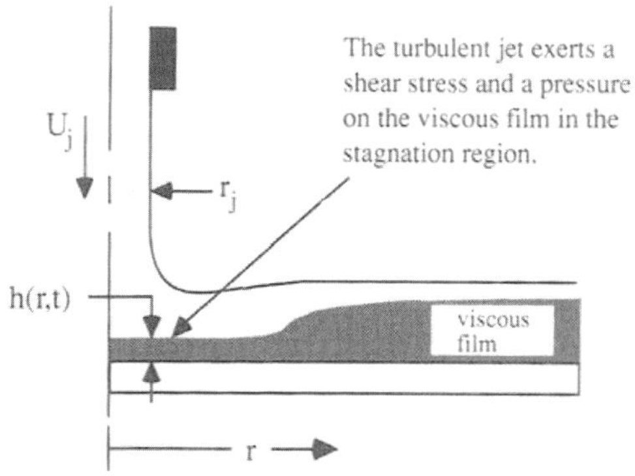


***FIGURE 5-4.1*** In the stagnation region of a turbulent jet, pressure and shear stress erode a viscous film.

and

$$\frac{\partial(ru_r)}{\partial r} + \frac{\partial(ru_z)}{\partial z} = 0. \tag{5-4.5}$$

(These are essentially the Reynolds lubrication equations for a thin viscous film.)
Boundary conditions take the form

$$\mu_f \frac{\partial u_r}{\partial z} = \tau \quad \text{at} \quad z = h(r,t) \tag{5-4.6}$$

and

$$u_r = u_z = 0 \quad \text{at} \quad z = 0. \tag{5-4.7}$$

The solutions for the velocity components may be written as

$$u_r = \frac{z}{h}\left[-\frac{h^2}{\mu_f}\frac{\partial p}{\partial r}\left(1 - \frac{z}{2h}\right) + \frac{\tau h}{\mu_f}\right] \tag{5-4.8}$$

and

$$u_z = -\frac{z^2}{2h^2}\left[-\frac{h^3}{\mu_f r}\frac{\partial}{\partial r}\left(r\frac{\partial p}{\partial r}\right)\left(1 - \frac{z}{3h}\right) + \frac{h^2}{\mu_f r}\frac{\partial(r\tau)}{\partial r}\right]. \tag{5-4.9}$$

A kinematic condition on the film thickness connects $h(r,t)$ to the velocity components at the interface:

$$\frac{\partial h}{\partial t} + [u_r]_{z=h}\frac{\partial h}{\partial r} = [u_z]_{z=h}. \tag{5-4.10}$$

After the velocity components are substituted into this kinematic condition we obtain an equation for the film thickness $h(r,t)$:

$$\frac{\partial h}{\partial t} + \left[-\frac{h^2}{2\mu_f}\frac{\partial p}{\partial r} + \frac{\tau h}{\mu_f}\right]\frac{\partial h}{\partial r} = \left[\frac{h^3}{3\mu_f r}\frac{\partial}{\partial r}\left(r\frac{\partial p}{\partial r}\right) - \frac{h^2}{2\mu_f r}\frac{\partial(r\tau)}{\partial r}\right]. \tag{5-4.11}$$

with initial and boundary conditions

$$\frac{\partial h}{\partial r} = 0 \quad \text{at} \quad r = 0,$$

$$h = h_0 \quad \text{at} \quad t = 0. \tag{5-4.12}$$

In order to solve Eq. 5-4.11 we need expressions for the pressure and shear stress distributions. In the stagnation region we have such expressions—Eqs. 5-4.1 and 2—and they are algebraically simple. As a consequence it is possible to solve Eq. 5-4.11 analytically, with the result

$$\frac{1}{s} - 1 - \frac{Fh_0}{G} \ln\left(\frac{1}{s}\,\frac{1 + (Fh_0/G)\,s}{1 + Fh_0/G}\right) = Gh_0 t, \tag{5-4.13}$$

where

$$s \equiv \frac{h}{h_0},$$

$$F \equiv -\frac{1}{3\mu_f r}\,\frac{\partial}{\partial r}\left(r\,\frac{\partial p}{\partial r}\right),$$

$$G \equiv \frac{1}{2\mu_f r}\,\frac{\partial}{\partial r}(r\,\tau). \tag{5-4.14}$$

In the stagnation region, where Eqs. 5-4.1 and 2 are presumed to hold, $F$ and $G$ are constants, and the film thickness $h$ is independent of radial position. It is not difficult to find that

$$\Phi \equiv \frac{Fh_0}{G} = \frac{0.302(a^*)^2 h_0\,\mathrm{Re}_j^{1/2}}{r_j},$$

$$\Psi \equiv Gh_0 t = \frac{0.656(a^*)^{3/2}\rho U_j^2 h_0 t}{\mathrm{Re}_j^{1/2} r_j \mu_f}, \tag{5-4.15}$$

and we may write Eq. 5-4.13 in the form

$$\frac{1}{s} - 1 - \Phi \ln\left(\frac{1}{s}\,\frac{1 + \Phi s}{1 + \Phi}\right) = \Psi. \tag{5-4.16}$$

Figure 5-4.2 shows a plot of film thickness in the stagnation region of the jet, as a function of time, in this dimensionless format. We see that as $s$ gets very small (very thin films) the model predicts that

$$s \to \Psi^{-1}. \tag{5-4.17}$$

We have carried out some experimental tests of the ability of Eq. 5-4.16 to mimic observations on film thinning under an impinging turbulent jet of water. The experiments were designed to produce very thin (submicron) viscous films on an extremely smooth planar surface. Thin films of silicone oil (polydimethylsiloxane) were spin-coated onto electronic grade silicon wafers. Turbulent jets of water were used to "remove" the film, and the residual film thickness at the stagnation point of the jet was measured as a function of time. Film thickness was measured directly on the wafer using an optical method based on ellipsometry. A Rudolph Research Auto El II autonulling ellipsometer was used. All experiments were carried out inside a Modulair MS-100 laminar flow hood in order to reduce the risk of particle contamination of these thin films. A more complete discussion of some of the issues associated with accurate film thickness measurements of this type can be found in Strong and Middleman (1989).

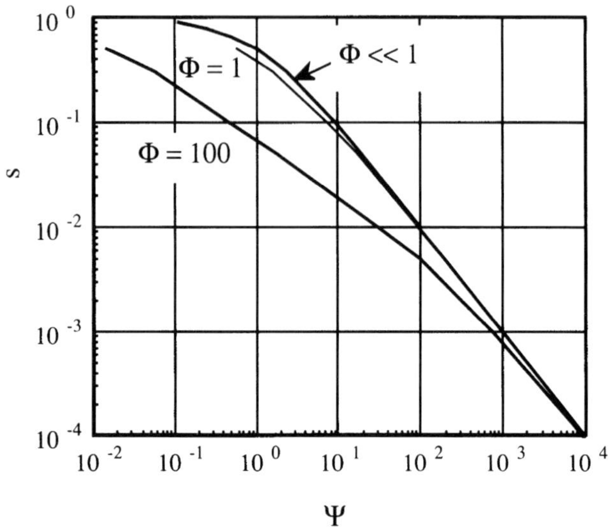


***FIGURE 5-4.2*** Film thinning in the stagnation region of a jet, according to Eq. 5-4.16.

In order for Eq. 5-4.16 to yield a predictive model for $h(t)$ it is necessary to have a value for the dimensionless hydrodynamic constant $a^*$. Experimental data cited by Chin and Tsang (1978) suggest that $a^*$ is of order unity, but it depends to some degree on the ratio of the capillary radius to the distance the jet travels from the capillary exit to the horizontal plane.

Nakoryakov *et al.* (1978) have made extensive measurements of the wall stress under an impinging jet, and they find that over a range of jet Reynolds numbers from 17,000 to 41,000, a single value of $a^* = 0.88$ permits their data to be fit by the single-fluid analysis (Eq. 5-4.2) very well.

An example of our data on film thinning (at a jet Reynolds number of 19,000) is shown in Fig. 5-4.3. With $a^* = 0.88$, Eq. 5-4.16 yields the line shown. The data are very well described by the approximate two-fluid model. We note that in the film thickness range of 20 to 40 nm the scatter in the measurements increases. We believe this is due to optical effects arising from silicon surface roughness of the order of 2 nm. This latter measurement follows from a series of studies of the surfaces of the bare silicon wafers, using a Sloan Dektac 3030 Surface Profilometer.

## *5-5 TWO-FLUID HOMANN FLOW*

One of the classical problems of fluid dynamics is the axisymmetric stagnation flow, or Homann flow, in which fluid in the region $z > 0$ approaches the plane $z = 0$ with symmetry about the axis ($r = 0$) normal to the plane. The flow at infinity (i.e., for $z \to \infty$) is simply $\mathbf{U} = (U_z, 0)$, and the solution seeks the velocity,

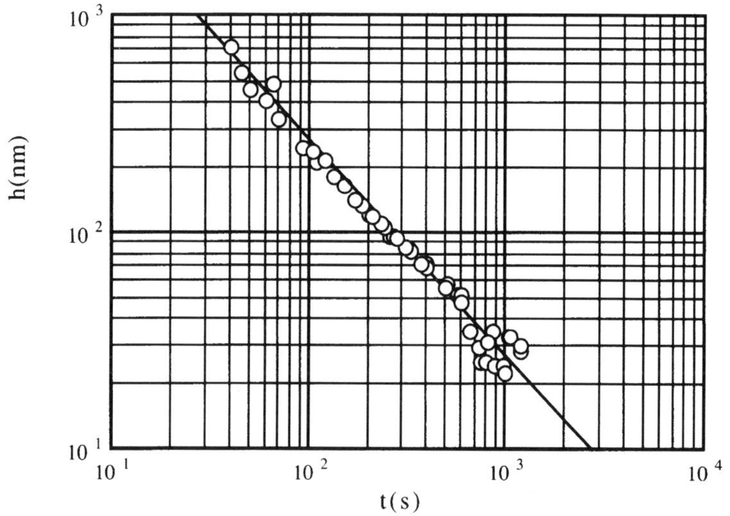


***FIGURE 5-4.3*** Thinning of a film of silicone oil in the stagnation region of a turbulent water jet. The silicone oil has a viscosity of 1000 centistokes. The jet issues from a capillary of inside diameter $D = 4.57$ mm, at a volume flow rate of 68 $cm^3/s$. (Data from Yeckel *et al.*, 1994.)

stress, and pressure fields in the neighborhood of the surface $z = 0$. The details of the solution to this problem are given, for example, in Schlichting (1960).

In the previous section we described the effect of an impinging liquid jet on an underlying viscous film. The pressure and shear stress in the neighborhood of the stagnation point of the jet were taken as given by Eqs. 5-4.1 and 2. This is not exactly the classical Homann flow because the jet is of a finite radius, and the ensuing flow from the jet spreads across a *mobile* viscous film, not a rigid plane. The applicability of Homann flow to the stagnation region of a jet has, however, been argued by Chin and Tsang (1978).

In what follows, we modify and extend the classical results for Homann flow, considering the effect of the mobile film, and show that under most conditions of interest the assumptions of the previous section, and hence the results of the analysis of film erosion, are valid to a good approximation.

The classical analysis for viscous Homann flow begins with the assumption that the velocity field may be written in the form

$$U_r = arf'(\eta), \quad U_z = -\sqrt{2\nu a}\ f(\eta), \tag{5-5.1}$$

where $\eta$ is defined as

$$\eta \equiv z \Big/ \sqrt{\frac{\nu}{2a}} \tag{5-5.2}$$

and $a$ is called the hydrodynamic constant. When the pressure field is written as

$$-\frac{1}{\rho}\frac{\partial p}{\partial r} = a^2 r, \tag{5-5.3}$$

the Navier–Stokes equations can be reduced to a single ordinary differential equation for $f(\eta)$ of the form

$$f''' + ff'' + \tfrac{1}{2}(1 - f'^2) = 0, \tag{5-5.4}$$

with boundary conditions

$$f = f' = 0 \quad \text{at } \eta = 0,$$
$$f' \to 1 \quad \text{at } \eta \to \infty. \tag{5-5.5}$$

The underlying viscous film is removed by the action of the jet through two mechanisms: the jet creates a radial pressure distribution in the region of the stagnation point, and the jet exerts a shear stress on the interface of the two liquids. It is necessary to have a model for these characteristics of the jet before a model for flow in the viscous layer can be developed. From the numerical solution to Eq. 5-5.4 the radial pressure distribution along the fixed plane $z = 0$ can be written as (see Eq. 5-4.1)

$$p(r) = \frac{1}{2}\rho U_j^2\left[1 - \left(\frac{a^* r}{2r_j}\right)^2\right], \tag{5-5.6}$$

where $a^*$ is a dimensionless form of the hydrodynamic constant, defined as

$$a^* \equiv \frac{2ar_j}{U_j}. \tag{5-5.7}$$

The shear stress is given by (see Eq. 5-4.2)

$$\tau(r) = 0.656 a^{*3/2} \rho U_j^2 \, \mathrm{Re}_j^{-1/2} \frac{r}{r_j}. \tag{5-5.8}$$

The Reynolds number is defined as

$$\mathrm{Re}_j = \frac{2U_j r_j}{\nu}. \tag{5-5.9}$$

In all of the above, $U_j$ is the velocity of the liquid jet at the stagnation point. The jet radius $r_j$ appears, but always as $a^*/r_j$, which in fact is independent of $r_j$.

### *5-5.1 The Modified Homann Flow over a Mobile Film*

It is not difficult to modify the classical Homann analysis and obtain a theory for stagnation flow over a viscous but mobile film. Once this is available it is possible to determine the conditions under which the simpler theory (Section 5-4) is a good approximation to the more complete analysis. Taking note of Fig. 5-5.1, we begin by rewriting Eqs. 5-5.1 and 2 for the jet flow as

$$U_r^j = arf'(\eta_j), \quad U_z^j = -\sqrt{2\nu_j a}\, f(\eta_j), \tag{5-5.10}$$

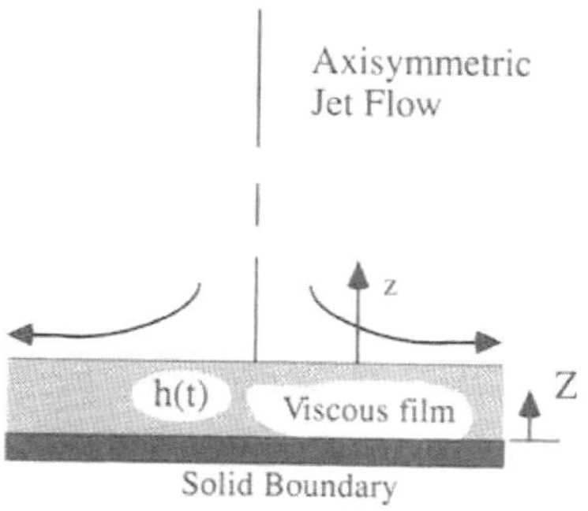


***FIGURE 5-5.1*** Geometry for the analysis of flow over a mobile film.

where $\eta_j$ is defined as

$$\eta_j \equiv z \Big/ \sqrt{\frac{\nu_j}{2a}}. \tag{5-5.11}$$

The Navier–Stokes equations for the jet still reduce to Eq. 5-5.4, but for $f(\eta_j)$:

$$f''' + ff'' + \tfrac{1}{2}(1 - f'^2) = 0 \tag{5-5.12}$$

and the boundary condition at infinity is

$$f' \to 1 \quad \text{at} \quad \eta \to \infty. \tag{5-5.13}$$

In the viscous liquid film ( $0 < Z < h$ ) the velocity field is also expressed in the form

$$U_r^{\mathrm{f}} = arF'(\eta_f), \quad U_z^{\mathrm{f}} = -\sqrt{2\nu_f a}\; F(\eta_f), \tag{5-5.14}$$

where $\eta_f$ is defined as

$$\eta_f \equiv Z \Big/ \sqrt{\frac{\nu_f}{2a}}. \tag{5-5.15}$$

The Navier–Stokes equations for the viscous film also reduce to Eq. 5-5.4, but for $F(\eta_f)$:

$$F''' + FF'' + \tfrac{1}{2}(1 - F'^2) = 0. \tag{5-5.16}$$

The no-slip condition at $Z = 0$ requires that

$$F = F' = 0 \quad \text{at} \quad \eta_f = 0. \tag{5-5.17}$$

The remaining two boundary conditions are at the interface, where the velocities and shear stresses are matched. Velocity matching requires that

$$F' = f' \tag{5-5.18}$$

and

$$F = \sqrt{\frac{\nu_j}{\nu_f}}\, f = \sqrt{\frac{\mathsf{M}}{\mathsf{P}}}\, f, \tag{5-5.19}$$

while the matching of the shear stresses requires that

$$F'' = \sqrt{\mathsf{MP}}\, f''. \tag{5-5.20}$$

All of these conditions are at the interface, defined by

$$\eta_j = 0, \quad \eta_f = h/\sqrt{\nu_f/2a}. \tag{5-5.21}$$

In these equations two dimensionless parameters appear:

$$\mathsf{M} \equiv \frac{\mu_j}{\mu_f} \quad \text{and} \quad \mathsf{P} \equiv \frac{\rho_j}{\rho_f}. \tag{5-5.22}$$

It is convenient to rename the dimensionless film thickness as

$$\mathsf{h} \equiv h/\sqrt{\nu_f/2a}. \tag{5-5.23}$$

Once a solution for $F$ is available from the system of equations written above, the dynamics of the film thickness follow from the kinematic condition

$$\frac{dh}{dt} = U_z^f(h) = -\sqrt{2\nu_f a}\; F(\mathsf{h}) \tag{5-5.24}$$

or

$$\int_{\mathsf{h}}^{\mathsf{h}_0} \frac{d\eta_f}{F} = \mathsf{T} \equiv 2at. \tag{5-5.25}$$

In order to evaluate the validity of the approximate lubrication analysis presented earlier (Section 5-4) we first note that $\Phi$ and $\Psi$ of that analysis may be written as

$$\Phi = 0.359\, \mathsf{h}_0\, \mathsf{M}^{-1/2},$$

and

$$\Psi = 0.464\, \mathsf{PH}_0\, \mathsf{M}^{1/2}\, \mathsf{T}. \tag{5-5.26}$$

Thus we will examine the behavior of $s$ vs. $\Psi$, where now we can write $\Psi$ as

$$\Psi = 0.464\, \mathsf{PH}_0\, \mathsf{M}^{1/2} \int_{\mathsf{h}}^{\mathsf{h}_0} \frac{d\eta_f}{F}. \tag{5-5.27}$$

Equations 5-5.12 and 16 are a coupled set of third-order ordinary differential equations, with the coupling arising through the matching conditions at the interface between the two fluids. The equations are integrated numerically. Details of the numerical procedure are presented in Yeckel *et al.* (1994).

Figure 5-5.2 shows a plot of the shear stress at the wall as a function of the film thickness $\Phi$, defined in Eq. 5-5.23. The dimensionless shear stress is expressed as $F''/\sqrt{\mathsf{MP}}$. Note in particular that the *jet* viscosity is used in this nondimensionalization. Figure 5-5.3 shows the same thing for the shear stress $f''$ at the jet–film interface. Curves are shown at three values of the viscosity ratio

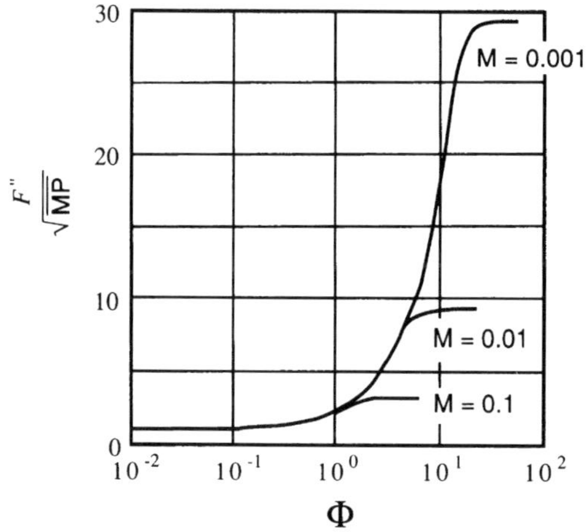


*FIGURE 5-5.2* Dimensionless wall shear stress.

M. The density ratio is set at the value P=1 in all cases, to reflect our interest in liquid–liquid systems. In both figures the stress approaches the value obtained from the Homann analysis as Φ tends toward zero, an expected result.

The approximate analysis given earlier (Section 5-4) is based on the assumption that the shear stress at the interface of the two liquids is virtually identical to the Homann value for a single fluid. If the film is sufficiently thin and viscous that the decrease in shear stress from the interface to the wall is small, relative to the Homann value, we expect that this approximation will be a good one. A comparison of Figs. 5-5.2 and 3 shows that this is the case at values of Φ up to about 0.1. A fortuitous result (Fig. 5-5.3) is that the interface stress remains close to the Homann value at values of Φ an order of magnitude greater, up to Φ = 1, at which point the wall stress is more than twice the Homann value.

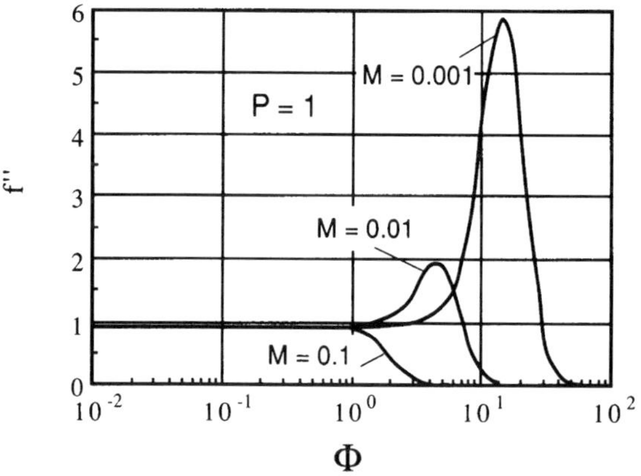


*FIGURE 5-5.3* Dimensionless interface shear stress.

Figure 5-5.3 shows that for $\Phi$ greater than one, the interface stress diverges sharply from the Homann value, in which case clearly the approximate model fails. At large enough $\Phi$ the interface stress falls to zero. That the interface stress has fallen to zero indicates that viscous effects are confined to the film, and the jet consists of a potential flow right up to the interface. Under these conditions, only the force caused by the radial pressure gradient acts to remove the viscous film. From this point on, the dimensionless *wall* stress is a constant, equal to the Homann stress (based on the jet viscosity) divided by the square root of $\mathsf{M}$, as seen in the large $\Phi$ asymptotes of Fig. 5-5.2. If the wall stress is made dimensionless by the *film* viscosity rather than the jet viscosity, the dimensionless wall stress equals the Homann value. This indicates that the jet transfers its momentum without viscous loss to the film, which itself is in a state of Homann flow.

The use of the shear stress at the solid boundary according to the *single-fluid* Homann analysis, as an approximation to the shear stress at the mobile interface in the two-fluid case, is seen to be a good approximation if the initial film thickness is so small that $\Phi < 0.03$. Under these conditions we would expect that the simple analytical lubrication analysis will be an accurate model of flow in the viscous film. It turns out that the approximate lubrication model for the dynamics of the film thinning is accurate to even higher values of $\Phi$, as Fig. 5-5.4 shows.

In Fig. 5-5.4 we see a comparison of the approximate lubrication analysis (the solid curves) and the mobile-film two-fluid Homann analysis for the rate of thinning of the mobile film. Results are shown for selected values of the parameters $\mathsf{P}$ and $\mathsf{M}$. With these choices, selection of an $\mathsf{h}_0$ value fixes the value of $\Phi_0$, at the values shown. The approximate model coincides nearly perfectly with the more exact model, even for short times (small $\Psi$) where the film is relatively

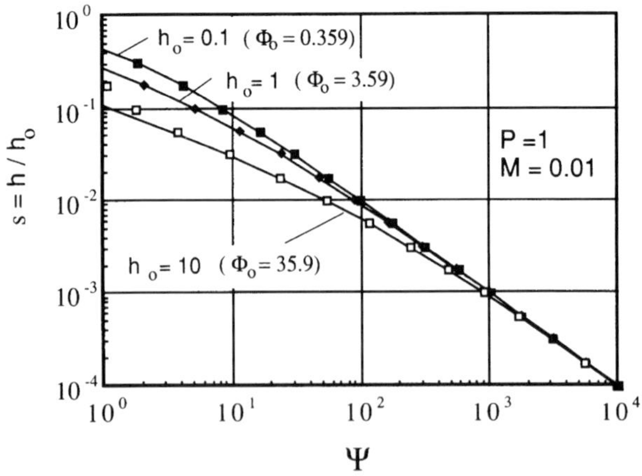


***FIGURE 5-5.4*** Comparison of the simple approximate model to the two-fluid Homann analysis, for the rate of thinning of the mobile film.

thick, when $h_0$ is smaller than unity. For the relatively large value of $h_0 = 10$ ($\Phi_0 = 35.9$) we see a significant deviation of the two models of the film thickness history at short times. However, as the initially thick film grows thinner, the behavior of $s$ vs. $\Psi$ becomes independent of $h_0$, the two models coincide, and both models exhibit the expected approach of $s$ to $1/\Psi$.

A key assumption of the two-fluid model outlined here is that the mobile interface is stable. It is well known that parallel flows of this type have the capacity to display several modes of instabilities. [See, for example, the works of Joseph *et al.* (1984), Papageorgiou *et al.* (1990), and Preziosi *et al.* 1989).] No such instabilities have been observed in our experimental studies.

## 5-6 LAMINAR FLOW OVER A DISK

Watson's analysis (Section 5-2) is a boundary layer theory, presumably valid for laminar jets but at high enough Reynolds numbers that the retarding effects of viscosity are balanced by the inertia of the radial flow. It is interesting to look at this type of flow (i.e., radial flow across a disk) when inertial effects disappear. Figure 5-6.1 suggests the flow field of interest.

If we assume that somewhere downstream of the stagnation region the flow is nearly parallel to the disk, the axial momentum equation reduces to

$$0 = -\frac{\partial p}{\partial z} - \rho g. \tag{5-6.1}$$

The solution is

$$p = -\rho g z + \rho g H(r). \tag{5-6.2}$$

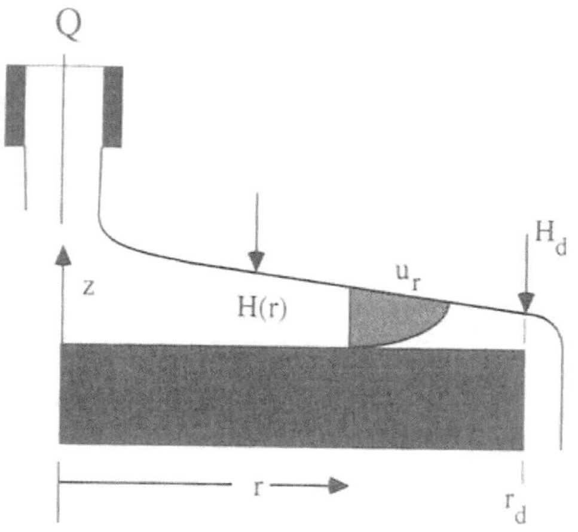


***FIGURE 5-6.1*** Viscous flow over a disk.

The radial momentum equation takes the form

$$0 = -\frac{\partial p}{\partial r} + \mu\frac{\partial^2 u_r}{\partial z^2}. \tag{5-6.3}$$

(We have used the continuity equation and the assumption that the flow is nearly parallel to the disk in simplifying the viscous terms.) Global conservation of mass can be written in the form

$$Q = \int_0^{H(r)} 2\pi r u_r \, dz. \tag{5-6.4}$$

When Eqs. 5-6.2 and 3 are combined we find

$$0 = -\rho g\frac{dH}{dr} + \mu\frac{\partial^2 u_r}{\partial z^2}. \tag{5-6.5}$$

This may be integrated to find the velocity profile, and when the boundary conditions

$$\mu\frac{\partial u_r}{\partial z} = 0 \text{ at } z = H(r) \tag{5-6.6}$$

and

$$u_r = 0 \text{ at } z = 0 \tag{5-6.7}$$

are used the result is

$$u_r = \frac{\rho g}{\mu}\left(\frac{dH}{dr}\frac{z^2}{2} - H\frac{dH}{dr}z\right). \tag{5-6.8}$$

When Eq. 5-6.4 is used we find

$$\frac{\mu Q}{2\pi r\rho g} = -\frac{H^3}{3}\frac{dH}{dr}. \tag{5-6.9}$$

This is a differential equation for $H(r)$, and the solution is

$$H^4 - H_d^4 = -\frac{6\mu Q}{\pi\rho g}\ln\frac{r}{r_d}. \tag{5-6.10}$$

The problem that we have to deal with now is how we obtain a value for $H_d$, the film thickness at the edge of the disk. A second issue is the behavior of this solution near the jet axis, where the model predicts that $H$ grows unbounded. (This same problem arises in Watson's analysis. We ignored it.) The failure of the model in this region is not unexpected, since we assumed that the flow was nearly parallel to the disk downstream from the stagnation region. In effect, we excluded the stagnation region from the model.

We might expect on physical grounds that $H_d$ depends upon the flow rate $Q$ and the size of the disk. One way to deal with this is to introduce an *ad hoc* hypothesis and examine the degree to which the resulting model mimics observation. As an example, we will assume that there is no viscous dissipation of energy when the liquid "turns the corner" at the edge $r = r_d$ of the disk. If this is true, then we have a laminar film flowing down a vertical cylinder, with no change in speed and film thickness, since the energy flux is conserved. We know that for a laminar falling film along a vertical surface the film thickness $H_f$ and the flow rate are related by (Bird *et al.*, 1960)

$$H_f^3 = \frac{3\mu Q}{2\pi\rho g r_d} \quad \text{if} \quad \frac{H_f}{r_d} << 1. \tag{5-6.11}$$

(This is just the solution for the *planar* falling film.) With the assumption that the two film thicknesses $H_f$ and $H_d$ are the same we find

$$H^4 = \left(\frac{3\mu Q}{2\pi\rho g r_d}\right)^{4/3} - \frac{6\mu Q}{\pi\rho g} \ln \frac{r}{r_d} \tag{5-6.12}$$

or

$$\left[\frac{H(r)}{r_d}\right]^4 = \mathsf{A}^{4/3} - 4\mathsf{A} \ln\left(\frac{r}{r_d}\right), \tag{5-6.13}$$

where

$$\mathsf{A} \equiv \frac{3\mu Q}{2\pi\rho g r_d^4}. \tag{5-6.14}$$

Equation 5-6.13 is plotted in Fig. 5-6.2 for three values of the parameter A.

Regardless of our reservations in deriving this result, this is a testable model. In Fig. 5-6.3 we show a comparison of film profile data with the theory for one value of the parameter A. For the case shown the measured film thickness profile is in good agreement with the predicted shape from Eq. 5-6.13.

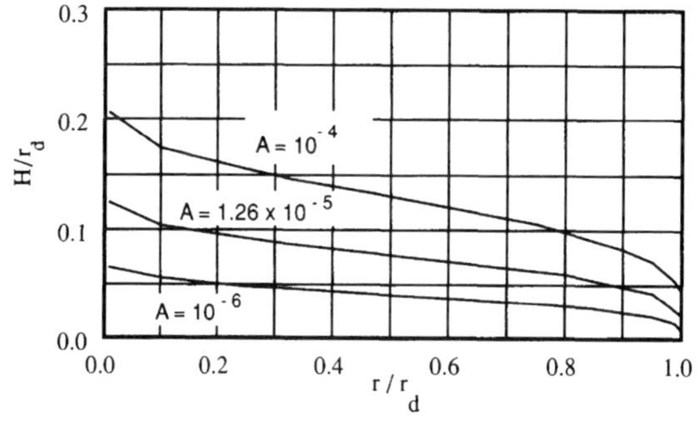


***FIGURE 5-6.2*** Film thickness across a disk: laminar flow.

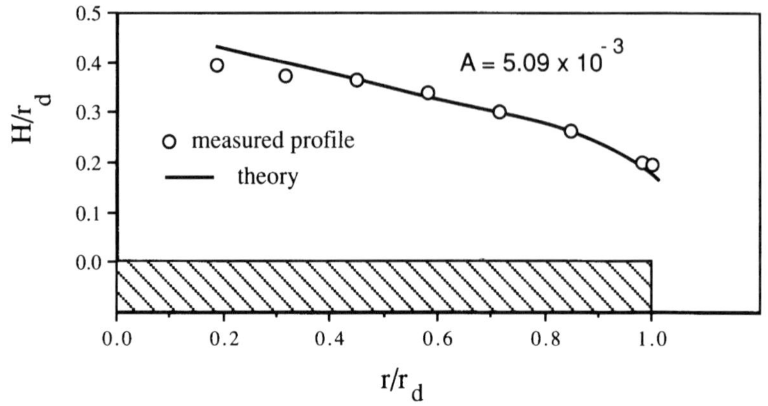


*FIGURE 5-6.3* Experimental test of Eq. 5-6.13. Re = 1.52; $r_d$ = 1.27 cm; $r_t$ = 0.05 cm. Liquid is shampoo, with $\mu$ = 1.48 Pa-s; $\sigma$ = 0.038 N/m. (Unpublished data of Chow and Middleman).

According to Eq. 5-6.13 we should find

$$\frac{H_d}{r_d} = A^{1/3}. \tag{5-6.15}$$

Data (see Fig. 5-6.4) obtained over a range of A values, and for jet Reynolds numbers from approximately 1 to 6, show that the film thickness at the edge of the disk agrees reasonably well with the prediction of Eq. 5-6.15.

Of course, the model still fails to address the issue of the behavior of $H(r)$ as $r$ approaches the jet axis. We should expect that a model for that region will require some assumptions about how the liquid is introduced along the axis. In the case of interest to us, where the liquid is supplied as a jet from a capillary, we would have to solve the problem of a free surface viscous stagnation flow.

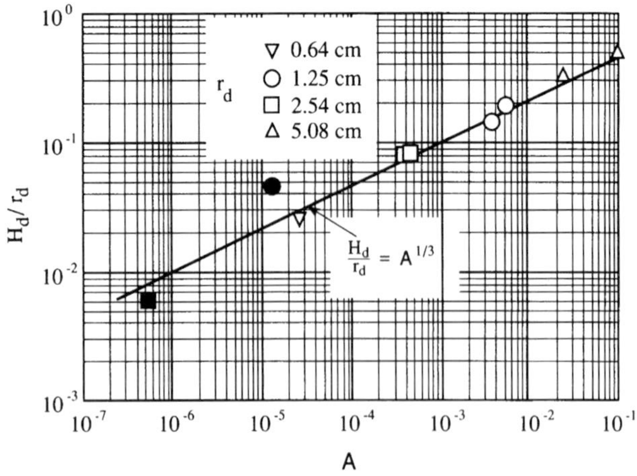


*FIGURE 5-6.4* Experimental test of $H_d = H_f$. (Chow and Middleman, unpublished data).

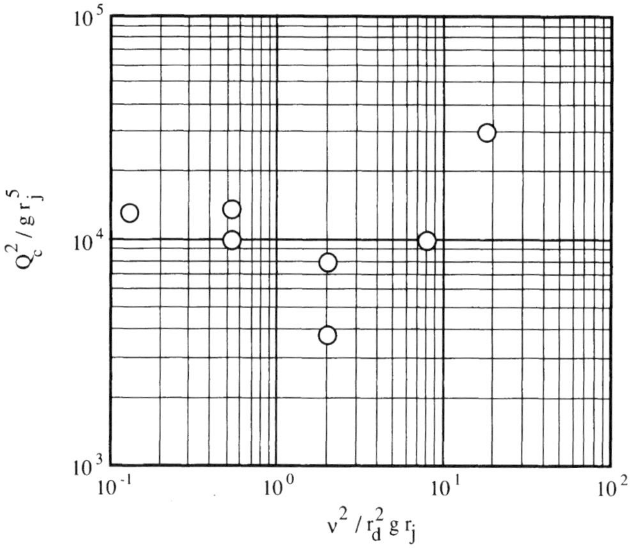


***FIGURE 5-6.7*** Correlation of the "critical dimple flow rate" for a laminar jet.

Certainly a numerical solution would be required.

Figure 5-6.5 (see color insert) shows several examples of the free surface shapes observed. In the viscous-dominated regime the surface shape is very sensitive to the flow rate, as well as to the standoff distance of the exit of the capillary from the disk. The "bell-shaped" surface is characteristic of very low flow rates and small standoff distances. It appears that under this combination of conditions "information" about the flow in the stagnation region can be propagated upstream to the jet, and under some conditions all the way to the capillary exit. At higher flow rates the surface is smoother and well described by the analysis of this section. Our limited experimental studies (see color insert, Fig. 5-6.6) indicate that a "dimpled" surface exists when the flow rate exceeds some critical value. An attempt at correlating data for the critical dimple flow rate is presented in Fig. 5-6.7 (see Problem P5-11). Additional work, both computational and experimental, will be required before this complex flow field is better understood. In particular, one would like to be able to predict the conditions that correspond to transitions from one surface profile to another.

## *Problems*

P5-1 Suppose the velocity profile across the liquid film in the radial flow region of Fig. 5-1.2 has the form

$$u_r = U(r)\,f\left(\frac{z}{\delta}\right) = \frac{4}{3}\,U_{\text{fd}}(r)\left[3\left(\frac{z}{\delta}\right) - 3\left(\frac{z}{\delta}\right)^2 + \left(\frac{z}{\delta}\right)^3\right]. \tag{P5-1.1}$$

Find the coefficient $K_{31}$ defined by Eq. 5-1.4.

P5-2 Derive Eq. 5-2.16.

P5-3 Derive Eqs. 5-2.22 and 24.

P5-4 Using Watson's model of the radial flow across a surface, find an expression for the shear stress distribution along the radius from the point of impact of the jet. Plot the result as a dimensionless shear stress

$$\varphi \equiv \frac{\nu(\partial u_r/\partial z)_{z=0}}{U_j^2} \tag{P5-4.1}$$

as a function of the radial position $r/r_j$, and the Reynolds number defined in Eq. 5-2.32. Show first that $\varphi$ is given by

$$\varphi = \frac{1.16r/r_j}{(r/r_j)^3 + 0.286\mathrm{Re}_j} \quad \text{for} \quad r < r_0 \tag{P5-4.2}$$

and

$$\varphi = \frac{0.141\mathrm{Re}_j\,(r/r_j)}{\left[(r/r_j)^3 + 0.286\mathrm{Re}_j\right]^2} \quad \text{for} \quad r > r_0 \tag{P5-4.3}$$

P5-5 Using Fig. 5-2.2, plot the film height at the edge of a disk of radius $r_d$, normalized to the jet radius, as a function of Reynolds number, with $r_d/r_j$ as a parameter. Choose parameters in the ranges $10^3 < \mathrm{Re}_j < 10^4$ and $10 < r_d/r_j < 100$.

P5-6 Modify the development of Eq. 5-3.12 in the following manner. Replace the assumption $u_r = U_1$ in Eq. 5-3.11 with a profile of the form of Eq. P5-1.1 in Problem P5-1:

$$u_r = \frac{4}{3}U_1\left[3\left(\frac{z}{H_\infty}\right) - 3\left(\frac{z}{H_\infty}\right)^2 + \left(\frac{z}{H_\infty}\right)^3\right]. \tag{P5-6.1}$$

Find the modified form of Eq. 5-3.12a.

P5-7 Equation 5-4.1 follows from the analysis of the flow of an inviscid jet impinging normal to a rigid surface. Show that the velocity components

$$U_r = ar, \qquad U_z = -2az \tag{P5-7.1}$$

satisfy the continuity equation and that when they are inserted into the radial component of the Navier–Stokes equations and the viscous terms are dropped, the result is Eq. 5-4.1, with $a^*$ defined as

$$a^* \equiv \frac{2ar_j}{U_j}. \tag{P5-7.2}$$

This velocity field (Eq. P5-7.1) is called inviscid Homann flow.

P5-8 Inviscid Homann flow (see Problem P5-7) does not yield any result for the shear stress in the neighborhood of the stagnation point. Show that a boundary layer analysis is possible which leads to a model for the shear stress. To obtain the boundary layer equations, define a function $f(\eta)$ such that

$$U_r = arf'(\eta), \qquad U_z = -\sqrt{2\nu a}\; f(\eta),$$
$$z = \sqrt{\frac{\nu}{2a}}\;\eta. \tag{P5-8.1}$$

Confirm that the continuity equation is satisfied for this velocity assumption. Show that with these definitions the Navier–Stokes equations can be reduced to a single ordinary differential equation of the form

$$f''' + ff'' + \tfrac{1}{2}(1 - f'^2) = 0 \tag{P5-8.2}$$

with boundary conditions

$$f = f' = 0 \quad \text{at } \eta = 0,$$
$$f' \to 1 \quad \text{at } \eta \to \infty. \tag{P5-8.3}$$

Find the shear stress predicted by this solution, and compare the form to that given in Eq. 5-4.2.

P5-9 Given a solution $f(\eta)$ to Eq. 5-5.4, show the relationship of $p(r,z = 0)$ and $\tau(r,z = 0)$ to $f(\eta)$.

P5-10 Consider the situation shown in Fig. P5-10.1 below. Suppose the flow rate is so slow that the liquid in the region at the edge of the disk is quasi-stationary. Find the maximum film thickness at the disk edge, $H_d$, if the determining factor is the balance between hydrostatic pressure and surface tension.

P5-11 We seek a rational suggestion for correlating data on the "critical dimple flow rate." Predict the critical dimple flow rate if the following were true:

> A "dimple" occurs at the flow rate such that the stagnation pressure of the jet just exceeds the hydrostatic head of the film downstream of the dimple.

P5-12 In an interesting series of papers by Bergink-Martens *et al.* (1990, 1992, 1994) a radial free-surface flow is described that corresponds to the sketch in Fig. P5-12.1.

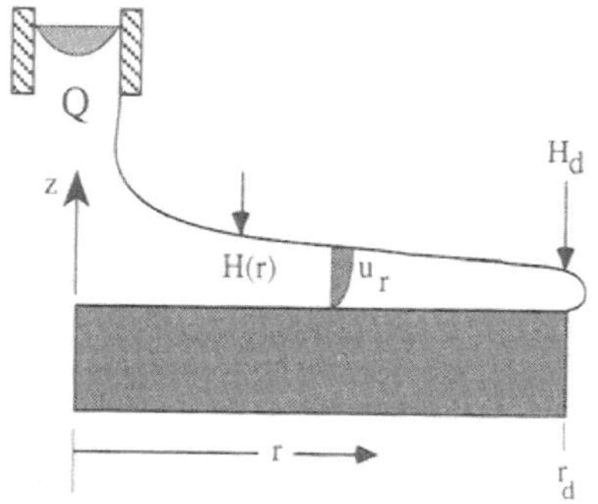


***FIGURE P5-10.1*** Definition sketch for Problem P5-10.

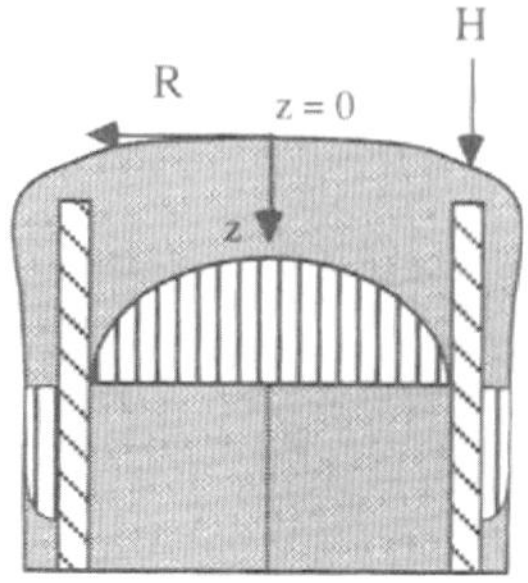


***FIGURE P5-12.1*** Definition sketch for Problem P5-12.

(a) Apply the quasi-steady analysis of Problem P5-10 and predict the height of the free surface above the rim. The measured values (Bergink-Martens *et al.*, 1990) for water are approximately 4 mm and independent of flow rate in the range $10 < Q < 40$ cm$^3$/s.

(b) Apply the transformation given in Problem P5-8 to this problem, and write the resulting form of the radial momentum equation. Specify the boundary conditions based on a reasonable physical picture of this flow. Discuss any difference between your result and that of Bergink-Martens *et al.* (1994). Why do you not obtain Eq. P5-8.2?

## References

Bergink-Martens, D. J. M. , H. J. Bos, A. Prins, and B. C. Schulte. (1990). Surface dilation and fluid-dynamical behavior of Newtonian liquids in an overflowing cylinder. I. Pure liquids. *J. Coll. Int. Sci.* **138**, 1.

Bergink-Martens, D. J. M., C. G. J. Bisperink, H. J. Bos, A. Prins, and A. F. Zuidberg. (1992). Surface dilational behavior of surfactant solutions. A comparison between the overflowing cylinder and the free falling film technique. *Coll. Surf.* **65**, 191.

Bergink-Martens, D. J. M., H. J. Bos, and A. Prins. (1994). Surface dilation and fluid-dynamical behavior of Newtonian liquids in an overflowing cylinder. II. Surfactant solutions. *J. Coll. Int. Sci.* **165**, 221.

Bird, R. B., W. E. Stewart, and E. N. Lightfoot. (1960). *Transport Phenomena.* Wiley, New York.

Bowles, R. I., and F. T. Smith. (1992). The standing hydraulic jump: Theory, computations and comparisons with experiments. *J. Fluid Mech.* **242**, 145.

Chaudhry, M. H. (1994). Computation of flows with shocks and bores. Ch. 3 in Computer Modeling of Free-Surface and Pressurized Flows, M. H. Chaudhry and L. W. Mays, eds., Kluwer Academic Publ., The Netherlands.

Chin, D-T., and C-H. Tsang. (1978). Mass transfer to an impinging jet electrode. *J. Electrochem. Soc.* **125**, 1461.

Craik, A. D. D., R. C. Latham, M. J. Fawkes, and P. W. F. Gribbon. (1981). The circular hydraulic jump. *J. Fluid Mech.* **112**, 347.

Errico, M. (1986). A study of the interaction of liquid jets with solid surfaces. Ph.D. thesis, Univ. of California, San Diego.

Higuera, F. J. (1994). The hydraulic jump in a viscous laminar flow. *J. Fluid Mech.* **274**, 69.

Joseph, D., M. Renardy, and Y. Renardy. (1984). Instability of the flow of two immiscible liquids with different viscosities in a pipe. *J. Fluid Mech.* **141** , 309.

Liu, X. and J. H. Lienhard V. (1993). The hydraulic jump in circular jet impingement and in other liquid films. *Exp. Fluids* **15**, 108.

Middleman, S. and J. Gavis. (1961). Expansion and contraction of capillary jets of Newtonian liquids. *Phys. Fluids* **4**, 355.

Nakoryakov, V. E., B. G. Pokusaev, and E. N. Troyan. (1978). Impingement of an axisymmetric liquid jet on a barrier. *Int. J. Heat Mass Transfer* **21**, 1175.

Olsson, R. G., and E. T. Turkdogan. (1966). Radial spread of a liquid stream on a horizontal plate. *Nature* **211**, 813.

Papageorgiou, D. T., C. Maldarelli, and D. S. Rumschitzki. (1990). Nonlinear interfacial stability of core-annular film flows. *Phys. Fluids A* **2**, 340.

Preziosi, L., K. Chen, and D. Joseph. (1989). Lubricated pipelining. stability of core-annular flow. *J. Fluid Mech.* **201**, 323.

Rahman, M. M., W. L. Hankey, and A Faghri. (1991). Analysis of the fluid flow and heat transfer in a thin liquid film in the presence and absence of gravity. *Int. J. Heat Mass Transfer* **34**, 103.

Schlichting, H. (1960). *Boundary Layer Theory*. McGraw–Hill, New York, pp. 81–83.

Strong, L., and S. Middleman. (1989). Lubricant retention on a spinning disk. *AIChE J.* **35**, 1753.

Watson, E. J. (1964). The radial spread of a liquid jet over a horizontal plane. *J. Fluid Mech.* **20**, 481.

Yeckel, A., and S. Middleman. (1987). Removal of a viscous film from a rigid plane surface by an impinging liquid jet. *Chem. Eng. Commun.* **50**, 165.

Yeckel, A., L. Strong, and S. Middleman. (1994). Removal of a thin viscous film from the stagnation region of an axisymmetric jet impinging on a planar surface. *AIChE J.* **40**, 1611.

Younus, M., and M. H. Chaudhry. (1994). A depth-averaged $k$–$\varepsilon$ turbulence model for the computation of free-surface flow. *J. Hydraulic Res.* **32**, 415.

# 6 *The Transfer of a Drop from a Capillary to an Impermeable Surface*

## *6-1 INTRODUCTION*

In the next chapter we will consider the fate of a drop sitting on a surface, under conditions where the drop is not in its equilibrium configuration. In this chapter we consider some problems associated with how the drop gets onto the surface. In particular, we examine the *transfer* of a pendent (hanging) drop from a vertical capillary to a horizontal surface. The simplest mode of transfer is through the growth of the drop to a size that can no longer be supported on the tip of the capillary. At some critical size (weight) the drop breaks off. However, only a portion of the drop leaves the capillary tip. Our goal is to learn how to describe the shape of a pendent drop and to determine the critical drop size and the fraction of the drop that falls away from the tip.

In another mode of transfer, we imagine that at some instant of time the drop has grown to the point that it touches the lower surface while still attached to the capillary, and thus it bridges the space between the two surfaces, as suggested in Fig. 6-1.1. In this case our goal is to determine the shape of the bridge as a function of the separation distance and ultimately to determine how the bridge partitions itself into drops between the two surfaces if the surfaces are separated

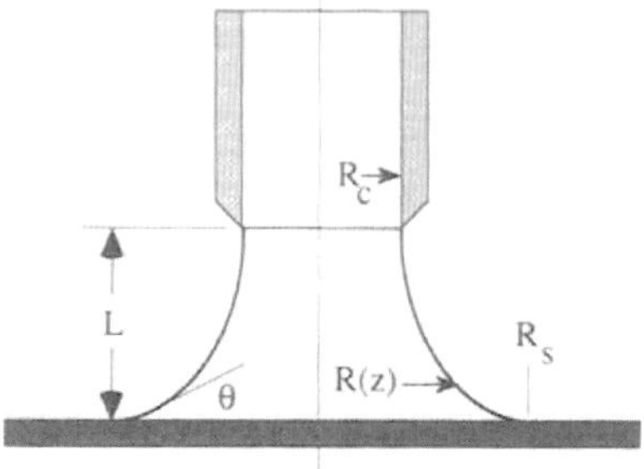


***FIGURE 6-1.1*** A static bridge between two surfaces.

while the volume of connecting liquid remains constant. Prior to attacking that dynamic problem, however, we have to understand the static case.

## 6-2 SHAPE OF A STATIC DROP

Figure 6-1.1 shows a drop which bridges the two surfaces and exhibits a contact angle $\theta$ less than $\pi/2$ radians on the lower surface. (Contact angle is measured from the lower surface to the tangent, from *within* the liquid.) We know from experience that such a configuration is stable for small values of $L$. Our experience also suggests that if $L$ is made large enough we cannot support the bridge, and the liquid breaks into two portions: a sessile (sitting) drop on the lower surface and a pendent drop on the capillary. An important task is to examine the stability of a pendular bridge of liquid—what factors determine the maximum stable $L$ for a given system? Then, for an unstable configuration, we would like to be able to predict the partitioning of the bridge—what fraction is transferred to the lower surface? Finally, we could return to the dynamic case—how do the dynamics affect the stability, and at what rate does a pendular bridge break down into the upper and lower portions? (In a practical sense this turns out to be an unimportant question—the bridge breaks so rapidly, under most conditions of practical interest, that the time to break is unmeasurably small.)

The first thing to note is that the shape of the bridge is determined, in part, by conditions imposed on the contact lines of the bridge. In Fig. 6-1.1, for example, we take the bridge to be constrained by the capillary at the upper boundary, so that

$$R(z) = R_c \quad \text{at} \quad z = L. \qquad (6\text{-}2.1)$$

(We take $z = 0$ to be the plane of the lower bounding surface.)

At the lower surface only the contact angle is specified:

$$dR/dz = -\cot\theta \quad \text{at} \quad z = 0. \qquad (6\text{-}2.2)$$

Another constraint on the shape is that the volume is specified, so

$$V = \int_0^L \pi R^2(z)\, dz. \tag{6-2.3}$$

These particular constraints (at least, the first two) are specific to the assumed contact geometry of the liquid bridge. For example, a bridge could be constrained between two disks, or between two infinite parallel planes. Figure 6-2.1 suggests these configurations. Others can be imagined as well. When necessary we will distinguish these cases by referring to radius-limited, or "R-bridges," and contact angle–limited, or "θ-bridges." In Fig. 6-1.1 we have a mixed case.

There is an enormous literature about the shapes of axisymmetric interfaces on or between solid surfaces. In order to calculate the shape of the interface, one begins with the Young–Laplace equation and an appropriate set of boundary conditions. Except under special conditions, the Young–Laplace equation is nonlinear and it must be solved by some numerical method. Let's begin by examining some issues associated with the formulation and solution of the Young–Laplace equation.

We start with the special case of a sessile drop, as shown in Fig. 6-2.2. We assume that gravity acts in the direction of the negative $z$ axis. Then the Young–Laplace equation takes the form

$$\sigma\left(\frac{1}{R_1} + \frac{1}{R_2}\right) + \rho g\left(H_0 - H\right) = 0, \tag{6-2.4}$$

where $H_0$ is a length parameter, to be determined later, but related to the size of the drop.

The shape of the drop is expressed as the *meridianal curve* $R(H)$. Equivalently, we can describe the shape as $H(R)$. The sum of the radii of curvature may be written in terms of $H(R)$ as

$$\frac{1}{R_1} + \frac{1}{R_2} = \frac{d^2H/dR^2}{\left[1 + \left(dH/dR\right)^2\right]^{3/2}} + \frac{dH/dR}{R\left[1 + \left(dH/dR\right)^2\right]^{1/2}}. \tag{6-2.5}$$

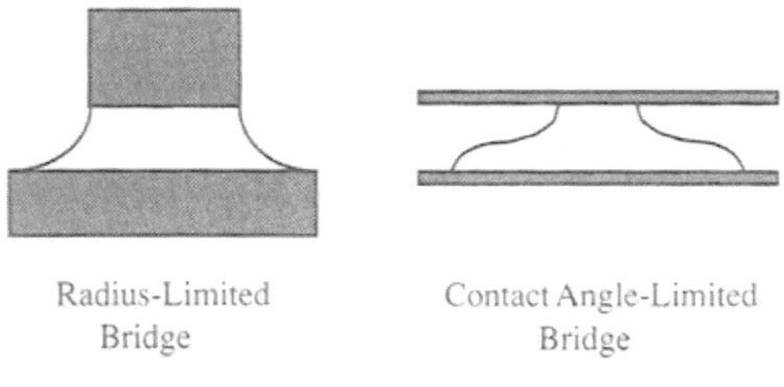


***FIGURE 6-2.1*** Two bridges of the same volume can have shapes that depend on the boundary constraints.

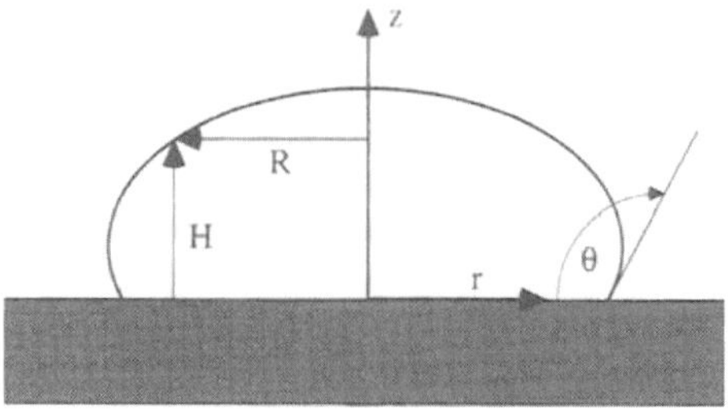


***FIGURE 6-2.2*** A static sessile drop. Note that the contact angle θ is measured within the liquid.

It follows that the shape of the drop $H(R)$ must be the solution to

$$\sigma \left\{ \frac{d^2H/dR^2}{\left[1 + (dH/dR)^2\right]^{3/2}} + \frac{dH/dR}{R\left[1 + (dH/dR)^2\right]^{1/2}} \right\} + \rho g(H_0 - H) = 0, \tag{6-2.6}$$

where, as previously noted, $H_0$ is a parameter characteristic of the size of the drop. The boundary conditions are usually taken as

$$\begin{aligned} (dH/dR) &= 0 \quad \text{for } R = 0, \\ (dH/dR) &= -\tan\theta \quad \text{for } R = R_0. \end{aligned} \tag{6-2.7}$$

As noted earlier, this is a nonlinear equation which must be solved numerically. One additional constraint on the solution (which gives the shape of the drop) is that the volume must satisfy

$$V = \int_0^{H_0} \pi R^2 \, dH. \tag{6-2.8}$$

If the contact angle is fixed by the nature of the solid–liquid thermodynamics, then specification of the drop volume $V$ leads to values of the contact or "footprint" radius $R_0$, through solution of Eq. 6-2.6. Alternatively, if θ and $R_0$ are specified, the solution yields the volume of the drop. In either event, the upper limit on the integral now identifies $H_0$ as the height of the drop at the axis of revolution.

Various strategies are described (see, for example, Fordham, 1948; Huh and Scriven, 1969; Boucher and Evans, 1975) for obtaining a numerical solution to Eq. 6-2.6. We will not describe these here. Instead, we take a look at two special cases for which *analytical* solutions are possible. We begin with the case of vanishing gravity (or a very small drop), and seek the solution to

$$\sigma \left\{ \frac{d^2H/dR^2}{\left[1 + (dH/dR)^2\right]^{3/2}} + \frac{dH/dR}{R\left[1 + (dH/dR)^2\right]^{1/2}} \right\} = 0. \tag{6-2.9}$$

It is not difficult to show that one solution to this equation corresponds to a spherical surface (see Problem P6-1). This is to be expected, since in the absence of gravity the pressure within the drop is everywhere uniform. Therefore the sum of the radii of curvature must be the same at any point on the surface. A surface with this property of uniform curvature is the sphere.

The other special case is that for which the contact angle is so small that the drop slope $dH/dR$ is everywhere small. This permits the *linearization* of Eq. 6-2.6 to the form

$$\sigma \left\{ \frac{d^2H}{dR^2} + \frac{dH/dR}{R} \right\} + \rho g\left(H_0 - H\right) = 0 \tag{6-2.10}$$

or, in nondimensional format,

$$\frac{d^2h}{dr^2} + \frac{1}{r}\frac{dh}{dr} + \left(h_0 - h\right) = 0, \tag{6-2.11}$$

where $H$, $R$, $H_0$, and $R_0$ are all nondimensionalized (to $h$, $r$, $h_0$, and $r_0$) with a "capillary length" $a^*$ defined as

$$a^* \equiv \sqrt{\frac{\sigma}{\rho g}}. \tag{6-2.12}$$

The boundary conditions given in Eqs. 6-2.7 now take the form

$$\begin{aligned} \left(dh/dr\right) &= 0 \text{ for } r = 0, \\ \left(dh/dr\right) &= -\tan\theta \text{ for } r = \sqrt{\mathrm{Bo}}, \end{aligned} \tag{6-2.13}$$

where the Bond number is defined as

$$\mathrm{Bo} \equiv \frac{\rho g R_0^2}{\sigma} = \frac{R_0^2}{a^{*2}}. \tag{6-2.14}$$

(Note that if $R_0$ is not specified, the Bond number is an unknown parameter.) Equation 6-2.11 is a Bessel equation, and the solution that satisfies the boundary conditions is

$$h(r) = h_0 - \frac{\tan\theta}{I_1(\sqrt{\mathrm{Bo}})}\left[1 - I_0(r)\right] \quad \theta\text{-limited.} \tag{6-2.15}$$

To determine $h_0$ we use the condition that

$$h = 0 \text{ at } r = \sqrt{\mathrm{Bo}},$$

which yields the result

$$h_0 = \frac{\tan\theta}{I_1(\sqrt{\mathrm{Bo}})}\left(1 - I_0(\sqrt{\mathrm{Bo}})\right). \tag{6-2.16}$$

The drop volume and the contact (footprint) radius are related through the volume constraint (Eq. 6-2.8), which takes the form

$$\frac{V}{a^{*3}} = \int_0^{h_0} \pi r^2\, dh = -\int_0^{\mathrm{Bo}} \pi r^2 \frac{dh}{dr}\, dr \tag{6-2.17}$$

with

$$\frac{dh(r)}{dr} = -\frac{\tan\theta \; I_1(r)}{I_1(\sqrt{\mathrm{Bo}})}. \tag{6-2.18}$$

If $R_0$ is given, the volume follows directly upon integration. If the volume is specified, then the indicated integration is performed for various (trial) values of $R_0$ until Eq. 6-2.17 is satisfied. In either case, Eq.6-2.17 is the connection between $h_0$ and $V$, for a given angle $\theta$.

The shape of a sessile drop is a relatively straightforward problem in capillary hydrostatics. Conceptually, the physics are quite simple—the drop sits on the surface, and if there is a *finite* contact angle the drop attains some equilibrium shape. (If the drop wets the surface, in the sense that the equilibrium contact angle is zero, then there is no equilibrium drop—the liquid spreads to wet the surface with an infinitesimally thin film.)

A somewhat more interesting and complicated problem is that of the hanging, or pendent drop. Two case of practical interest are possible and are illustrated in Fig. 6-2.3.

The pendent drop may hang from a large planar surface on which the liquid has an equilibrium contact angle $\theta_c$. For a given drop volume the shape and hence the contact radius are determined (limited) by the contact angle. On the other hand, the drop may hang from a rod (or capillary) of radius $R_0$. Then the drop is assumed to be "pinned" to this contact radius, and for a given volume the angle $\theta$ is not the *thermodynamic* contact angle $\theta_c$. Instead, $\theta$ adjusts itself in accordance with the requirements of the Young–Laplace equation. This is the "R-limited" drop mentioned earlier.

The $\theta$-limited cases of the sessile and pendent drops have one significant distinction. With the sessile drop of nonzero equilibrium contact angle we may imagine that if we continually increase the volume of the drop, the drop will spread to a larger contact radius in order to satisfy the Young–Laplace equation while maintaining a fixed contact angle. The pendent drop volume is limited, however. At some maximum volume the pendent drop falls away from the surface—in a finite gravitational field its weight is unstable above a certain volume.

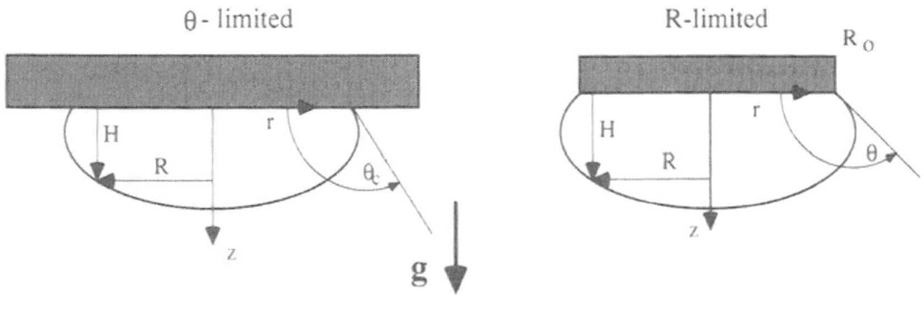


***FIGURE 6-2.3*** Pendent drops.

This is not the case with the θ-limited sessile drop. The same instability exists for the R-limited pendent case—there is a maximum drop size that will remain on a capillary of a given radius. The same is true of the *sessile* drop in the R-limited case. As the sessile drop grows in volume on an upward-pointed thin-walled capillary, the drop eventually overflows the boundary of the tip.

We can immediately find an *analytical* solution for the shape of a pendent drop in the θ-limited case, for liquids which "wet" the surface in the sense that the equilibrium contact angle is $\theta_c = 0$. These drops will attain shapes that satisfy the Young–Laplace equation and the volume constraint, for a given value of $a^*$ (Eq. 6-2.12). For the pendent drop, if the small-slope approximation is valid (this will require that the drop "size" be small in some sense), we may simply modify Eq. 6-2.11 by changing the sign on the term corresponding to the hydrostatic pressure:

$$\frac{d^2h}{dr^2} + \frac{1}{r}\frac{dh}{dr} - \left(h_0 - h\right) = 0. \tag{6-2.19}$$

The solution may be shown to have the form

$$\frac{h(r)}{h_0} = J_0(r) - J_0(\sqrt{\mathsf{Bo}}) \tag{6-2.20}$$

The length parameter $h_0$ and the contact radius $r_0$ (i.e., $\mathsf{Bo}$) are determined by the volume constraint (Eq. 6-2.17) and the condition of zero contact angle:

$$\frac{dh(r)}{dr} = 0 = J_1(\sqrt{\mathsf{Bo}}). \tag{6-2.21}$$

The smallest nonzero root of Eq. 6-2.21 yields

$$\mathsf{Bo} = 14.68. \tag{6-2.22}$$

For a fixed value of $a^*$ this yields the radius (and ultimately, the volume) of the drop that is supported on the underside of a planar surface, when the contact angle is zero.

Figure 6-2.4 shows a comparison of the shapes of two drops with the same Bond number. For the same fluid this would correspond to the same contact radius. The drops do not have the same volume, however. The Bond number of $\mathsf{Bo} = 14.68$ is chosen for the comparison. Obviously the pendent drop has a zero contact angle for $\mathsf{Bo} = 14.68$, in view of Eq. 6-2.21. The sessile drop has a finite contact angle, which is not the equilibrium contact angle for the fluid–solid boundary, because we are looking at the R-limited case for the sessile drop (see Problem P6-7).

We noted above that the small-slope approximation that permits us to find an analytical solution to the drop shape equation requires that the drop be "small." Since drop size has units of length, the statement that it is small is not well defined. We need a dimensionless criterion for the validity of the small-slope

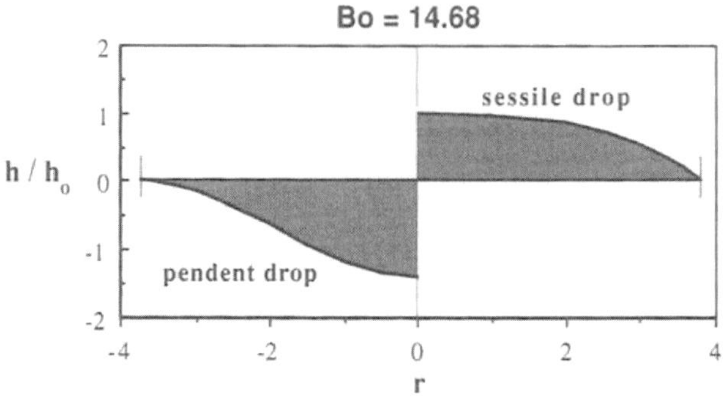


***FIGURE 6-2.4*** Comparison of the shapes of pendent and sessile drops at the same Bond number. R-limited case.

approximation. We begin with the statement that *dh/dr* be small, say

$$\left|\frac{dh}{dr}\right| \leq 0.2, \tag{6-2.23}$$

so that we may neglect the square of this quantity compared to one, in Eq. 6-2.6. From Eq. 6-2.20 we find

$$\left|\frac{dh(r)}{dr}\right| = h_0\left|J_1(r)\right|. \tag{6-2.24}$$

This Bessel function takes on its maximum value at $r = 1.8$, for which

$$\left|\frac{dh(r)}{dr}\right| = 0.58\ h_0. \tag{6-2.25}$$

Now we must connect $h_0$ to the volume of the drop. This follows from the volume constraint applied to Eq. 6-2.20:

$$\frac{V}{a^{*3}} = \int_0^{\text{Bo}} \pi r^2\left(-\frac{dh}{dr}\right)\ dr = \int_0^{\text{Bo}} \pi r^2 h_0 J_1(r)\ dr. \tag{6-2.26}$$

Performing the indicated integration we find

$$\frac{V}{a^{*3}} = 18.57 h_0. \tag{6-2.27}$$

Finally, a criterion for the small slope approximation takes the form

$$V \leq 6.4 a^{*3}. \tag{6-2.28}$$

For water, for example,

$$a^* \equiv \sqrt{\frac{\sigma}{\rho g}} = \sqrt{\frac{0.072}{1000 \times 9.8}} = 0.027\ \text{m}. \tag{6-2.29}$$

Hence this analytical model for the shape of a pendent drop should not be used for water drops of volume exceeding (from Eq. 6-2.28) 126 microliters. (This corresponds roughly to a spherical drop of radius 3 mm.)

If we begin with a pendent wetting drop of the shape given in Fig. 6-2.4, and slowly increase the volume of the drop, the shape will change and Eq. 6-2.20 will no longer hold. It will then be necessary to solve the nonlinear Young–Laplace equation (Eq. 6-2.6 with the appropriate sign change for the pendent case). Physically meaningful solutions are possible up to a limiting value of $V/a^{*3}$. This limitation represents the largest drop that can be supported on the underside of a planar surface that is completely wetted by the liquid. An example of such numerical results is shown in Fig. 6-2.5, based on the work of Boucher and Evans (1975). The largest supportable drop has a volume and contact radius given by

$$\left(\frac{V_{\max}}{a^{*3}}\right) = 18.96 \text{ and } \sqrt{\mathrm{Bo}_{\max}} = \frac{R_{0\max}}{\sqrt{\sigma/\rho g}} = 3.2196. \tag{6-2.30}$$

In the works cited at the end of this chapter, especially Padday and Pitt (1973), Boucher and Evans (1975), and Boucher *et al.* (1976), one can find very detailed computations of the shapes of pendent and sessile drops under R-limited and θ-limited conditions. We will not attempt to outline or even summarize these studies here. Our primary interest is in the relevance of this work to the central issue in this chapter, which is the transfer of a pendent drop to a surface. Hence we turn to that question now.

## 6-3 BREAKAWAY OF A PENDENT DROP

When a pendent drop grows too large to be supported on the tip of a capillary the drop breaks away from the tip. However, not all of the drop leaves the tip—some remains attached. In Chapter 4 we gave a brief discussion of the physics of this problem, including a very simple analysis for predicting the size of a drop. We come back to this problem now and provide a more detailed discussion.

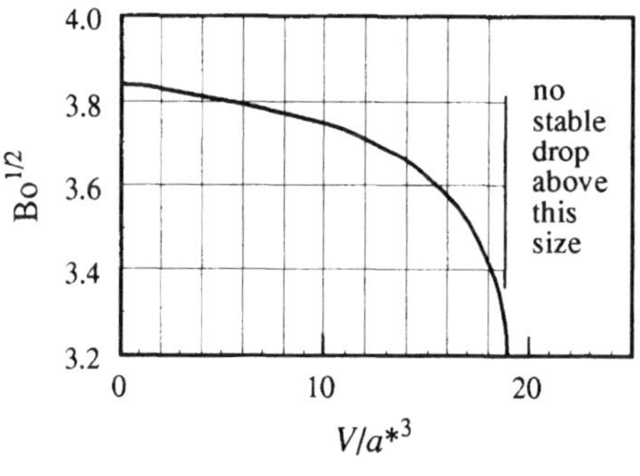


***FIGURE 6-2.5*** Dimensionless contact radius (as Bond number) as a function of dimensionless drop volume. (Based on Boucher and Evans, 1975.)

Figure 6-3.1 shows a montage of photographs of stages in the breakaway process. Clearly the dynamics can be quite complex.

Despite the fact that (or is it because?) the process of dripping is such a common occurrence, the study of dripping has created some classical literature in the field of capillary fluid dynamics. Early studies include those of Tate (1864) and Rayleigh (1899). Harkins and Brown (1919) reviewed the data of Rayleigh and presented their own large body of experimentation on the volume of drops falling from a capillary tip under gravity. Padday and Pitt (1973) then produced numerical solutions to the Young–Laplace equation for a variety of drop configurations, including predictions of the maximum stable drop volume on the tip of a capillary.

In Chapter 4 we derived a very simple model for the size of a drop falling from a capillary. From Eq. 4-1.3, for the case that the neck of the drop is parallel to the capillary axis at the point of departure, we may write (using the notation of this chapter)

$$\left(\frac{V_{\text{drop}}}{a^{*3}}\right) = \frac{2\pi R_0}{a^*} = 2\pi\sqrt{\text{Bo}}, \tag{6-3.1}$$

where the Bond number is based on the capillary radius. (In Chapter 4 we used a Bond number based on the *diameter*.) Equation 6-3.1 is often referred to as Tate's law for the volume of a drop.

In Fig. 6-3.2 we show the theoretical results of Padday and Pitt (1974) for the critical drop volume. *This is not the volume of the drop that falls.* It is the volume of the pendent drop just *before* a part of that drop falls away from the capillary tip. Note that the theory is represented fairly well by Tate's law. However, Tate's law is often used, incorrectly, to predict the volume of the *departing* drop. A selection of the data of Harkins and Brown on the volume of the falling drop is included on this plot. As expected, the falling drop has a smaller volume than that of the critical drop, since some of the drop remains on the tip after the pendent drop "splits." This is shown clearly in Fig. 6-3.3. The line labeled "Rayleigh" in Fig. 6-3.2 represents a fit to his data for the falling drop volume. His data agree very well with those of Harkins and Brown.

Harkins and Brown presented extensive data for the weight of falling drops and calculated a correction factor $f$, defined with respect to the "ideal weight" as predicted by the simplistic force balance that leads to Tate's law. Figure 6-3.4 shows the Harkins and Brown factor as a function of the dimensionless radius of the capillary tip. Using this factor, we may write the weight of a falling drop in the form

$$W = \rho g V = 2\pi R_0 \sigma f\left(\frac{R_0}{\sqrt{2}\,a^*}\right). \tag{6-3.2}$$

With these results we may predict the size (weight, volume, or equivalent spherical diameter) of a drop falling from a capillary. In all of this discussion we have assumed that the drop separates slowly, so that there are no dynamic phenomena,

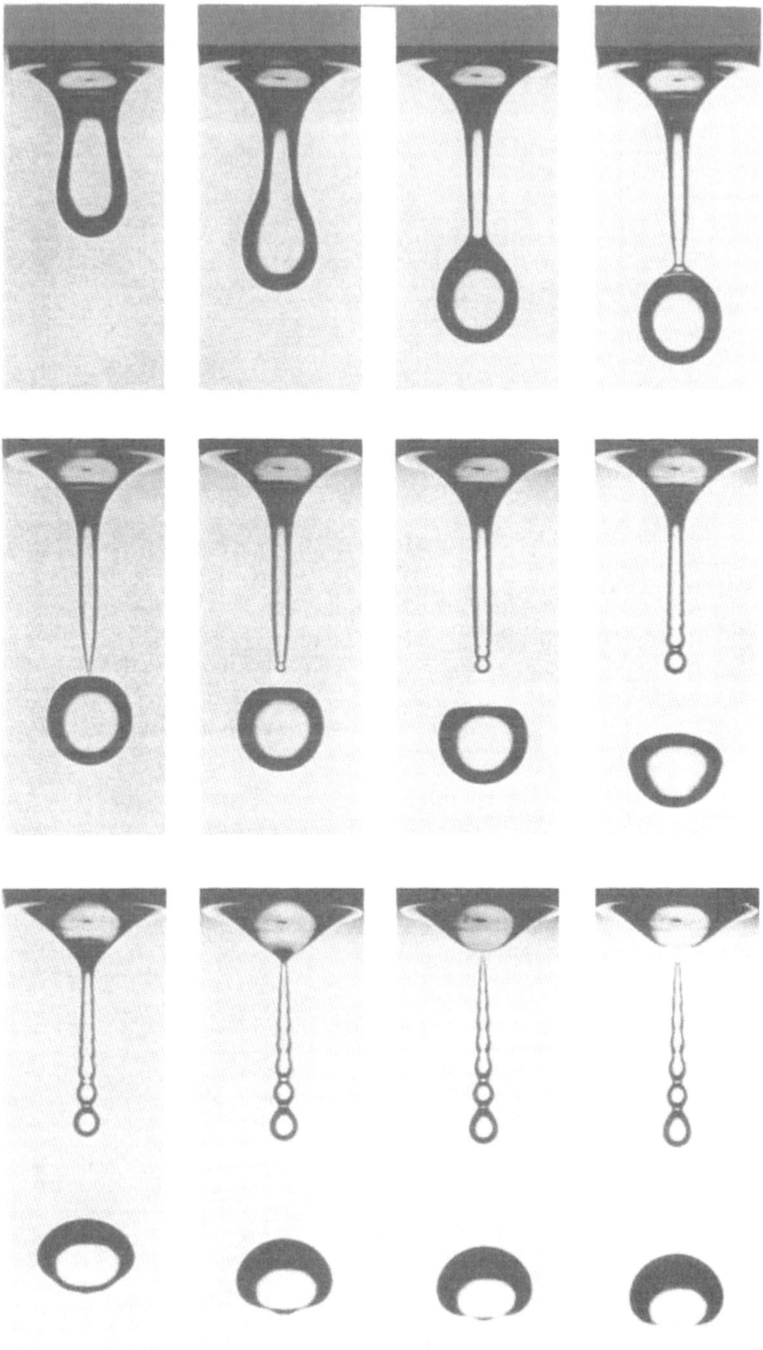

***FIGURE 6-3.1*** Slow dripping from a capillary (From Shi *et al.*, 1994. Reprinted with permission from the cover of *Science*, July 8, 1994. Copyright 1994 by the AAAS.)

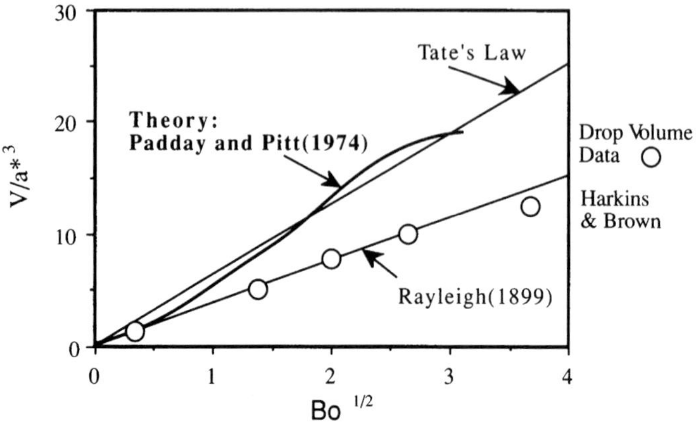


*FIGURE 6-3.2* Critical drop volume, and drop volume falling from a capillary.

such as viscous resistance, that affect the manner in which the pendent drop splits at its critical volume.

With all of the discussion of the previous three sections as background, we now turn to the question of the fate of a liquid bridge between two surfaces, the situation depicted at the beginning of this chapter, in Fig. 6-1.1.

*FIGURE 6-3.3* Drop falling from a capillary. Note the significant volume of liquid remaining attached to the tip.

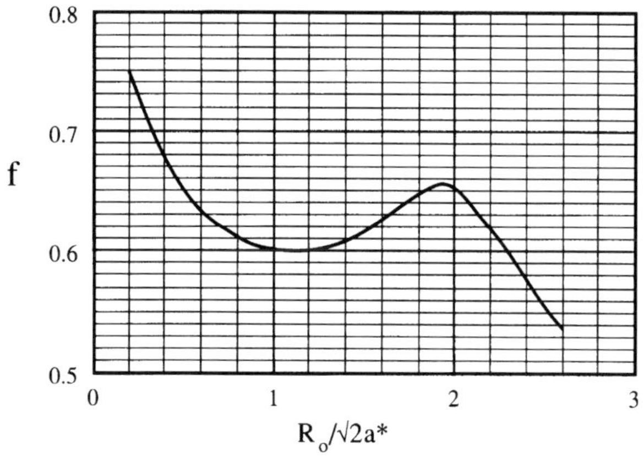


***FIGURE 6-3.4*** The Harkins and Brown correction factor.

## *6-4 Stability of an Axisymmetric Bridge*

In principle, the topic of this section is an extension of the discussion of the shape and stability of a pendent drop. The difference is that the "drop" is now constrained from below, as well as from above. Figure 6-4.1 shows the geometry of an axisymmetric pendular bridge between two disks of different radii, in a gravitational field. The contact radius of the liquid is assumed to be constrained at the two disks, and so we refer to this case as "doubly R-limited."

The first issue is with respect to the stability of this bridge: what is the maximum stable length for such a bridge? We may already have come across an answer to this question, for the special case when there is no gravity, the two disks have the same radius $R_d$, and the initial bridge is a cylinder of volume $V = \pi R_c^2 L$. In Chapter 4 we looked at the stability of an *unbounded* cylinder of liquid, subject to surface tension, and we found that the cylinder is unstable to all periodic disturbances of wavelengths $\lambda$ that satisfy Eq. 4-3.24, which we write now as

$$\frac{\lambda}{2\pi R_c} \geq 1. \tag{6-4.1}$$

If the cylinder is shorter than $2\pi R_c$, it can only be subjected to *periodic* disturbances of wavelengths shorter than the axial length of the cylinder, and so Eq. 6-4.1 would not be satisfied. We conclude that for the special case cited here, a cylinder of liquid would be stable so long as $L/R_c < 2\pi$. However, in a gravitational field acting along the axial direction, the cylinder could be unstable at shorter lengths, and it is necessary to examine the effect of gravity on stability.

A series of papers by Meseguer and coworkers provides discussion of analytical and numerical methods of solving the stability problem of the axisymmetric

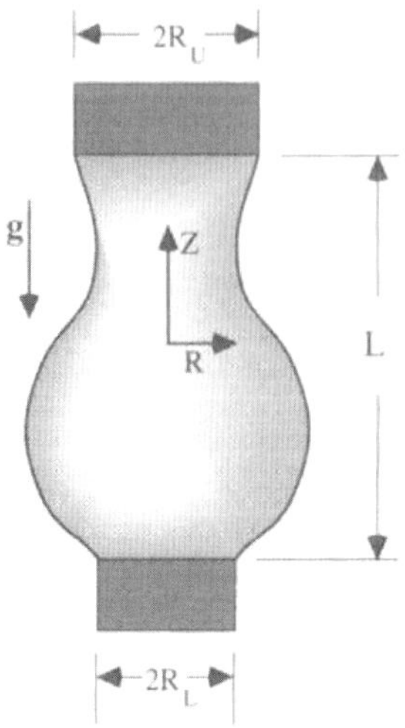


*FIGURE 6-4.1* The axisymmetric pendular liquid bridge.

liquid bridge. Experimental data are also provided. We will attempt to summarize the key methods and results of these works here, but the original papers warrant close study. The analysis is organized about several classes of approximations related to the physics, having to do with the relative roles of viscosity and gravity. In all cases, of course, it is surface tension which drives the instability. Our intuition should suggest that in the problem of interest to us in this chapter—the transfer of liquid to a surface from a capillary—gravity should be an important factor. Although this is true, there are some situations of practical interest in which gravity may be neglected. One arises if the length scales are so small that the weight of the liquid zone is very small in comparison to the forces associated with surface tension (and possibly viscosity). We would expect that a Bond number would determine conditions under which gravity is unimportant. In another case, which has motivated much of the work to be described here, large cylinders of single-crystal semiconductor material are grown, or purified, in the molten state. Because gravity limits the length of liquid cylinder which would remain stable, there is considerable interest in growing large crystals in space, i.e., in a microgravity environment. Hence much of the theoretical work neglects the effects of gravity.

If we assume that the fluid dynamics are maintained axisymmetric, that the external fluid has no influence on the dynamics (other than through the presence of the interfacial tension), and if we take all physical properties to be constant, the stability is governed by the Navier–Stokes and continuity equations in cylindrical coordinates:

**Navier–Stokes:**

*r* component

$$\rho\left(\frac{\partial u_r}{\partial t} + u_r\frac{\partial u_r}{\partial r} + u_z\frac{\partial u_r}{\partial z}\right) = -\frac{\partial p}{\partial r} + \mu\left\{\frac{\partial}{\partial r}\left[\frac{1}{r}\frac{\partial}{\partial r}(r\,u_r)\right] + \frac{\partial^2 u_r}{\partial z^2}\right\}. \tag{6-4.2}$$

*z* component

$$\rho\left(\frac{\partial u_z}{\partial t} + u_r\frac{\partial u_z}{\partial r} + u_z\frac{\partial u_z}{\partial z}\right) = -\frac{\partial p}{\partial z} + \mu\left[\frac{1}{r}\frac{\partial}{\partial r}\left(r\frac{\partial u_z}{\partial r}\right) + \frac{\partial^2 u_z}{\partial z^2}\right] + \rho g_z. \tag{6-4.3}$$

**Continuity:**

$$\frac{\partial u_z}{\partial z} + \frac{1}{r}\frac{\partial(r\,u_r)}{\partial r} = 0. \tag{6-4.4}$$

We will nondimensionalize these equations in the following manner.

All lengths are divided by the disk radius $R_d$. We take the upper and lower radii to be identical in this analysis.

All velocities are made dimensionless by dividing by $(\sigma/\rho R_d)^{1/2}$.

Time is nondimensionalized by dividing by $(\rho R_d^3/\sigma)^{1/2}$.

Pressure is made dimensionless by dividing by $(\sigma/R_d)$.

We avoid using the viscosity $\mu$ in the nondimensionalization because we are going to want to examine the limiting case of the inviscid liquid.

The nondimensional form of Eqs. 6-4.2, 3, and 4 is

$$\frac{\partial U}{\partial \tau} + U\frac{\partial U}{\partial s} + W\frac{\partial U}{\partial \zeta} = -\frac{\partial P}{\partial s} + \mathrm{C}\left\{\frac{\partial}{\partial s}\left[\frac{1}{s}\frac{\partial}{\partial s}(s\,U)\right] + \frac{\partial^2 U}{\partial \zeta^2}\right\}, \tag{6-4.5}$$

$$\frac{\partial W}{\partial \tau} + U\frac{\partial W}{\partial s} + W\frac{\partial W}{\partial \zeta} = -\frac{\partial P}{\partial \zeta} + \mathrm{C}\left(\frac{1}{s}\frac{\partial}{\partial s}\left(s\frac{\partial W}{\partial s}\right) + \frac{\partial^2 W}{\partial \zeta^2}\right) - \mathrm{Bo}, \tag{6-4.6}$$

$$\frac{\partial W}{\partial \zeta} + \frac{1}{s}\frac{\partial(s\,U)}{\partial s} = 0. \tag{6-4.7}$$

The radial and axial velocity components are now $(U,W)$, respectively, and the radial and axial coordinates are $(s,\zeta)$. The dimensionless time is denoted $\tau$. Gravity is taken as acting in the $-z$ direction. Two dimensionless groups appear here—the Bond number and the Ohnesorge number:

$$\mathrm{Bo} \equiv \frac{R_d^2}{\sigma/\rho g} \quad \text{and} \quad \mathrm{C} \equiv \frac{\mu}{\sqrt{\rho\sigma R_d}} = \mathrm{Oh}. \tag{6-4.8}$$

Next we must write appropriate boundary conditions. We impose a no-slip condition at the upper and lower disks:

$$U(s, \pm\Lambda, \tau\ ) = W(s, \pm\Lambda, \tau\ ) = 0, \tag{6-4.9}$$

where the so-called "slenderness" of the bridge is $\Lambda = L/2R_d$.

Axial symmetry requires that

$$U(0, \zeta, \tau\ ) = 0,$$
$$\frac{\partial W(0, \zeta, \tau\ )}{\partial s} = 0. \tag{6-4.10}$$

Dynamic boundary conditions on the interface reflect statements about the shear and normal stresses. We denote the surface by the dimensionless radial coordinate $F(\zeta,\tau)$. A detailed description of the formulation of free surface boundary conditions, in Cartesian coordinates, was given in Chapter 2, Section 2-5. The analogous equations in the axisymmetric case become the following:

Normal stress condition

$$P - \frac{2\mathrm{C}}{1 + (\partial F/\partial\zeta)^2}\left[\frac{\partial U}{\partial s} + \left(\frac{\partial F}{\partial\zeta}\right)^2\frac{\partial W}{\partial\zeta} - \frac{\partial F}{\partial\zeta}\left(\frac{\partial W}{\partial s} + \frac{\partial U}{\partial\zeta}\right)\right]$$
$$= \frac{1}{\left[1 + (\partial F/\partial\zeta)^2\right]^{3/2}}\left[\frac{1 + (\partial F/\partial\zeta)^2}{F} - \frac{\partial^2 F}{\partial\zeta^2}\right] - \mathrm{Bo}\ \zeta. \tag{6-4.11}$$

Shear stress condition

$$\mathrm{C}\left\{2\frac{\partial F}{\partial\zeta}\left(\frac{\partial U}{\partial s} - \frac{\partial W}{\partial\zeta}\right) + \left[1 - \left(\frac{\partial F}{\partial\zeta}\right)^2\right]\left(\frac{\partial W}{\partial s} + \frac{\partial U}{\partial\zeta}\right)\right\} = 0. \tag{6-4.12}$$

The fact that there is no mass flow across the interface is expressed in the form of a kinematic boundary condition (see Problem P6-8)

$$\frac{\partial F}{\partial\tau} - U_F + W_F\frac{\partial F}{\partial\zeta} = 0, \tag{6-4.13}$$

where the subscript F on $U$ and $W$ implies that these are the velocity components at the free surface.

The interface is pinned to the edges of the disks, so

$$F(\pm\ \Lambda, \tau\ ) = 1. \tag{6-4.14}$$

Initial conditions require that there be no motion anywhere within the liquid bridge:

$$U(s,\zeta,0) = W(s,\zeta,0) = 0 \tag{6-4.15}$$

and that the initial distortion of the surface be of the form

$$F(\zeta,0) = F_0(\zeta), \tag{6-4.16}$$

subject to the volume conservation statement:

$$\int_{-\Lambda}^{+\Lambda} F_0^2(\zeta)\, d\zeta = 2\Lambda. \qquad (6\text{-}4.17)$$

Even if this set of partial differential equations is linearized it is not possible to write an analytical solution, although the problem resembles one studied earlier—Rayleigh instability of a capillary jet. The difference lies in the fact that the jet problem is unbounded and periodic in the axial direction, while the liquid bridge is bounded by the upper and lower disks. Mathematically, the complication arises from the need to satisfy Eqs. 6-4.9 and 14. Hence we seek simplifications to this formulation that still retain the essential physics of the problem.

We begin the stability analysis by dropping the viscous terms. Of course this yields, if at all, a model valid only for the case where surface tension dominates the dynamics. We then average the dynamic equations across the cross section of the bridge. In effect, we make the approximation that the axial velocity $W$ and the pressure $P$ are not functions of radial position. Equation 6-4.6 becomes

$$\frac{\partial W}{\partial \tau} + W\frac{\partial W}{\partial \zeta} = -\frac{\partial P}{\partial \zeta} - \mathrm{Bo}. \qquad (6\text{-}4.18)$$

This is the primary momentum balance that must be satisfied by the flow. Central to the physics is the effect of the interface shape on the internal pressure $P$, through the boundary condition on normal stress, Eq. 6-4.11. Setting $\mathrm{C} = 0$ in that equation we find

$$P = \frac{1}{\left[1 + (\partial F/\partial\zeta)^2\right]^{3/2}}\left[\frac{1 + (\partial F/\partial\zeta)^2}{F} - \frac{\partial^2 F}{\partial \zeta^2}\right] - \mathrm{Bo}\,\zeta. \qquad (6\text{-}4.19)$$

Equation 6-4.18 is an axial momentum equation. Equation 6-4.19 accounts for the role of surface forces. The third essential part of the physics is conservation of mass. The continuity equation is converted just as in our earlier discussion of the capillary jet, and Eq. 4-3.54 is obtained. In the present notation this is

$$\frac{\partial F^2}{\partial \tau} + \frac{\partial\left(WF^2\right)}{\partial \zeta} = 0. \qquad (6\text{-}4.20)$$

We now have three equations in the variables $F$, $W$, and $P$. One variable ($U$) has been eliminated, and the dimensionality of the mathematics has been reduced to one space dimension and time. The primary requirement for the accuracy of this approach is that the bridge be thin; that is, it must have a large slenderness ratio $\Lambda$.

These three nonlinear coupled partial differential equations may be solved numerically, and such solutions are presented by Meseguer (1983), in the case of the absence of gravity. However, we first want to extract from this formulation a criterion of stability of the bridge to small disturbances. Hence we proceed to carry out a linearized stability analysis. We begin by writing the functions $F, W$, and $P$ in the form of a small perturbation about the steady-state (in this case, static) solution:

$$F = 1 + \varepsilon f, \quad P = (1 - \mathrm{Bo}\,\zeta) + \varepsilon p, \quad W = \varepsilon w. \tag{6-4.21}$$

These three equations are substituted into Eqs. 6-4.18, 19, and 20, and only linear terms in $\varepsilon$ are retained. The result is the following set of equations for the new unknowns—$f$, $w$, and $p$:

$$2\frac{\partial f}{\partial \tau} + \frac{\partial w}{\partial \zeta} = 0, \tag{6-4.22}$$

$$\frac{\partial w}{\partial \tau} + \frac{\partial p}{\partial \zeta} = 0, \tag{6-4.23}$$

$$p = -f - \frac{\partial^2 f}{\partial \zeta^2}. \tag{6-4.24}$$

We can eliminate $p$ and $w$ and write a single equation for the shape function $f$:

$$2\frac{\partial^2 f}{\partial \tau^2} + \frac{\partial^4 f}{\partial \zeta^4} + \frac{\partial^2 f}{\partial \zeta^2} = 0. \tag{6-4.25}$$

Boundary conditions on $f$ are

$$\frac{\partial f}{\partial \zeta} + \frac{\partial^3 f}{\partial \zeta^3} = 0 \quad \text{at} \quad \zeta = \pm\Lambda,$$

$$f = 0 \quad \text{at} \quad \zeta = \pm\Lambda. \tag{6-4.26}$$

(The first of these is from the statement that $w = 0$ on the disk surfaces.) The solution to Eq. 6-4.25 may be written in the form

$$f = e^{\gamma\tau}\,\Theta(\zeta), \tag{6-4.27}$$

where the function $\Theta$ satisfies an ordinary differential equation of the form

$$\frac{d^4\Theta}{d\zeta^4} + \frac{d^2\Theta}{d\zeta^2} + 2\gamma^2\Theta = 0. \tag{6-4.28}$$

The general solution to this differential equation is

$$\Theta = e^{\theta\zeta}. \tag{6-4.29}$$

There are four values of $\theta$ corresponding to the four roots to

$$\theta^4 + \theta^2 + 2\gamma^2 = 0. \tag{6-4.30}$$

These roots are $\pm\theta_1$, $\pm\theta_2$ from the solutions to

$$\theta_{1,2}^2 = -\frac{1}{2} \pm \frac{1}{2}\sqrt{1 - 8\gamma^2}. \tag{6-4.31}$$

As far as criteria for stability of the cylinder, the only physically meaningful solutions of interest to us are those for the case that $\gamma$ is real and positive. This requires that

$$\gamma \leq \frac{\sqrt{2}}{4} = 0.3536. \tag{6-4.32}$$

Then we may write $\theta$ as $\pm\,\beta_1$, $\pm\beta_2$ where the $\beta$'s are real numbers. From Eq. 6-4.31 the $\beta$'s satisfy

$$\beta_1\,\beta_2 = \sqrt{2}\,\gamma,$$
$$\beta_1^2 + \beta_2^2 = 1. \tag{6-4.33}$$

Going back to Eq. 6-4.27 we will write the shape function as the superposition of the four independent solutions, in the trigonometric form

$$f = e^{\gamma\tau}\left[A\cos\beta_1\zeta + B\sin\beta_1\zeta + C\cos\beta_2\zeta + D\sin\beta_2\zeta\right]. \tag{6-4.34}$$

Application of the four boundary conditions given in Eqs. 6-4.26 yields a set of four linear algebraic equations that are homogeneous in the constants $A$, $B$, $C$, and $D$. The condition that there be a nontrivial solution for these constants is that the determinant of the coefficients of the constants in this set of four equations vanish. Actually, symmetry about the midplane transverse to the cylinder axis reduces the number of constants to two. After performing the required algebra (and quite a bit of it!) we find that the determinant equation reduces to an *eigenvalue* equation of the form

$$0 = \sin 2\beta_1\Lambda\,\sin 2\beta_2\Lambda - 2\sqrt{2}\,\gamma\left(1 - \cos 2\beta_2\Lambda\,\cos 2\beta_1\Lambda\right). \tag{6-4.35}$$

From Eq. 6-4.31 we may find that $\beta_1$ and $\beta_2$ are the positive roots of

$$\beta_1^2 = \frac{1 + \sqrt{1 - 8\gamma^2}}{2},$$
$$\beta_2^2 = \frac{1 - \sqrt{1 - 8\gamma^2}}{2}. \tag{6-4.36}$$

Hence we may regard Eq. 6-4.35 as an algebraic equation for $\gamma$ as a function of $\Lambda$. The equation must be solved numerically, and the result is shown in Fig. 6-4.2. The graph confirms the expected result that for $\Lambda < \pi$ the cylinder is stable (there is no positive value of $\gamma$.)

Meseguer (1983) goes on to examine details of the solutions for the evolution of the interface as a function of time. The effect of a finite viscosity is included. As expected, the rate of growth of an instability is slowed by viscosity. However, Meseguer shows that $\Lambda = \pi$ is still the stability limit, even for the case of finite viscosity.

The stability method illustrated here, particularly the expansion of the variables $F$, $P$, and $W$ as shown in Eqs. 6-4.21, leads to the conclusion that the results are independent of the presence of gravity, since the Bond number contribution drops out of the analysis. This is not correct. The presence of finite gravity does alter the critical slenderness ratio. The fault lies in the assumption that Eqs. 6-4.21 are correct. There is no *a priori* reason that the power of $\varepsilon$ should be the same in all three functions. A different approach to the stability analysis is given in a subsequent paper by Rivas and Meseguer (1984). We will not illustrate the details here. A key result is that for the case of vanishing viscosity the critical slenderness ratio *does* depend on gravity, and the critical value is predicted to be given by

$$\Lambda = \pi\left[1 - \frac{3}{2}\left(\frac{3}{2}\,\mathrm{Bo}^2\right)^{1/3}\right]. \tag{6-4.37}$$

According to this result, finite gravity reduces the stability of slender cylindrical bridges. The critical slenderness ratio is reduced to values below $\pi$. Experimental demonstration of this fact is shown in Fig. 6-4.3, based on data given by Coriell *et al.* (1977). The dashed line is Eq. 6-4.37, which deviates from the observations when the Bond number is no longer small compared to unity.

It is important to keep in mind that the theory shown here (Eq. 6-4.37, for example) is a *linearized* stability theory that assumes that the initial perturbation to the shape is small. It turns out that the stability limit of $\Lambda = \pi$ is not observed or predicted when the shape perturbation is large enough.

While the types of stability analyses described here shed some light on the expected behavior of liquid bridges, the information that can be obtained by this analytical methodology is limited. In particular, one would like to be able to predict the final partitioning of the bridge into volumes attached to the upper and lower disks. A nonlinear analysis of the dynamic equations, including viscous and gravitational effects, is required in order to produce useful predictions. The set of equations that must be solved, if we maintain the thin filament approxima-

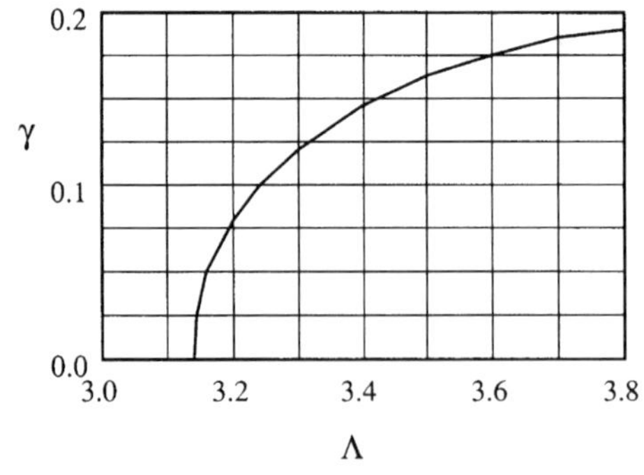


***FIGURE 6-4.2*** Growth rate parameter as a function of the slenderness ratio. Gravity and viscosity negligible.

tion, is as follows:

$$\frac{\partial Q}{\partial \tau} + \frac{\partial}{\partial \zeta}\left(\frac{Q^2}{S}\right) = -S\frac{\partial P}{\partial \zeta} - S\,\text{Bo}, \tag{6-4.38}$$

$$\frac{\partial S}{\partial \tau} + \frac{\partial Q}{\partial \zeta} = 0, \tag{6-4.39}$$

and

$$P = \frac{S + \frac{1}{2}(\partial S/\partial \zeta)^2 - \frac{1}{2}\,S(\partial^2 S/\partial \zeta^2)}{\left[S + \frac{1}{4}(\partial S/\partial \zeta)^2\right]^{3/2}} - \text{Bo}\,\zeta. \tag{6-4.40}$$

where

$$S = F^2 \quad \text{and} \quad Q = F^2 W. \tag{6-4.41}$$

Boundary conditions on $S$ and $Q$ are

$$S(\pm\Lambda, \tau) = 1, \qquad Q(\pm\Lambda, \tau) = 0, \tag{6-4.42}$$

and the initial condition, written in terms of the cross-sectional area $S$ instead of with respect to $F$, is

$$S(\zeta, 0) = 1 + \varepsilon(\varepsilon - 2)\sin\frac{\pi\zeta}{\Lambda}, \qquad Q(\zeta, 0) = 0. \tag{6-4.43}$$

The initial shape function is defined so that the distorted volume is identical to that of an undistorted cylinder. Some aspects of the numerical procedure are described by Meseguer (1983), but few solutions are presented. Figure 6-4.4 shows some calculated results (Meseguer, 1983) for the partitioning of the

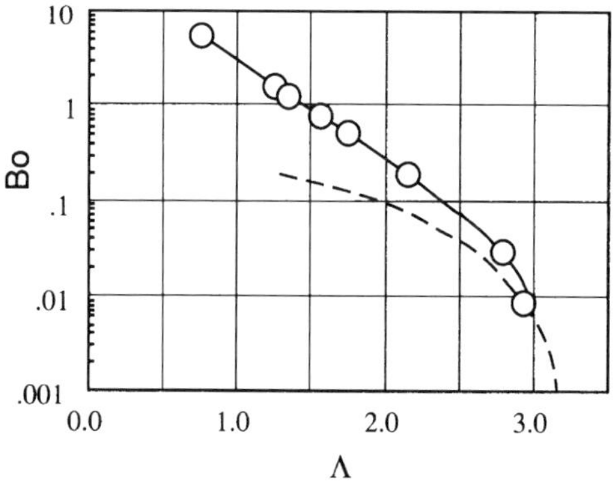


***FIGURE 6-4.3*** Effect of gravity on critical value of $\Lambda$.

bridge in the absence of gravity and viscosity. For large values of $\varepsilon$ instability occurs even though $\Lambda$ is less than $\pi$.

Insufficient numerical work is available to permit one to draw conclusions with regard to the effect of parameters on the manner in which the bridge is partitioned into the "upper" and "lower" portions. This is especially so in the presence of significant gravitational effects. Hence, at least at the time of writing, there is a need for additional studies—both theoretical and experimental—of the partitioning of a liquid bridge between two planar surfaces.

It is important to keep in mind that the boundary condition $S(\pm\,\Lambda,\,\tau) = 1$ expressed in Eq. 6-4.42 corresponds to the doubly R-limited case. We would expect that the case of drop partitioning between larger surfaces would depend upon the wetting characteristics of the liquid for each surface (contact angle). Evidence along these lines is provided in a set of experiments reported by Yakhnin and Chadov (1983). They placed small drops ($V \approx 0.005$ cm$^3$) between a pair of parallel solid surfaces. Initially the drop was spread into a thin "pancake" with a slenderness ratio of about 0.03. Then the two surfaces were separated at some rate, and the partitioning of the drop was observed as a function of the separation rate and as a function of the surface materials. Typical data are reproduced in Fig. 6-4.5. Keep in mind that these data represent partitioning of the drop under conditions that the bounding surfaces are moving apart, whereas the discussion of partitioning prior to this point has been for a bridge between *stationary* boundaries.

Several points should be noted in the data of Fig. 6-4.5. First of all, the parti-

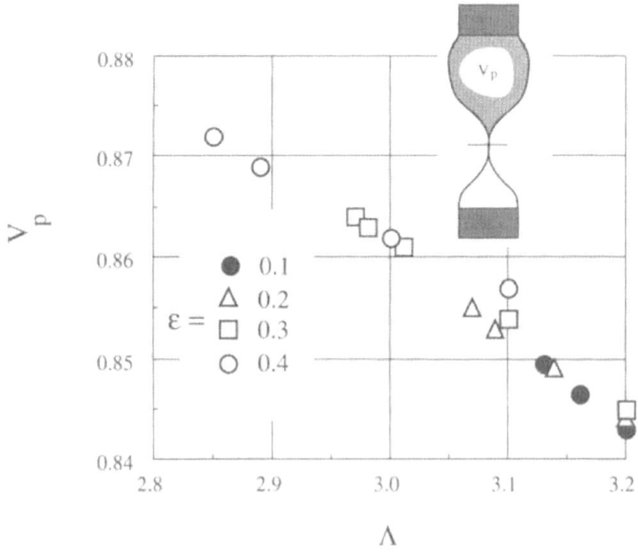


***FIGURE 6-4.4*** Volume fraction of the larger drop after the cylinder partitions between two equal disks. (No gravity or viscous forces.)

tioning depends upon the separation rate, but in all cases when the rate is high enough, the drop splits in half. At lower rates the partitioning depends upon the nature of the surfaces (presumably, contact angle), and for a given pair of surfaces the partitioning depends upon the viscosity and (though not shown here) the surface tension of the liquid. In order to model such results it would be necessary to account for the dynamics of the receding contact line. This is a difficult problem to which no solution is currently known. However, the *equal* partitioning that is observed at high separation rates, independent of the nature of the surfaces, is not unexpected. At high separation rates the liquid–solid contact line cannot move fast, compared to the elongation rate of the liquid, and as a consequence the dynamics near the solid surfaces occurs far from the contact boundaries, in the radial direction. Hence the nature of the contact line dynamics—the receding dynamic contact angle and the rate of retraction of the footprint radius—will have little effect on the flow within the filament as it approaches separation into two drops.

A numerical study of filament splitting by Bousfield (1988) suggests that the velocity distribution in the liquid, along its axis, plays an important role in determining how the drop partitions. Since filament splitting of the type studied by Yakhnin and Chadov is under conditions of a rapid transient extensional flow, the rate at which the moving surface induces motion in the liquid will vary with the viscosity of the liquid. As Bousfield points out, this problem is analogous to the classic transient shear flow problem. In a low-viscosity liquid the filament will be largely stationary except very near the moving surface, and we would expect the liquid to partition in such a way that most of the drop remains on the

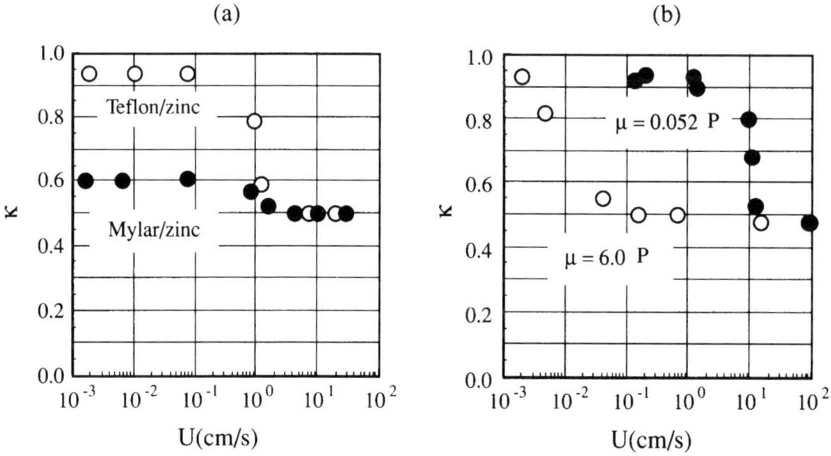


***FIGURE 6-4.5*** Partitioning of drops under dynamic separation conditions (Data from Yakhnin and Chadov, 1983). (a) A glycerol solution between two different pairs of surfaces. (b) Two different viscosity liquids between a Teflon and a zinc surface. $\kappa$ is the fraction of the mass left on the zinc surface.

stationary surface. For a high-viscosity liquid the velocity field in the filament will quickly reach the fully developed steady flow, which corresponds to a nearly linear velocity profile along the axis. Symmetric splitting should be expected in this case. We note that the data of Fig. 6-4.5 indicate that the symmetric splitting is achieved at lower separation velocities as the viscosity of the liquid increases, consistent with this idea.

## 6-5 IMPACT OF A DROP WITH A SOLID SURFACE

So far we have examined several aspects of the transfer of a pendent drop to a solid surface. A somewhat related problem is that in which the transfer takes place through the impact of a spherical drop, free of any attachment to a boundary, onto a solid surface. The simplest case is that in which the drop falls freely to the surface. Figure 6-5.1 shows a sequence of photographs of the stages in the impact of a small heptane drop on a stainless steel surface (Chandra and Avedisian, 1991).

The drop spreads radially into a "pancake" shape. The rate of spreading is "driven" by the inertia of the drop, and it is slowed by viscous effects. When the inertial energy is dissipated the drop reaches its maximum radius. If the liquid does not wet the surface, surface tension forces will cause the drop to retract to an equilibrium sessile drop shape. If the drop does wet the surface there could be a second stage of slow spreading as the drop moves toward its equilibrium configuration. We would like to be able to predict the maximum radius achieved under the inertial force of impact. A rough model for this can be worked out from a simple but approximate energy balance.

The kinetic energy of the drop, prior to impact, is given by

$$E_k = \left(\frac{1}{2}\rho U^2\right)\left(\frac{1}{6}\pi D^3\right). \tag{6-5.1}$$

The drop also has a surface energy, before impact, given by

$$E_s = \pi D^2 \sigma. \tag{6-5.2}$$

After contact with the surface we consider the drop to be a disk of height $h$ and diameter $d$ such that

$$V = \frac{\pi D^3}{6} = \frac{\pi d^2 h}{4}. \tag{6-5.3}$$

This disk has a surface energy arising from its surface area and from its contact area with the solid surface. The former is given by

$$E_{s1} = \left(\frac{\pi d^2}{4} + \pi d h\right)\sigma. \tag{6-5.4}$$

The latter can be expressed in terms of the contact angle:

$$E_{s2} = -\frac{\pi d^2}{4}\sigma \cos \theta. \tag{6-5.5}$$

The viscous dissipation term is difficult to assess in any simple way, with any

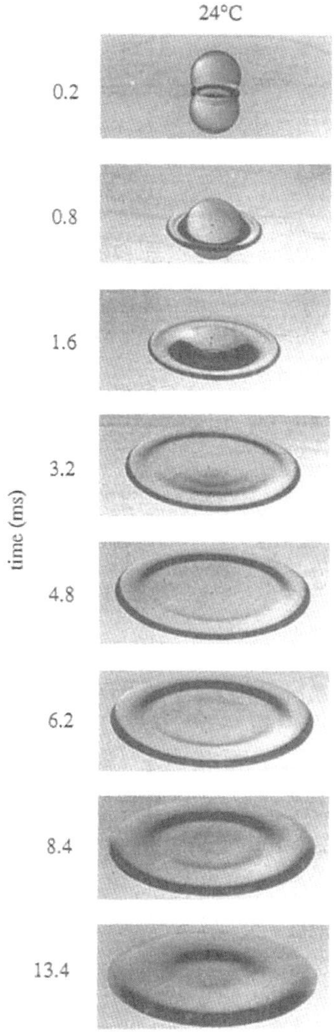


***FIGURE 6-5.1*** Impact of an $n$-heptane drop on a stainless steel surface at 24°C (From Chandra and Avedisian, 1991. Reprinted with kind permission of The Royal Society.)

precision. A crude approximation, suggested by Chandra and Avedisian, takes the rate of dissipation per unit of volume to be that for a simple shear flow:

$$\Phi_v = \mu \left(\frac{U}{h}\right)^2. \tag{6-5.6}$$

The time over which the drop spreads on the surface is of the order of $D/U$. As the spreading drop slows down the dissipation will be less than that given in Eq. 6-5.6. Hence we will write the energy loss due to dissipation as

$$W = \Phi_v V \frac{D}{U} = \alpha\, \mu \left(\frac{U}{h}\right)^2 \frac{\pi D^3}{6} \frac{D}{U}, \tag{6-5.7}$$

where $\alpha$ is an unknown coefficient that accounts for our uncertainty in the dissipation function and the spreading time. We now write an energy balance that states that the maximum spreading of the drop corresponds to the point where the initial energy has been converted to surface energy and heat through dissipation:

$$E_k + E_s = E_{s1} + E_{s2} + W. \tag{6-5.8}$$

Using the terms written above, we find that this takes the form

$$\frac{3\alpha\,\mathrm{Ca}}{2}\beta^4 + (1 - \cos\theta)\,\beta^2 - \left(\frac{\mathrm{We}}{3} + 4\right) = 0. \tag{6-5.9}$$

(In Eq. 6-5.4 we neglect the $\pi dh$ term.) The capillary and Weber numbers are defined here as

$$\mathrm{Ca} = \frac{\mu U}{\sigma}, \quad \mathrm{We} = \frac{\rho U^2 D}{\sigma}, \tag{6-5.10}$$

and the maximum diameter of the spreading drop is defined by

$$\beta_{max} \equiv \frac{d_{max}}{D}. \tag{6-5.11}$$

The solution of Eq. 6-5.9 is

$$\beta_{max}^2 = \frac{-(1 - \cos\theta) + \sqrt{(1 - \cos\theta)^2 + 4\left(\frac{\mathrm{We}}{3} + 4\right)\frac{3\alpha\,\mathrm{Ca}}{2}}}{3\alpha\,\mathrm{Ca}}. \tag{6-5.12}$$

The data of Chandra and Avedisian correspond to the following parameter values:

$$\theta = 32°, \qquad \mathrm{Ca} = 0.019, \qquad \mathrm{We} = 43,$$

from which it follows that

$$\beta_{max} = 4.8. \tag{6-5.13}$$

Figure 6-5.2 shows measured values of the "pancake" diameter as a function of time. Considering the simplicity of this model, the predicted maximum spreading is in good agreement with the data. Note that the time scale of the spreading is about 6 ms. For the data shown, $D = 1.5$ mm and $U = 0.93$ m/s. This yields a value of $D/U = 1.6$ ms. Hence we see that the estimate used for the spreading time in Eq. 6-5.7 in this model is poor. We could select a value of the coefficient $\alpha$ larger than unity to improve the fit to the data, but we have no other data with which to test the model further.

We would like to be able to model the *dynamics* of the spreading, and obtain an estimate of the $d(t)$ behavior. Harlow and Shannon (1967) give a numerical solution to the Navier–Stokes equations for a drop impacting on a planar rigid surface. However, they neglect viscous, interfacial, and gravitational effects in their simulation. Hence their drop spreads continuously, since there is no dissipative (viscous) force to retard the motion on the rigid surface. Although this seems quite unrealistic, we can examine the possibility that these assumptions are reasonable in the early stages of drop spreading. This might be the case for small values of the capillary number, for example. In Fig. 6-5.2 we show the curve corresponding to the behavior of $\beta$ predicted by the numerical solution. The failure of this model as the spreading proceeds is not surprising. However, it would appear that in the early stages of spreading, at least for the single set of data shown, viscous effects are not important.

Harlow and Shannon's (1967) numerical solution represents one of the early attempts at computational fluid dynamics in the presence of free boundaries. In the intervening 25 years great strides have been made in numerical analysis, and a more recent study of the same problem is presented by Fukai *et al.* (1993). Among their many interesting results they show (their Fig. 7) that the early stages of drop spreading are only weakly affected by the Reynolds number. Hence the viscous dissipation of the initial kinetic and surface energy is a secondary phenomenon, a point in agreement with Fig. 6-5.2.

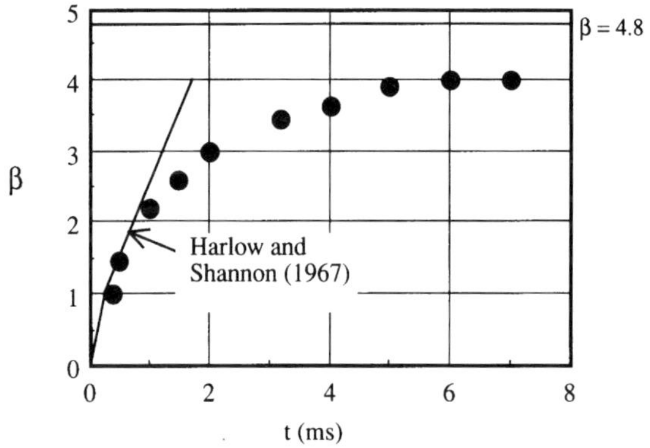


***FIGURE 6-5.2*** Data of Chandra and Avedisian (1991) for drop spreading on impact with a solid surface.

Two sequences of simulations from Fukai *et al.* (1993) are shown in Fig. 6-5.3. For one combination of parameters the drop spreads and then retracts with sufficient energy that the height of the retracting column of liquid exceeds slightly the initial diameter of the drop prior to impact. The drop shape proceeds to oscillate thereafter until viscosity damps the motion. In the second sequence, the same qualitative phenomena occur, but the dynamic shape of the drop is quite different. In particular, the drop spreads very rapidly, leaving (temporarily)

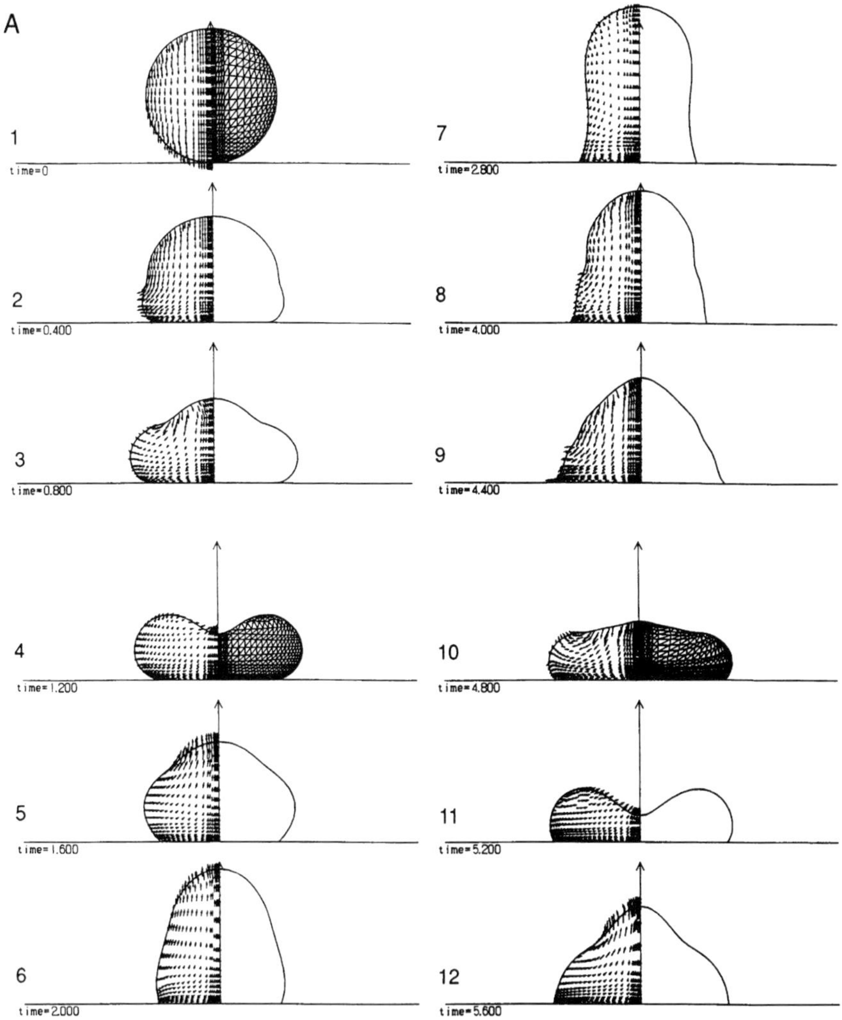


***FIGURE 6-5.3A*** Simulated drop impacts (From Fukai *et al.*, 1993). Sequence A corresponds to Re = 100, We = 1.4, Fr = 1020. Sequence B corresponds to Re = 1200, We = 100, Fr = $5.6 \times 10^4$. All dimensionless groups use the preimpact radius $R_0$ as the length scale. Time is normalized to $R_0/V_0$.

a very thin film surrounded by a circumferential thickened rim. Fukai *et al.* present experimental confirmation of this simulation.

Another paper worthy of study, but with too much detail for us to summarize here, is that of Stow and Hadfield (1981). Of special importance in their work is their study of the critical conditions that lead to "splashing," that is, the ejection of secondary drops subsequent to the impact of the primary drop. This is a very interesting, important, and challenging problem but we will not address the problem here.

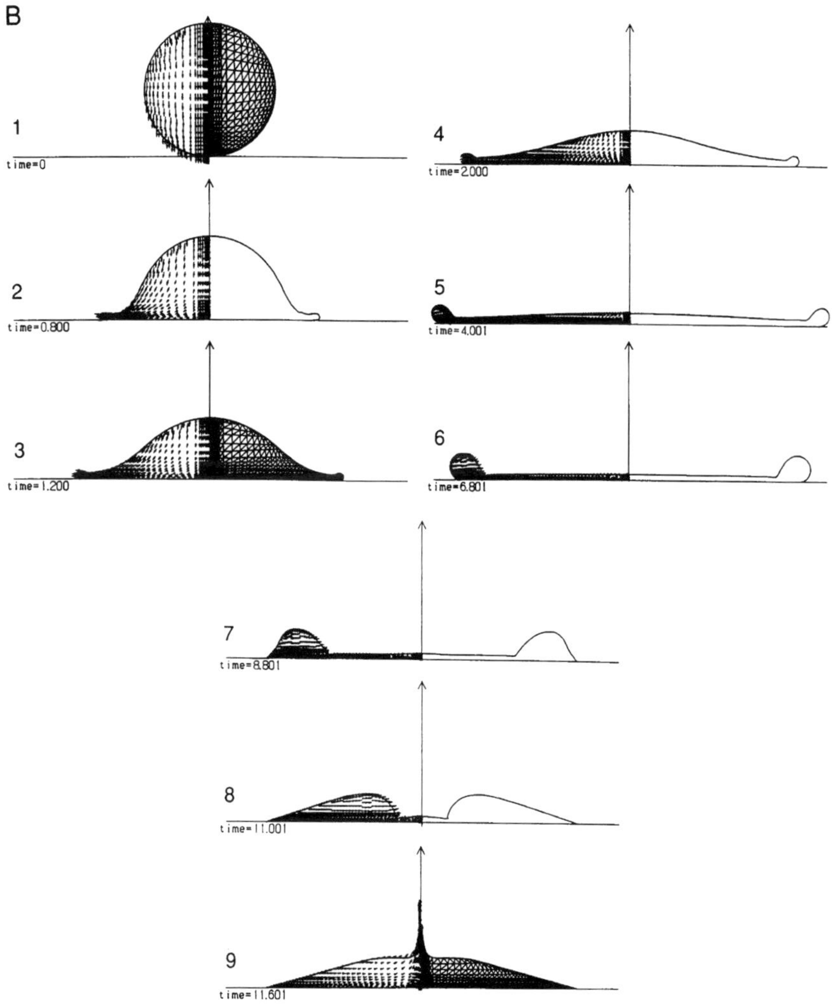


***FIGURE 6-5.3B*** *Continued.*

## Problems

P6-1 Prove that a spherical surface satisfies Eq. 6-2.9.

P6-2 The volume of a spherical segment of height $H_0$ and footprint radius $R_0$ (see Fig. P6-2.1) is found to be

$$V = \frac{1}{6}\pi H_0^3\left(1 + 3\frac{R_0^2}{H_0^2}\right). \tag{P6-1.1}$$

For the sessile drop in the absence of gravity, find a relationship between the contact angle and $H_0$ and $R_0$.

P6-3 For the sessile drop in the absence of gravity, find a relationship between the contact (footprint) area $A_0$ and the drop volume $V$, with the contact angle as a parameter.

P6-4 For the drops shown in Fig. 6-2.4, if $R_0 = 0.2$ mm in each case, find the volume of each drop, given a contact angle of $\theta = 10°$ for the sessile drop. What is the maximum slope $dh/dr$ for the pendent drop? Assume that the liquid is water.

P6-5 Obtain Eq. 6-2.27 by using the following alternative relationship for the drop volume:

$$\frac{V}{a^{*3}} = \int_0^{\mathrm{Bo}} 2\pi r\, h(r)\; dr$$
$$= \int_0^{\mathrm{Bo}} 2\pi r\, h_0\left[J_0(\sqrt{\mathrm{Bo}}) - J_0(r)\right]\; dr. \tag{P6-5.1}$$

P6-6 What boundary conditions have been imposed in solving Eq. 6-2.19?

P6-7 Figure 6-2.4 shows the R-limited sessile drop shape, based on the analytical solution to the Young–Laplace equation (Eq. 6-2.4). Show that the solution satisfying the equation and boundary conditions is

$$\frac{h}{h_0} = \frac{I_0(r) - I_0(\sqrt{\mathrm{Bo}})}{1 - I_0(\sqrt{\mathrm{Bo}})}. \tag{P6-7.1}$$

Does $h_0$ have the same meaning in this case as it does in the case of the

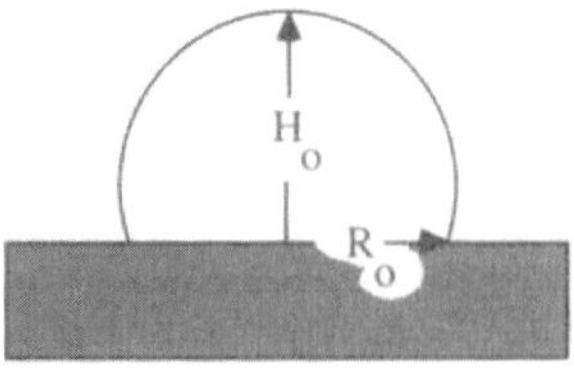


***FIGURE P6-2.1*** Sketch for Problem P6-2.

pendent drop? Derive a relationship between $h_0$ and the (dimensionless) drop volume $V$.

P6-8 Why is Eq. 6-4.13 not simply $\partial F/\partial \tau = U_F$? In what coordinate system might this be true?

P6-9 A water bridge of radius 1 mm and slenderness $\Lambda = 3$ is in a gravitational field of $g = 9.8$ m/s$^2$. What is the Bond number? Is the bridge stable?

P6-10 Compare Eq. 6-2.15 to Eq. 6-2.6. Is the difference in format due to anything other than choice of parameters for the nondimensionalization?

P6-11 Garandet *et al.* (1994) have made experimental determinations of the Harkins and Brown correction factor (see Fig. 6-3.4). Test their data against Fig. 6-3.4. Before doing so you will have to plot the Harkins and Brown correction factor for smaller values of $R_0/\sqrt{2}\, a^*$ than shown in Fig. 6-3.4. Convert the theory presented by Garandet *et al.* to a form suitable for plotting on Fig. 6-3.4. Discuss any reservations you have about the theory.

P6-12 Do you think the spelling "pendant" is preferable to "pendent" for the pendular drop?

P6-13 Look at Fig. 6-2.4. For the sessile drop we find $h/h_0 = 1$, consistent with the interpretation of $h_0$ as the height of the drop at the axis of symmetry. This is apparently not so for the pendent drop solution, given as Eq. 6-2.20. Why? Pitts (1974) gives an analytical solution for the small slope approximation which may be written in the form

$$\frac{h(r)}{h_0} = \frac{1 - J_0\left(r/a^*\right)}{1 - J_0\left(R/a^*\right)}. \qquad \text{(P6-13.1)}$$

What is the relationship of this solution to our Eq. 6-2.20?

P6-14 Boucher and Evans (1975) have presented numerical solutions for the shape of pendent drops and an analysis of the maximum stable drop size that can hang from a capillary. Their results are plotted in Fig. P6-14.1. Boucher and Evans define a dimensionless capillary radius as $X = R_c/a$ and a dimensionless maximum drop volume as $V = V_{max}/a^3$. The length scale $a$ is defined as

$$a \equiv \sqrt{\frac{2\sigma}{\rho g}}. \qquad \text{(P6-14.1)}$$

Compare these numerical results to our simplistic model (Eq. 6-3.1) which asserts that the maximum volume corresponds to $\theta = 0$. Is the Boucher and Evans result compatible with the Harkins and Brown correction factor shown in Fig. 6-3.4?

P6-15 Garandet *et al.* (1994) have obtained data for the volume of molten metal drops that fall from a cylindrical support when their weight exceeds the surface tension force. Compare the data below to the prediction of Eq. 6-3.1.

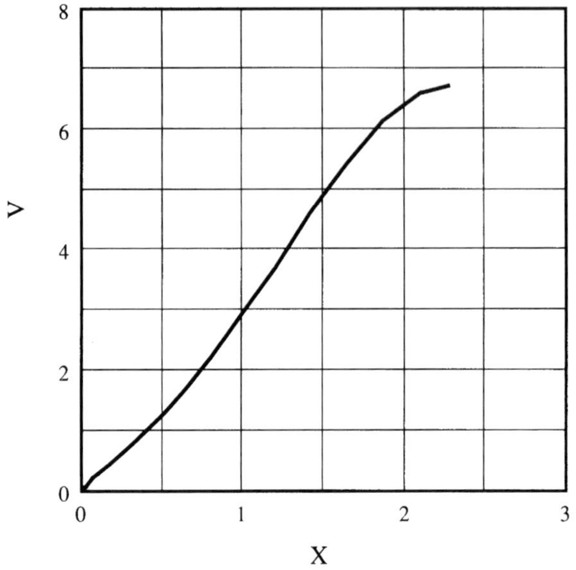


***FIGURE P6-14.1*** Theoretical prediction of maximum stable drop size for a pendent drop on a capillary.

Rhenium: $\rho = 18{,}000\ \mathrm{kg/m^3}$, $\sigma = 2.52\ \mathrm{N/m}$

| $D_c$ (mm) | $V$ ($10^{-8}$ m$^3$) |
|---|---|
| 0.31 | 1.29 |
| 0.5 | 1.96 |
| 1 | 3.69 |
| 1.77 | 6.13 |
| 4 | 12.1 |

Zirconium: $\rho = 6050\ \mathrm{kg/m^3}$, $\sigma = 1.435\ \mathrm{N/m}$

| $D_c$ (mm) | $V$ ($10^{-8}$ m$^3$) |
|---|---|
| 0.76 | 5.16 |
| 1 | 6.54 |

P6-16 Mori (1990) presents an equation that curve-fits the Harkins–Brown correction factor (see Eq. 6-3.2) in the form

$$f = 0.6 + 0.4\left[1 - 0.778\left(\frac{\rho g R_0^2}{\sigma}\right)^{1/3}\right]^{2.2} \qquad \text{(P6-16.1)}$$

when $\mathrm{Bo} = \rho g R_0^2/\sigma \le 2.2$. Using this equation, plot $1/\mathrm{Bo}$ as a function of $V_{\mathrm{drop}}/R_0^3$.

## References


Boucher, E. A., and M. J. B. Evans. (1975). Pendent drop profiles and related capillary phenomena. *Proc. R. Soc. Lond. A* **346**, 349.

Boucher, E. A., M. J. B. Evans, and H. J. Kent. (1976). Capillary phenomena. II. Equilibrium and stability of rotationally symmetric fluid bodies. *Proc. R. Soc. Lond. A* **349**, 81.

Bousfield, D. W. (1988). Filament splitting between separating plates. *Chem. Eng. Comm.* **73**, 19.

Chandra, S., and C. T. Avedisian. (1991). On the collision of a droplet with a solid surface. *Proc. R. Soc. Lond. A* **432**, 13.

Chandra, S., and C. T. Avedisian. (1992). Observations of droplet impingement on a ceramic porous surface. *Int. J. Heat Mass Transf.* **35**, 2377.

Coriell, S. R., S. C. Hardy, and M. R. Cordes. (1977). Stability of liquid zones. *J. Coll. Interface Sci.* **60**, 126.

Fordham, S. (1948). On the calculation of surface tension from measurements of pendant drops. *Proc. R. Soc. Lond. A* **194**, 1.

Fortes, M. A. (1982). Axisymmetric liquid bridges between parallel plates. *J. Coll. Interface Sci.* **88**, 338.

Fukai, J., Z. Zhao, D. Poulikakos, C. M. Megaridis and O. Miyatake. (1993). Modeling of the deformation of a liquid droplet impinging upon a flat surface. *Phys. Fluids A* **5**, 2588.

Garandet, J. P., B. Vinet and P. Gros. (1994). Considerations on the pendant drop method. A new look at Tate's law and Harkins' correction factor. *J. Coll. Interface Sci.* **165**, 351.

Harkins, W. D., and F. E. Brown. (1919). The determination of surface tension (free surface energy), and the weight of falling drops. *J. Am. Chem. Soc.* **41**, 499.

Harlow, F.H., and J.P. Shannon. (1967). The splash of a liquid drop. *J. Appl. Phys.* **38**, 3855.

Huh, C. and L. E. Scriven. (1969). Shapes of axisymmetric fluid interfaces of unbounded extent. *J. Coll. Int. Sci.* **30**, 323.

Meseguer, J. (1983). The breaking of axisymmetric slender liquid bridges. *J. Fluid Mech.* **130**, 123.

Meseguer, J., and A. Sanz. (1985). Numerical and experimental study of the dynamics of axisymmetric slender liquid bridges. *J. Fluid Mech.* **153**, 83.

Mori, Y. H., and T. Mochizuki. (1992). Explicit expressions for volume of drops released from submerged nozzles: Their derivation from semiempirical implicit correlations. *Int. J. Multiphase Flow* **18**, 141.

Orr, F. M., L. E. Scriven, and A. P. Rivas. (1975). Pendular rings. Meniscus properties and capillary force. *J. Fluid Mech.* **67**, 723.

Padday, J. F., and A. R. Pitt. (1974). The stability of axisymmetric menisci. *Phil. Trans. Roy. Soc. Lond.* **275**, 489.

Perales, J. M., and J. Meseguer. (1992). Theoretical and experimental study of the vibration of axisymmetric viscous liquid bridges. *Phys. Fluids A* **4**, 1110.

Pitts, E. (1974). The stability of pendent liquid drops. Part 2. Axial symmetry. *J. Fluid Mech.* **63**, 487.

Rayleigh, Lord. (1899). Investigations in capillarity. *Phil. Mag.* **48**, 321.

Rivas, D., and J. Meseguer. (1984). One-dimensional self-similar solution of the dynamics of axisymmetric slender liquid bridges. *J. Fluid Mech.* **138**, 417.

Shi, X. D., M. P. Brenner and S. Nagel. (1994). A cascade of structure in a drop falling from a faucet. *Science* **265**, 219.

Stow, C. D., and M. G. Hadfield. (1981). An experimental investigation of fluid flow resulting from the impact of a water drop with an unyielding dry surface. *Proc. R. Soc. Lond. A* **373**, 419.

Tate, T. (1864). On the magnitude of a drop of liquid formed under different circumstances. *Phil. Mag.* **27**, 176.

Yakhnin, E. D., and A. V. Chadov. (1983). Investigation of the transfer of a liquid from one solid surface to another. 2. Dynamic transfer. *Koll. Zhur.* **45**, 1183.

# 7 The Isolated Drop on an Impermeable Surface

When *a liquid* jet strikes a surface the radial flow in the ensuing axisymmetric film on the surface is quite complex, as our discussion in Chapter 5 suggests. When the jet drips onto the surface, as a series of *discrete* drops, rather than as a continuous jet, the result is an unsteady and even more complicated flow field. We begin a discussion of the interaction of drops with a surface by considering a single, isolated drop. The drop is initially at rest, but not necessarily in its equilibrium configuration on the surface.

## *7-1 SPREADING OF AN ISOLATED DROP WITH NO SURFACE TENSION*

The axisymmetric spreading of a drop of an isothermal incompressible Newtonian fluid on an impermeable surface that is wetted by the liquid is a complex flow field under all but the simplest conditions. We begin by consideration of the simplest case of interest—a viscous drop spreading slowly under the action of gravity. Figure 7-1.1 aids the discussion.

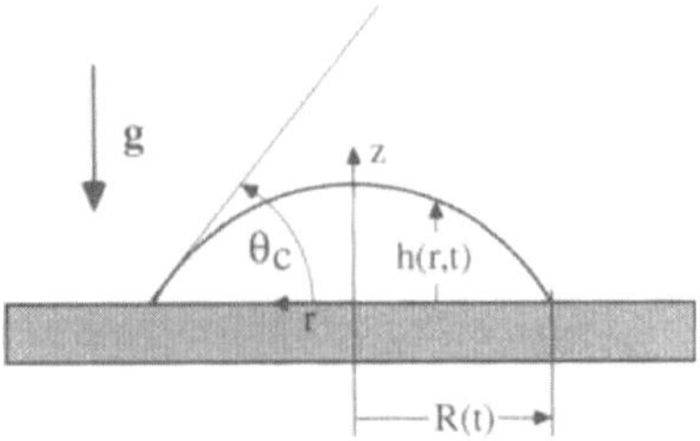


*FIGURE 7-1.1* An axisymmetric drop spreads across the plane $z = 0$.

We will neglect surface tension forces in this analysis. This will turn out to be a reasonable assumption for large drops of high viscosity.

An axisymmetric drop of volume $V$ is somehow formed with a circular "footprint" of radius $R_0$ at time $t = 0$. Our goal is to develop a dynamic model for the radius of the drop as a function of time: $R(t)$.

We begin with the assumption that the flow is quasi-steady (in some sense—slow) and nearly parallel to the surface $z = 0$, which is taken to be horizontal. Inspection of the radial component of the Navier–Stokes equations, and use of the continuity equation, leads us to begin with the following equation for the radial velocity:

$$\frac{1}{\rho}\frac{\partial p}{\partial r} = \nu \frac{\partial^2 u_r}{\partial z^2}. \tag{7-1.1}$$

In effect, we are neglecting any influence—in the momentum balance—of a velocity component $u_z$ as the drop spreads. This is essentially a lubrication analysis.

There is a pressure distribution within the drop that arises only from gravity if the effect of surface tension is ignored. We may write the pressure, for an axisymmetric drop, as

$$p = p_0 + \rho g(h - z) \tag{7-1.2}$$

and Eq. 7-1.1 becomes

$$g \frac{\partial h}{\partial r} = \nu \frac{\partial^2 u_r}{\partial z^2}. \tag{7-1.3}$$

The continuity equation for this incompressible axisymmetric flow is simply

$$\frac{\partial u_z}{\partial z} + \frac{1}{r}\frac{\partial}{\partial r} r u_r = 0. \tag{7-1.4}$$

If we integrate this over the definite limits $z = 0$ to $z = h(r,t)$ we find

$$\left[u_z(h) - u_z(0)\right] + \frac{1}{r}\frac{\partial}{\partial r} r \int_0^h u_r \, dz = 0. \tag{7-1.5}$$

At the lower limit, since the surface is assumed impermeable, $u_z(0) = 0$. At the upper limit, we may replace $u_z(h)$ with $\partial h/\partial t$. Hence we find an expression for conservation of mass in the form

$$\frac{\partial h}{\partial t} + \frac{1}{r}\frac{\partial}{\partial r}\left[r\int_0^h u_r\, dz\right] = 0. \tag{7-1.6}$$

Note that the existence of $u_z$ is essential to the mass balance. If there were no $u_z$, there would be no $\partial h/\partial t$. (See Problem P7-1.) Another statement of conservation of mass arises from the fact that whatever the shape of the drop, its volume remains fixed:

$$2\pi\int_0^{R(t)} rh\, dr = V. \tag{7-1.7}$$

Now we may integrate Eq. 7-1.3 twice with respect to $z$ and impose the boundary conditions

$$u_r = 0 \quad \text{on } z = 0 \quad \text{(no slip on the solid surface)} \tag{7-1.8}$$

and

$$\tau_{rz} = -\mu\frac{\partial u_r}{\partial z} = 0 \quad \text{(no shear at the upper surface).} \tag{7-1.9}$$

This permits us to find the radial velocity in the form

$$u_r = -\frac{g}{2\nu}\frac{\partial h}{\partial r}z(2h - z). \tag{7-1.10}$$

When this result is introduced into Eq. 7-1.6 and the indicated integration is performed, we obtain a differential equation for $h(r,t)$:

$$\frac{\partial h}{\partial t} - \frac{g}{3\nu}\frac{1}{r}\frac{\partial}{\partial r}\left(rh^3\frac{\partial h}{\partial r}\right) = 0. \tag{7-1.11}$$

This is a nonlinear partial differential equation in $h(r,t)$. It turns out that it can be solved analytically by introducing a similarity transformation. We begin by defining a new independent variable

$$\xi \equiv \left(\frac{gV^3}{3\nu}\right)^{-1/8} r\, t^{-1/8}. \tag{7-1.12}$$

In addition, a new dependent variable $\psi$ is defined such that the drop height $h(r,t)$ is expressed in the form

$$h(r,t) = \xi_R^{2/3}\left(\frac{3V\nu}{g}\right)^{1/4} t^{-1/4}\,\psi\left(\frac{\xi}{\xi_R}\right). \tag{7-1.13}$$

$\xi_R$ is the value of $\xi$ at $r = R(t)$. When Eqs. 7-1.13 and 12 are substituted into Eqs.

7-1.11 and Eq. 7-1.7 is used, we find that $\psi$ satisfies an ordinary differential equation of the form

$$\left(\zeta\psi^3\psi'\right)' + \frac{1}{8}\zeta^2\psi' + \frac{1}{4}\zeta\psi = 0, \tag{7-1.14}$$

where

$$\zeta \equiv \xi/\xi_R \tag{7-1.15}$$

and the prime means $d/d\zeta$.

Equation 7-1.7, after some algebraic manipulation, leads to the result

$$\xi_R = \left[2\pi\int_0^1 \zeta\,\psi(\zeta)\,d\zeta\right]^{-3/8}. \tag{7-1.16}$$

It can be shown (Pattle, 1959) that the solution to Eq. 7-1.14 is

$$\psi(\zeta) = \left(\frac{3}{16}\right)^{1/3}\left(1-\zeta^2\right)^{1/3} \tag{7-1.17}$$

and

$$\xi_R = \left(\frac{2^{10}}{3^4\pi^3}\right)^{1/8} = 0.894 \tag{7-1.18}$$

or

$$h = 0.531\left(\frac{3\nu V}{gt}\right)^{1/4}\left[1-\left(\frac{r}{R}\right)^2\right]^{1/3} \tag{7-1.19}$$

and

$$\left(\frac{3\nu}{gV^3}\right)^{1/8} R(t) = 0.894t^{1/8}. \tag{7-1.20}$$

Thus we predict that the drop spreads slowly, in the sense that $R$ is proportional to the 1/8 power of time.

Experimental data on the spreading of large silicone oil drops across a flat horizontal acrylic sheet were presented by Huppert (1982), who also presented the theoretical analysis outlined here. The data for the radial spreading, $R(t)$, are shown in Fig. 7-1.2, and the agreement with the theory is excellent.

The role of surface tension has been ignored in this analysis. Especially in the neighborhood of the moving contact line, that is, near $R(t)$, we might expect surface phenomena to be important. An important question is whether a drop of a particular liquid would spread without bound, or whether it could reach an equilibrium size (i.e., a specific final value of $R$) and then remain static after that time. After all, we *do* observe finite-sized drops sitting on surfaces. Hence we need to discuss the role of surface tension in affecting the spreading of a drop in some detail.

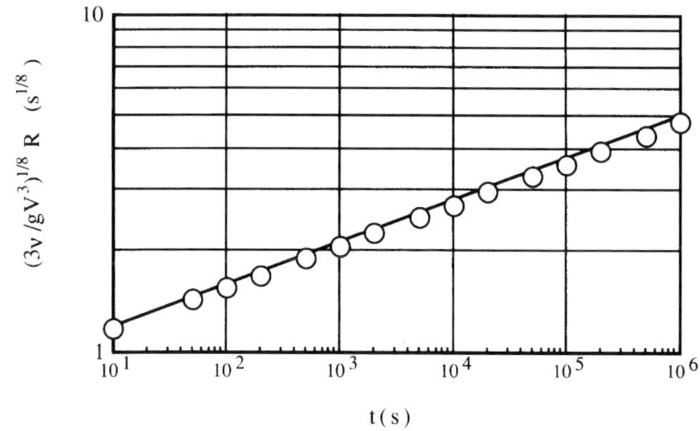


*FIGURE 7-1.2* Huppert's data for the spreading of silicone oil drops. The line is Eq. 7-1.20.

## 7-2 SPREADING OF A DROP WITH SURFACE TENSION

Observation convinces us that some drops do not spread continuously, but instead reach an equilibrium size and shape on a surface. In this section we consider the role of surface tension. We begin with an analysis of the static drop and then consider how the drop spreads to the static shape. Figure 7-2.1 shows the key features for the analysis.

The drop has a surface tension $\sigma$ and its interaction with the solid surface that it sits on, in the static case, is completely defined by the equilibrium contact angle $\theta_c$ (measured within the liquid drop, as in Fig. 7-2.1). It is possible that the liquid completely wets the surface, by which we imply that the liquid spreads spontaneously and continuously across the surface. This would correspond to a contact angle $\theta_c = 0$. It is common to differentiate between the cases that $\theta_c$ is greater, or less than, $\pi/2$. The case $\theta_c > \pi/2$ is often referred to as a nonwetting drop, and the case $\theta_c < \pi/2$ is often called a wetting drop. This latter nomenclature is then easily confused with what we might call the "completely wetting" case of $\theta_c = 0$.

The goal of an analysis of a static drop is the drop shape function $h(r)$. The static force balance is determined from a balance of hydrostatic pressure and the pressure that arises from surface curvature given by the Young-Laplace equation. This takes the form

$$p = p_0 + \rho g(h-z) - \sigma\left(\frac{\partial^2 h}{\partial r^2} + \frac{1}{r}\frac{\partial h}{\partial r}\right), \tag{7-2.1}$$

where we have assumed that the slope of the drop $(\partial h/\partial r)$ is small enough that the exact expression for the curvature can be linearized.

For the static drop Eq. 7-1.1 requires that

$$\frac{\partial p}{\partial r} = 0 = \frac{\partial}{\partial r}\left[\rho g h - \sigma\left(\frac{\partial^2 h}{\partial r^2} + \frac{1}{r}\frac{\partial h}{\partial r}\right)\right]. \tag{7-2.2}$$

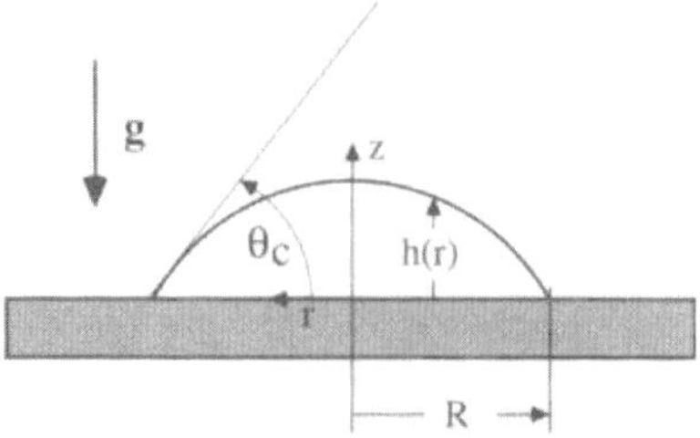


*FIGURE 7-2.1* A static drop on a surface.

This equation has the general solution, subject to the condition that $h$ be bounded at $r = 0$,

$$h(r) = A + BI_0(br), \tag{7-2.3}$$

where $I_0$ is a modified Bessel function. The coefficients $A$ and $B$ are determined from the boundary conditions

$$h = 0 \quad \text{and} \quad \frac{dh}{dr} = -\tan\theta_c \approx -\theta_c \quad \text{at } r = R. \tag{7-2.4}$$

The approximation of the slope is based on the assumption of a small slope, consistent with Eq. 7-2.1. The constant $b$ is determined from the requirement that the volume of the drop be given by

$$V = \int_0^R 2\pi r h(r)\, dr. \tag{7-2.5}$$

After these coefficients are determined, the solution may be recast into a nondimensional format

$$H(s) = \frac{2\left[I_0(\mathsf{B}^{1/2}S) - I_0(\mathsf{B}^{1/2}s)\right]}{S^2 I_2(\mathsf{B}^{1/2}S)} \tag{7-2.6}$$

by defining a new radial coordinate $s$ and a reduced drop height $H(s)$ as

$$s \equiv \frac{r}{a_0}, \quad S \equiv \frac{R}{a_0}, \quad \text{and} \quad H(s) \equiv a_0\theta_c h(r). \tag{7-2.7}$$

This format introduces a new Bond number $\mathsf{B}$ defined as

$$\mathsf{B} \equiv \frac{\rho g a_0^2}{\sigma} \tag{7-2.8}$$

and a length scale $a_0$ as

$$a_0 \equiv \left(\frac{V}{2\pi\theta_c}\right)^{1/3}. \tag{7-2.9}$$

To find the static drop radius we impose the boundary condition on the slope of the drop at $r = R$ (Eq. 7-2.4) and find

$$S^3 = \frac{2\mathrm{B}^{1/2}S\left[I_1(\mathrm{B}^{1/2}S)\right]}{I_2(\mathrm{B}^{1/2}S)}. \qquad (7\text{-}2.10)$$

The dimensionless drop radius is a function of the Bond number.

Having completed the analysis of the static drop, we can now examine the dynamics of spreading to the static configuration. We assume that a volume $V$ of liquid is deposited with an initial axisymmetric shape and with an initial contact angle greater than $\theta_c$. One is tempted to take the dynamic equation presented earlier, Eq. 7-1.11, but modified to account for the effect of surface tension. The result would be

$$\frac{\partial h}{\partial t} - \frac{1}{r}\frac{\partial}{\partial r}\left(\frac{rh^3}{3\mu}\frac{\partial}{\partial r}\left[\rho g h - \sigma\left(\frac{\partial^2 h}{\partial r^2} + \frac{1}{r}\frac{\partial h}{\partial r}\right)\right]\right) = 0. \qquad (7\text{-}2.11)$$

This may be nondimensionalized, as before (Eqs. 7-2.7, 8, and 9), with the addition of a reduced time variable

$$\tau \equiv \frac{\sigma\theta_c^3}{3\mu a_0}t. \qquad (7\text{-}2.12)$$

No simple analytical solution to this nonlinear partial differential equation is available. Worse than this, however, there is the widely held view that Eq. 7-2.11 does not really reflect the physics of drop spreading. The problem lies in two somewhat related issues. First of all, the usual hydrodynamic assumption of no slip between the liquid and the underlying solid boundary cannot hold along the moving contact line, since the boundary between the liquid and the solid is obviously in motion. The additional complication arises from the observation that when a drop spreads the apparent slope of the free surface, at the contact boundary, does not follow from the static contact angle constraint—the second half of Eqs. 7-2.4. Instead, there appears to be a *dynamic* contact angle which depends upon the rate of spreading. In fact, all of these notions are controversial, and the primary problem is that the physics in the neighborhood of the moving contact line is extremely complex. The consensus is that macroscopic concepts of fluid dynamics must be, in a sense, abandoned in order to get the physics "right" in the neighborhood of the moving contact line. Detailed discussions of these issues, and approaches to their resolution, may be found in the works of de Gennes (1985), Teletzke *et al.* (1987), and Dussan V (1979).

We are going to put these reservations aside for the moment. A quasi-steady hydrodynamic analysis of drop spreading, due to Tanner (1979), leads to an explicit model for the dynamic contact angle, and subsequently to a prediction of the drop radius $R(t)$ which is in good agreement with experimental data. We begin with Eq. 7-1.1 in the form

$$\frac{\partial p}{\partial r} = \mu \frac{\partial^2 u_r}{\partial z^2}. \tag{7-2.13}$$

This equation is integrated twice with respect to $z$, and the boundary conditions of no slip at $z = 0$ and no shear at $z = h$ are imposed. This leads (see Problem P7-3) to the mean radial velocity in the form

$$\overline{u_r} = \frac{dR}{dt} = -\frac{h^2}{3\mu}\frac{dp}{dr}. \tag{7-2.14}$$

For the pressure gradient we return to Eq. 7-2.2, neglect gravity for a very thin film, and neglect the term $(1/r)\partial h/\partial r$ with respect to the second derivative of $h$, with the result that

$$\frac{dR}{dt} = \frac{\sigma h^2}{3\mu}\frac{d^3h}{dr^3}. \tag{7-2.15}$$

or

$$h^2 \frac{d^3h}{dr^3} = 3\ \mathrm{Ca}, \tag{7-2.16}$$

where we define a capillary number based on the spreading rate $dR/dt$. Tanner (1979) studied the properties of solutions to this differential equation. It was first reduced to a nondimensional form by defining

$$d\xi = -\frac{dr}{x_0}, \qquad \eta = \frac{h}{y_0}, \qquad y_0 = x_0(3\ \mathrm{Ca})^{1/3}, \tag{7-2.17}$$

from which it follows that Eq. 7-2.16 takes the form

$$\eta^2 \frac{d^3\eta}{d\xi^3} = -1. \tag{7-2.18}$$

This equation has the property that there is a range of values of the parameter $y_0$ such that the curve $\eta(\xi)$ has an inflection point. The existence of this inflection point is a requirement for drops with small apparent contact angles, as suggested in Fig. 7-2.2.

We now identify the apparent contact angle with the slope of the drop height at the inflection point $x_i$, where the height of the drop surface is $h_i$. For small angles the slope is the angle (in radians), so we find

$$\theta \equiv \frac{y_0}{x_0}\left(\frac{d\eta}{d\xi}\right)_i = (3\ \mathrm{Ca})^{1/3}\left(\frac{d\eta}{d\xi}\right)_i. \tag{7-2.19}$$

For different values of $y_0$ that yield an inflection point it is possible to integrate Eq. 7-2.18 numerically. Tanner's results indicate that the coefficient of the capillary number term in Eq. 7-2.19 is nearly constant (with a value of approximately $2.2 \pm 0.3$ ) over a wide range of parameters. Hence the conclusion of Tanner's

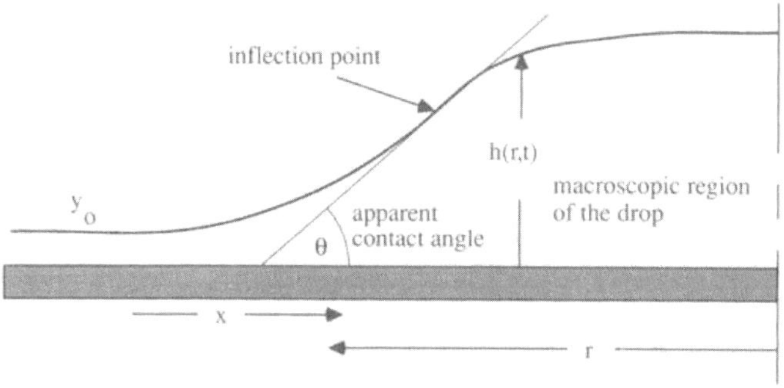


***FIGURE 7-2.2*** Drop shape near the inflection point.

analysis is expressed in the form

$$\theta = C(\mathrm{Ca})^{1/3}. \tag{7-2.20}$$

In an independent study of a related problem, Friz (1965) finds Eq. 7-2.20, with $C = 3.4$. (We note that $\theta$ is in units of radians here.)

Data consistent with Eq. 7-2.20 are presented by Chen and Wada (1992) and are replotted in our Fig. 7-2.3. With the contact angle transformed to radians, the line shown on the plot corresponds to Eq. 7-2.20 with $C = 3.6$.

It is now possible to link this result with a simple geometrical argument that will lead to a model for the spreading rate $dR/dt$. Since gravity has been neglected, we can expect the arguments presented in this section to hold only for very small drops. For small drops we can expect that the drop shape, on a macroscopic scale, is that of a spherical cap of radius $R$ and height $H$ and apparent contact angle $\theta$. For $H/R \ll 1$ it follows that

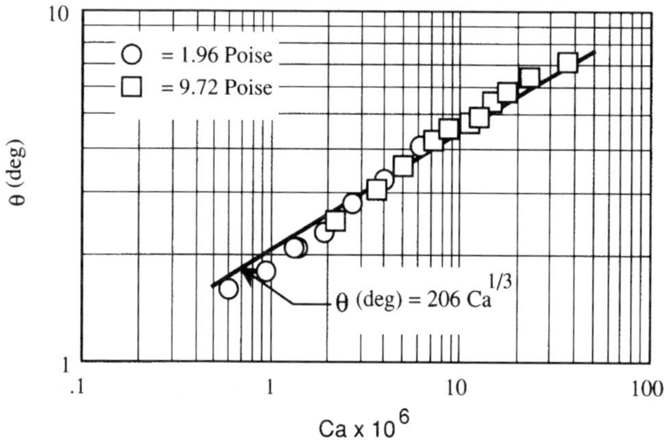


***FIGURE 7-2.3*** Data of Chen and Wada (1992) as a test of Eq. 7-2.20. Silicone oils with $\sigma = 21$ dyn/cm, spreading on glass.

$$V = \frac{\pi R^3}{4}\theta = \frac{\pi H R^2}{2}, \tag{7-2.21}$$

where $V$ is the drop volume. When Eq. 7-2.20 is introduced, and $U$ in the capillary number is replaced by $dR/dt$, we find

$$V = \frac{\pi C R^3}{4}\left(\frac{\mu}{\sigma}\frac{dR}{dt}\right)^{1/3}. \tag{7-2.22}$$

This may be integrated to give $R(t)$:

$$R(t) = \left(\frac{4V}{\pi C}\right)^{0.3}\left(\frac{10\sigma t}{\mu}\right)^{0.1} + R_0. \tag{7-2.23}$$

For $R \gg R_0$, and taking $C = 3.4$, we find

$$R(t) \approx 0.94\left(\frac{\sigma V^3 t}{\mu}\right)^{0.1}. \tag{7-2.24}$$

Data for the spreading of a silicone oil on glass are presented by Chen (1988), and one set of those data is shown in Fig. 7-2.4. The agreement between the data and the model given in Eq. 7-2.24 is excellent.

Just before presenting Tanner's analysis we noted a number of reservations with respect to application of a simple hydrodynamic analysis of drop spreading. The reader should study the works of de Gennes (1985), Teletzke *et al.* (1987), and Dussan V (1979). These references are much too detailed and complex to review in a meaningful manner in a short section on this topic. However, two notions are worth mentioning. One is the observation that the spreading drop is actually preceded by a thin "precursor film" of submicron dimensions. The other is that in the neighborhood of the contact line, where the liquid film thickness varies from the order of molecular dimensions to a few hundred angstroms, van der Waals forces come into play and introduce additional interfacial forces

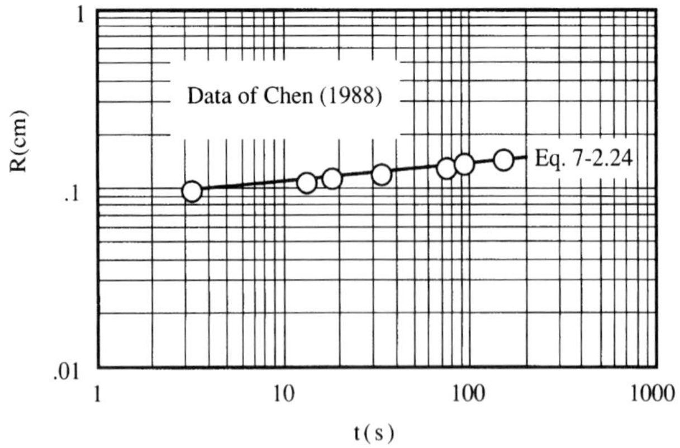


***FIGURE 7-2.4*** Data for spreading of a drop of silicone oil on glass ($V = 1.7 \times 10^{-4}$ cm$^3$, $\mu = 1.95$ poise, $\sigma = 21$ dyn/cm).

(through something called a "disjoining pressure," which will be discussed in Chapter 9) above that associated with surface tension.

Figure 7-2.5 suggests some of the details in the neighborhood of the moving contact boundary. To the naked eye it appears that the drop has a radius $R(t)$ and an apparent contact angle $\theta$, which may differ from the static angle $\theta_c$. There is a very thin precursor film of thickness $\zeta_0$ (not necessarily the same, conceptually, as the thickness $y_0$ of Tanner's analysis). It is believed that when normal macrophotographic techniques are used to view the drop, the apparent radius $R$ is the inflection point of the surface.

In a paper by Vavkushevskii et al. (1989) we find a solution (but not a clear derivation of it) for the radius of a spreading drop as a function of time, accounting for the role of the hydrodynamics in the precursor film, wherein a significant amount of viscous dissipation occurs. The final result may be written in the form

$$\rho^* \equiv \frac{R(t)}{(3V/2\pi)^{1/3}} = \left(1 + \frac{497\sigma\zeta_0}{\mu V^{2/3}} t\right)^{1/8}. \tag{7-2.25}$$

In this model it is assumed that the drop completely wets the surface and spreads without bound. No account of van der Waals forces is taken, and the length parameter $\zeta_0$ is assumed to be a property of the liquid–solid pair. Experimental confirmation of this model is given in Fig. 7-2.6 (Mickaily, 1992), for three low-molecular-weight oils spreading on smooth aluminum. The parameter $\zeta_0$ is obtained by fitting the data to the theory, and the value obtained (1.45 μm) is consistent with the observations of Vavkushevskii *et al.* for silicone oils on glass and aluminum. In the figure shown, $t'$ is proportional to time but it is not dimensionless.

$$t' \equiv \frac{497\sigma}{V^{2/3}\mu} t. \tag{7-2.26}$$

In plotting Fig. 7-2.6, cgs units were used for $t'$.

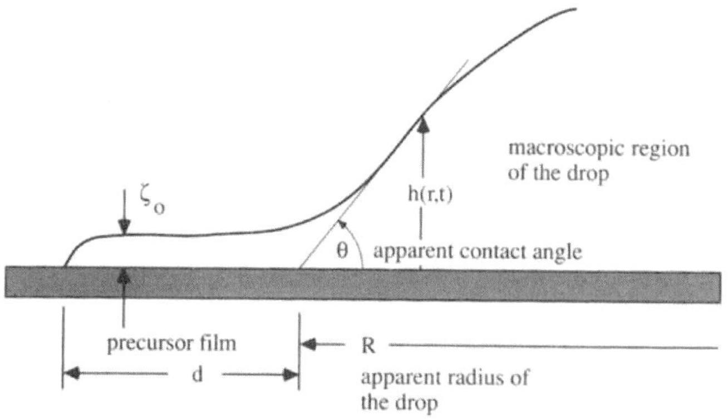


***FIGURE 7-2.5*** Microscopic view of a drop spreading on a solid.

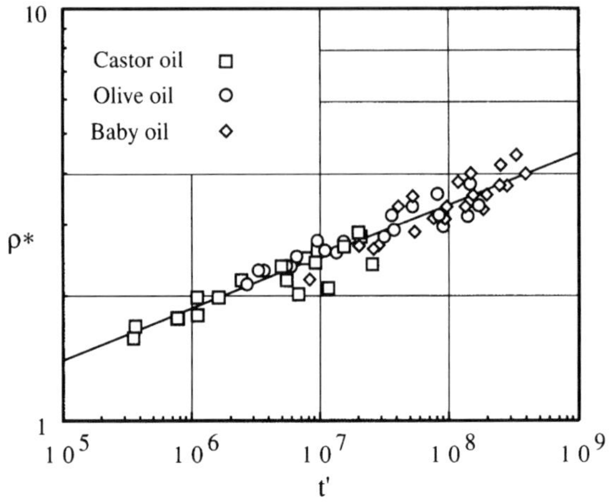


***FIGURE 7-2.6*** Data for horizontal spreading of drops of various liquids on smooth aluminum. (From Mickaily, 1992.)

## *Problems*

P7-1: Derive an equation of the form of (but perhaps not exactly the same as) Eq. 7-1.6 from an unsteady mass balance on a differential volume element. Do not assume that $u_z$ is zero.

P7-2. Levinson *et al.* (1988) present an equation of the form of Eq. 7-1.20 by equating the viscous and gravitational forces acting on a small drop. They give expressions for these forces in their Eqs. 6a and 6b, but the expressions they write do not have the dimensions of forces. Clarify their derivation by writing the terms proportional to the gravitational and viscous forces. On p. 1010 they define something called "self-similarity" of the drop shape, and state that "it is easy to see that self-similarity automatically implies" that $hR^2$ is independent of time as the drop spreads. ($h$ is the height of the drop at its center.) Write a brief explanation of this assertion, to someone to whom this is not so self-evident, such as the author of this text.

P7-3. Show that Eq. 7-2.10 leads to the predictions

$$S = 2 \quad \text{for} \quad \mathsf{B} \ll 1,$$

$$S = (4\mathsf{B})^{1/4} \quad \text{for} \quad \mathsf{B} \gg 1. \qquad \text{(P7-3.1)}$$

P7-4. For long spreading times we may put Eq. 7-2.25 in the form

$$R(t) \approx 1.7\zeta_0^{0.125}\left(\frac{\sigma V^2 t}{\mu}\right)^{0.125}. \tag{P7-4.1}$$

Compare this model to Eq. 7-2.24. It is difficult to distinguish $R \sim t^{0.1}$ from $R \sim t^{0.125}$. Suggest some experiments that could differentiate one model from the other.

P7-5. The procedure that leads to Eq. 7-2.24 is rather simplistic. Diez *et al.* (1994) present an analysis that appears to be more rigorous and that leads to the following result for the radius of the drop front:

$$\frac{R(t)}{V^{1/3}} = 0.944\left(\frac{\sigma t}{3\mu V^{1/3}}\right)^{0.1}. \tag{P7-5.1}$$

Compare the *result* of Diez *et al.* to Eq. 7-2.24. How different is it? Compare the *physical approach* that Diez *et al.* take to that of Tanner. How different is it?

P7-6. Some of Mickaily's data for spreading of oil drops on aluminum are given in Fig. P7-6. Test the ability of Eq. 7-2.24 to fit these data. The drop volume was 0.024 $cm^3$. Physical properties are $\rho = 0.841$ g/$cm^3$, $\mu =$ 0.225 poise, $\sigma = 35$ dyn/cm.

P7-7. Combine Eqs. 7-2.21 and 7-2.24 to yield a model for the height $H$ at the center of a spreading drop as a function of time. Compare the model to the data of Diez *et al.* (1994), shown here in Fig. P7-7.

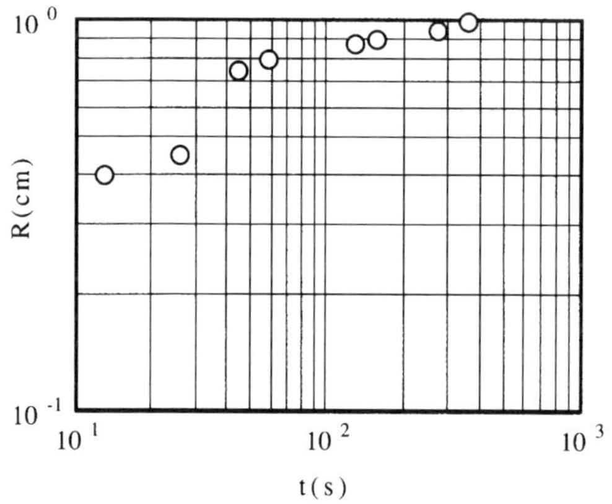


***FIGURE P7-6*** Drop spreading data of Mickailey (1992). Baby oil on aluminum.

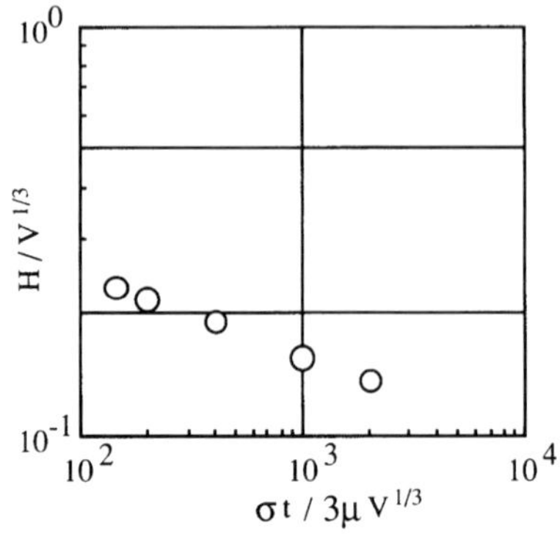


***FIGURE P7-7*** Data for height at the center of a spreading drop, from Diez *et al.* (1994). The liquid is polydimethylsiloxane (PDMS), for which $\sigma = 20$ dyn/cm, $\nu = 10.3\ \text{cm}^2/\text{s}$, $V = 0.12\ \text{mm}^3$.

## *References*


Chen, J.-D. (1988). Experiments on a spreading drop and its contact angle on a solid. *J. Coll. Interf. Sci.* **122**, 60.

Chen, J.-D., and N. Wada. (1992). Edge profiles and dynamic contact angles of a spreading drop. *J. Coll. Interf. Sci.* **148**, 207.

de Gennes, P. G. (1985). Wetting, statics and dynamics. *Rev. Mod. Phys.* **57**, 827.

Diez, J. A., R. Gratton, L. P. Thomas, and B. Marino. (1994). Laplace pressure driven drop spreading. *Phys. Fluids* **6**, 24.

Dussan V. E. B. (1979). On the spreading of liquids on solid surfaces. Static and dynamic contact lines. *Annu. Rev. Fluid Mech.* **11**, 371.

Friz, G. (1965). Uber den dynamischen randwinkel im fall der voll standigen benetzung. *Zeit. fur angew. Physik* **19**, 374.

Hocking, L. M. (1983). The spreading of a thin drop by gravity and capillarity. *Q. J. Mech. Appl. Math.* **36**, 55.

Hocking, L. M., and A. D. Rivers. (1982). The spreading of a drop by capillary action. *J. Fluid Mech.* **121**, 425.

Hocking, L. M. (1992). Rival contact angle models and the spreading of drops. *J. Fluid Mech.* **239**, 671.

Huppert, H. E. (1982). The propagation of two-dimensional and axisymmetric viscous gravity currents over a rigid horizontal surface. *J. Fluid Mech.* **121**, 43.

Kalliadasis, S. and H.-C. Chang. (1994). Apparent dynamic contact angle of an advancing gas-liquid meniscus. *Phys. Fluids* **6**, 12.

Levinson, P., A. M. Cazabat, M. A. Cohen Stuart, F. Heslot, and S. Nicolet. (1988). The spreading of macroscopic drops. *Rev. Phys. Appl.* **23**, 1009.

Mickaily, E. S. (1992). Hydrodynamics of cleaning. Ph.D. dissertation, Univ. of California San Diego.

Neogi, P., and C. A. Miller. (1982). Spreading kinetics of a drop on a smooth solid surface. *J. Coll. Interf. Sci.* **86**, 525.

Pattle, R. E. (1959). Diffusion from a point source with a concentration-dependent coefficient. *Q. J. Mech. Appl. Math.* **12**, 407.

Starov, V. M., V. V. Kalinin, and J.-D. Chen. (1994). Spreading of liquid drops over dry surfaces. *Adv. Coll. Interface Sci.* **50**, 187.

Tanner, L. H. (1979). The spreading of silicone oil drops on horizontal surfaces. *J. Phys. D Appl. Phys.* **12**, 1473.

Teletzke, G. F., H. T. Davis, and L. E. Scriven. (1987). How liquids spread on solids. *Chem. Engg. Comm.* **55**, 41.

Vavkushevskii, A. A., V. V. Arslanov, V. Y. Stepanenko, and V. A. Ogarev. Spreading of viscous liquid drop on solid horizontal surface. *Koll. Zh.* **51**, 439.

# 8 Capillary Penetration Dynamics

## 8-1 INTRODUCTION

We often speak of phenomena associated with surface tension as "capillary phenomena," although there may be no capillary tube in sight. The reason for this lies in the observation that most liquids will be spontaneously and rapidly drawn into small-diameter tubes upon contact of one end of the empty (open) tube with the liquid. The driving force for this phenomenon is surface tension, and the term "capillarity" as a synonym for surface tension effects was born of this observation. Of course, the liquid is drawn into the capillary only if it wets the material of which the capillary is made. Here, "wetting" would be defined by the observation of an equilibrium contact angle less than 90°.

In this chapter we examine the dynamics of the flow of a liquid into a capillary, or into the region between closely spaced parallel disks, under the action of surface tension. A number of important and interesting phenomena involve such dynamics. In addition, an understanding of this relatively simple geometry permits us to examine more complex problems, such as the wetting and drying of porous media.

## 8-2 CAPILLARY WICKING

"Wicking" is the term that we will use for the spontaneous penetration of a liquid into a capillary under the action of surface tension. The simplest model of wicking is based upon flow into a long straight capillary, as suggested in Fig. 8-2.1. A number of models of the wicking process are available, and much of this chapter is derived from these works. See, for examples, Joos *et al.* (1990), Batten (1984), and Good and Lin (1976).

The driving force for wicking is the capillary pressure difference $p_c$, which, for a small diameter capillary, is well approximated by using the Young–Laplace equation. Assuming that the meniscus is a section of a spherical surface that intersects the capillary at a finite (i.e., nonzero) contact angle $\theta_d$, we have (see Fig. 8-2.1)

$$p_c = -\frac{2\sigma}{R}\cos\theta_d, \tag{8-2.1}$$

where $\sigma$ is the surface tension of the liquid. Note the sign on the pressure. The pressure is reduced by the (negative) curvature of this meniscus as one passes from the ambient medium just above the meniscus *into the liquid.* In this analysis we allow for a dynamic contact angle $\theta_d$ that may be different from the equilibrium contact angle $\theta_c$. It is known (Joos, *et al.*, 1990) that the dynamic contact angle depends upon the rate of flow of the liquid through the capillary.

We assume that the capillary is vertical and that the flow is opposed by gravity. (The zero-gravity case is considered later.) Hence the net pressure driving force is given by

$$\Delta p = \frac{2\sigma}{R}\cos\theta_d - \rho g x, \tag{8-2.2}$$

where $x$ is the height of rise measured from the surface of the reservoir.

If we assume that the liquid is Newtonian and that the flow is very slow (i.e.,

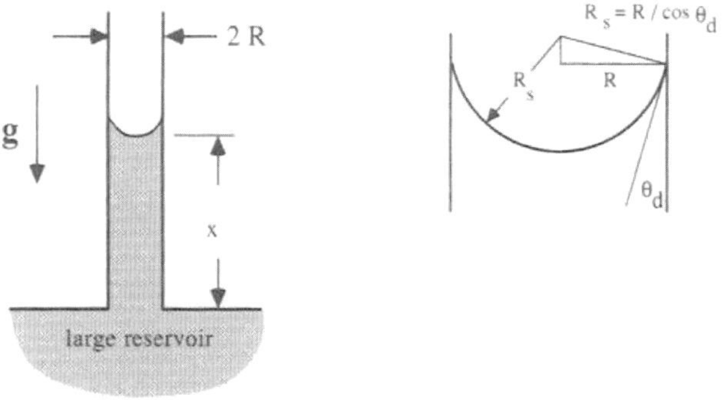


***FIGURE 8-2.1*** Wicking into a single capillary. The meniscus is a section of a sphere of radius $R_s$ which is related to the capillary radius $R$ and the contact angle as shown.

at low Reynolds number) then we may write the average velocity of liquid rising in the capillary from Poiseuille's law as

$$U = \frac{dx}{dt} = \frac{R^2 \, \Delta p}{8\mu x}. \tag{8-2.3}$$

(Here we have assumed that the only portion of the capillary that is wetted is the length $x$.) Since the velocity $U$ is just $dx/dt$, we are led to a differential equation in the form

$$\frac{dx}{dt} = \frac{\sigma R \cos \theta_d}{4\mu x} - \frac{\rho g R^2}{8\mu}. \tag{8-2.4}$$

First we regard the dynamic contact angle as a constant, independent of velocity. In that case we may solve the dynamic equation for capillary rise *analytically* and find the solution in the form

$$T = -H^2\left[\frac{X}{H} + \ln\left(1 - \frac{X}{H}\right)\right], \tag{8-2.5}$$

where the following parameters and variables have been defined:

$$T = \frac{t}{t'}, \qquad t' = \frac{\mu R}{4\sigma \cos \theta_d},$$
$$X = \frac{4x}{R}, \qquad H = \frac{4h}{R}, \qquad h = \frac{2\sigma \cos \theta_d}{\rho g R}. \tag{8-2.6}$$

We note that $T$, $X$, and $H$ are dimensionless. $t'$ is not dimensionless—it is a characteristic time scale for wicking, and $h$ is easily seen to be the equilibrium capillary rise. $H$ is just the inverse of a Bond number except for a numerical factor. Figure 8-2.2 shows the $X(T)$ behavior according to Eq. 8-2.5, which is often called the Washburn, Rideal, Lucas, or WRL, equation (Washburn, 1921). (Many authors refer to this simply as the Washburn equation.)

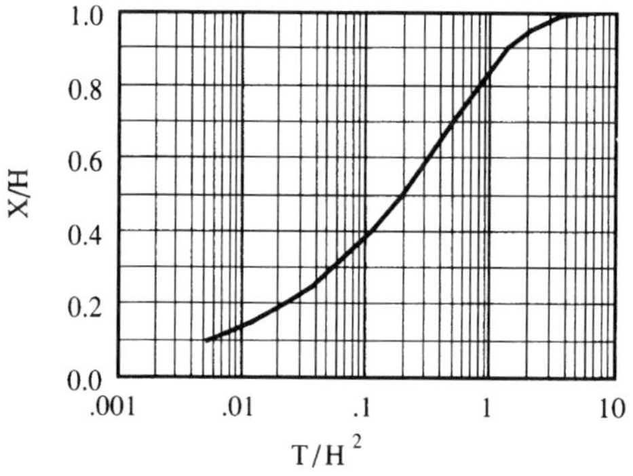


***FIGURE 8-2.2*** Wicking into a single capillary (Eq. 8-2.5).

If gravity is negligible, a simpler analytical model can be derived from Eq. 8-2.4 by dropping the term with $g$, with the result that

$$\frac{x}{R} = \left(\frac{\sigma}{2\eta R}\, t\right)^{1/2} \tag{8-2.7}$$

or, in dimensionless terms,

$$X = (2T)^{1/2}. \tag{8-2.8}$$

Equation 8-2.8 has the property that $X$ increases continuously with time. If the capillary is in contact with a *finite* volume of fluid, such as when a single drop is in contact with the end of the capillary, an unbounded increase in $X$ does not occur physically because the liquid is eventually depleted. Hence Eq. 8-2.8 holds until the drop disappears into the capillary. (At some point the radius of curvature of the original "feed" drop will be small enough to contribute to the dynamics of the flow. See, for example, Marmur, 1988.)

Equation 8-2.5 has been tested against data on the rate of capillary rise into single capillaries. The data reveal a number of discrepancies relative to this simple model. One problem is that the *dynamic* contact angle is an unknown function of the rate of rise itself. Another is that the results depend upon whether the capillary has been prewet with the liquid, or instead is dry prior to the intrusion of the liquid into the capillary. (On this point, see the evidence presented by Rillaerts and Joos, 1980.)

An example of experimental data on dynamic contact angle is presented in Fig. 8-2.3, which shows the data of Joos *et al.* cited earlier. They used silicone oils of varying viscosities. Their data consistently lie below the prediction of Eq. 8-2.5. This is largely a result of the fact that the dynamic contact angle is much larger than the assumed value. (Since silicone oil completely wets glass, $\theta_c = 0$ ). This lowers the capillary pressure driving the flow. Nevertheless, Eq. 8-2.5 provides a model that is useful for estimation of the time required to achieve wicking into a capillary.

In order to improve the dynamic model, it is necessary to introduce some

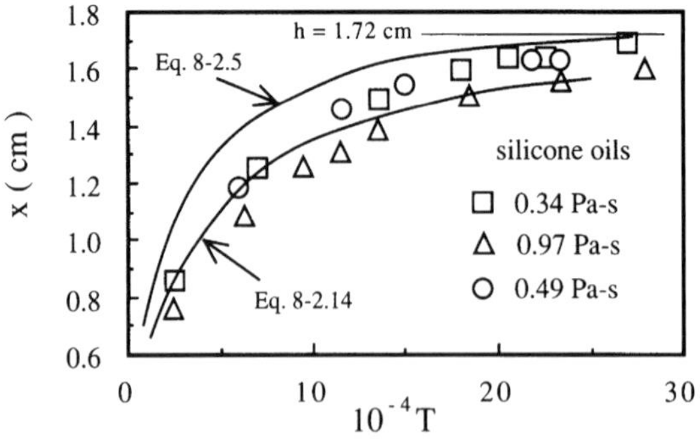


***FIGURE 8-2.3*** Experimental test of Eq. 8-2.5.

dependence of the contact angle on the rate of rise of the liquid in the capillary. Joos *et al.* (1990) present an experimental relationship for the dynamic contact angle, which they write in the form

$$\cos\theta_d = \cos\theta_c - 2\left(1 + \cos\theta_c\right)\mathrm{Ca}^{1/2}, \tag{8-2.9}$$

where the capillary number is defined as usual as

$$\mathrm{Ca} = \frac{\mu U}{\sigma} \tag{8-2.10}$$

and $U = dx/dt$ is the average velocity of the liquid rising in the capillary. If we now introduce Eq. 8-2.9 into Eq. 8-2.4 and consider the case of a liquid that completely wets the capillary, so that $\theta_c = 0$, we may solve for the rate of capillary rise from a differential equation of the form

$$\frac{dx}{dt} = \frac{\sigma R}{4\mu x}\left[1 - 4\left(\frac{\mu}{\sigma}\frac{dx}{dt}\right)^{1/2}\right] - \frac{\rho g R^2}{8\mu} \tag{8-2.11}$$

or, in dimensionless terms,

$$X\frac{dX}{dT} + 4\left(\frac{dX}{dT}\right)^{1/2} = 1 - \frac{X}{H}. \tag{8-2.12}$$

The dimensionless variables and parameters are defined as in Eqs. 8-2.6 except that $\theta_d$ is set to zero in those definitions. We may write Eq. 8-2.12 as a quadratic equation in $(dX/dT)^{1/2}$ and solve the quadratic equation exactly, with the result that

$$\frac{dX}{dT} = \frac{\left[-2 + \sqrt{4 + X\left(1 - X/H\right)}\right]^2}{X^2}. \tag{8-2.13}$$

The solution may be evaluated numerically from

$$T = \int_0^X \frac{X^2\, dX}{\left[-2 + \sqrt{4 + X\left(1 - X/H\right)}\right]^2}. \tag{8-2.14}$$

Figure 8-2.3 shows $X(T)$ for $H = 275$, which corresponds to the data shown on the figure. This modified theory fits these rise data better at short times than does the WRL theory, but is not especially good at long times, except for the most viscous liquid. However, more extensive testing of Eq. 8-2.13 by Joos *et al.* indicates that the data are well described by this equation.

### *8-2.1 Inertial Effects*

The model described here and the experimental test correspond to conditions where viscous effects permit us to use the steady Poiseuille flow assumption with some confidence. Because most capillary wicking flows involve capillaries of very small radius, the Reynolds numbers are generally small and there is no issue of the possibility of turbulent flow. However, in the early stages of wicking, where the liquid has penetrated only over an axial length of several capillary diameters, it is quite possible that the inertia of the liquid retards the development of the quasi-steady velocity field. For example, an order of magnitude estimate of the time scale for acceleration of the liquid to the quasi-steady flow is given by

$$t_\infty = \frac{R^2}{\nu}. \tag{8-2.15}$$

(This follows approximately from Eq. 4-2.10. We will shortly make a stronger case for this result.) We may ask: How far does the liquid move toward its equilibrium height during the time it is accelerating toward the Poiseuille velocity? Examination of Fig. 8-2.2 indicates that the value of $T/H^2$ determines the approach of $x$ to its final value. We may write $T/H^2$ as

$$\frac{T}{H^2} = \frac{\rho^2 g^2 R^3}{16\mu\sigma \cos\theta_d} t. \tag{8-2.16}$$

At the time given by Eq. 8-2.15 we find

$$\frac{T}{H^2} = \frac{\rho^3 g^2 R^5}{16\mu^2\sigma \cos\theta_d}. \tag{8-2.17}$$

If we choose the physical properties of water, and a capillary radius of 0.5 mm, this time is approximately $T/H^2 = 0.3$. From Fig. 8-2.2 we find that the water has traveled more than halfway toward its equilibrium height. We conclude that for water wicking into capillaries larger than a half-millimeter in radius, inertial effects are important. Because of the strong dependence on $R$ observed in Eq. 8-2.17, we expect that capillaries of radius smaller than 0.5 mm will not exhibit strong inertial effects.

When we talk about inertial effects here we are accounting for the fact that when a pressure gradient is imposed on a liquid, a finite time is required for the liquid to accelerate to (nearly) its steady velocity. Imagine that a long capillary is brought to a reservoir filled with liquid. At time $t < 0$ the liquid in the reservoir is at rest—there is no flow. The capillary is empty. Suddenly, at time $t > 0$, the tip of the capillary comes in contact with the liquid in the reservoir. If the liquid wets the tube, a constant pressure difference will be imposed across the ends of the tube, and the liquid begins to move down the tube axis. We have solved this problem, in a different context, in Section 4-2.

The summation in Eq. 4-2.7 gives the transient part of the velocity profile. A measure of how rapidly the steady-state profile is achieved is given by the rate at which the velocity along the tube axis ($s = 0$) builds up to its maximum value, which is unity in dimensionless terms. We examine the behavior of the infinite series term in Eq. 4-2.7 and observe how quickly this function becomes *small* compared to unity.

Setting $s = 0$ in the series in Eq. 4-2.7, we find

$$\Phi(0,\tau) \equiv -8 \sum_{n=1}^{\infty} \frac{J_0(0)}{\alpha_n^3 J_1(\alpha_n)} e^{-\alpha_n^2 \tau}. \tag{8-2.18}$$

Figure 8-2.4 shows the decay of $\Phi(0,\tau)$, and we conclude that the steady profile is achieved for $\tau \approx 1$, since $\Phi(0,\tau)$ falls well below 0.01 by that time.

The conclusion that we draw is that the transient disappears in a time of the order of $\tau = 1$. In terms of real time, the transient disappears, or steady flow is achieved, by a time given by

$$t_\infty = \frac{\rho R^2}{\mu}. \tag{8-2.19}$$

This is identical to Eq. 8-2.15.

Now let's use this model for the transient start-up of Poiseuille flow to modify the equation we have for the dynamics of wicking. In essence, we begin with Eq. 8-2.3, in the form

$$\frac{dx}{dt} = \overline{u_z^s} + \overline{U}. \tag{8-2.20}$$

The average of the steady-state velocity is

$$\overline{u_z^s} = \frac{\Delta \wp R^2}{8\mu x}. \tag{8-2.21}$$

For the average of the transient part of the velocity we must find

$$\overline{U} = 2u_z^{\max} \int_0^1 s\Phi(s,\tau)\, ds. \tag{8-2.22}$$

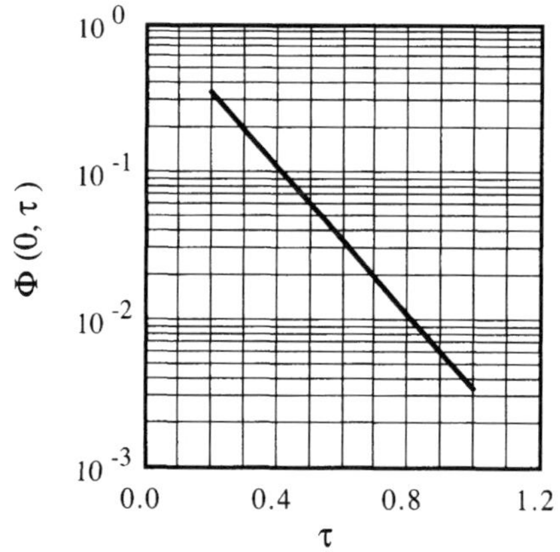


***FIGURE 8-2.4*** The function $\Phi(0, \tau)$ of Eq. 8-2.18.

Introducing Eq. 4-2.7, we can show that

$$\frac{\overline{U}}{u_z^{\max}} = -16 \sum_{n=1}^{\infty} \frac{e^{-\alpha_n^2 \tau}}{\alpha_n^4}. \tag{8-2.23}$$

Then the rate of penetration is given by the solution to

$$x\frac{dx}{dt} = \frac{\Delta \wp R^2}{8\mu}\left(1 - 32 \sum_{n=1}^{\infty} \frac{e^{-\alpha_n^2 \tau}}{\alpha_n^4}\right). \tag{8-2.24}$$

This is a good time to recall that the pressure is defined in such a way that

$$\Delta \wp = \frac{2\sigma \cos \theta_d}{R} - \rho g x. \tag{8-2.25}$$

Our goal is to get an idea of the importance of the transient start-up of this flow on the rate of penetration. We can achieve this by taking the case of a horizontal capillary (i.e., no gravity) for which we can write an analytical solution for $x(t)$. Using the nondimensionalization of Eqs. 8-2.6, we may write Eq. 8-2.24 in the form

$$X\frac{dX}{dT} = 1 - 32 \sum_{n=1}^{\infty} \frac{e^{-\alpha_n^2 \Omega T}}{\alpha_n^4}, \tag{8-2.26}$$

where the dimensionless group $\Omega$ is defined as

$$\Omega \equiv \frac{\mathrm{Oh}^2}{4 \cos \theta_d} \tag{8-2.27}$$

and Oh is the Ohnesorge number:

$$\mathrm{Oh} = \frac{\mu}{\sqrt{\rho \sigma R}}. \tag{8-2.28}$$

The solution to Eq. 8-2.26 may be written in the form

$$X^2 = 2T + 64 \sum_{n=1}^{\infty} \frac{e^{-\alpha_n^2 \Omega T} - 1}{\Omega \alpha_n^6}. \tag{8-2.29}$$

We may now plot the $X(T)$ function, for several values of the parameter $\Omega$, which apparently is the parameter that determines the importance of inertia on penetration rate in a horizontal capillary. The results are shown in Fig. 8-2.5.

As expected, significantly longer times are required to achieve the same extent of penetration when fluid inertia is accounted for. The smaller value of $\Omega$ shown on the graph corresponds to water in a capillary of 0.5 mm radius.

Although the effect of inertia appears to be important, the results are in a sense misleading. We may replot these results in *real* variables (i.e., not in dimensionless variables) in Fig. 8-2.6. What we see here, for the case of water in a capillary of 0.5 mm radius, is that at any time or extent of penetration *of practical interest* the effect of the transient flow is really quite small.

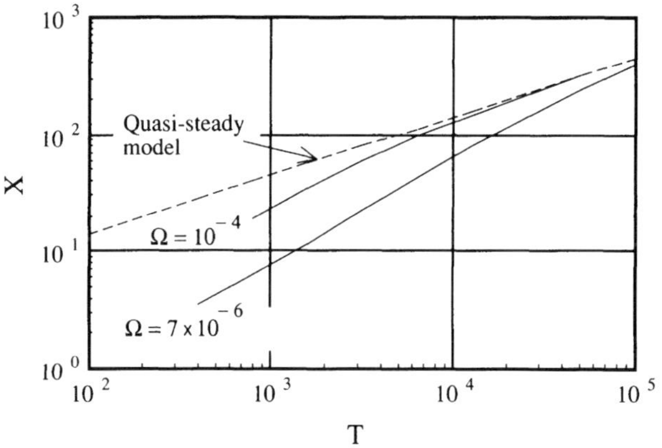


***FIGURE 8-2.5*** Effect of inertia on penetration rate in a horizontal capillary. (The quasi-steady model is Eq. 8-2.8.)

Similar analyses could be carried out for the effect of inertia on penetration rate under conditions where gravity is important. In addition, it is possible to include the effect of a dynamic contact angle in Eq. 8-2.25. Furthermore, entrance effects (Szekely *et al.* 1971; Batten, 1984) may be significant when the extent of penetration is very small, since most of the viscous friction would then lie within the region where the parabolic velocity profile is being developed from the entrance condition. An important "bottom line" to keep in mind is the following. Extreme care is required in the experimental study of the dynamics of capillary penetration. Cleanliness of the liquid and of the interior of the capillary must be maintained. In an interesting and relevant study, Good and Lin (1976) showed that the rate of penetration of organic liquids into a glass capillary depends upon whether the glass is outgassed in a vacuum to remove adsorbed

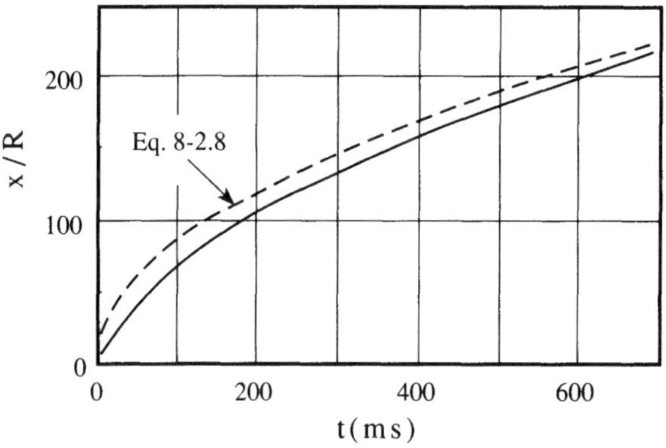


***FIGURE 8-2.6*** Effect of inertia on penetration rate in a horizontal capillary ($\Omega = 7 \times 10^{-6}$). The solid line is from Eq. 8-2.29.

contaminants. Rillaerts and Joos (1980) demonstrated the importance of prewetting of the capillary on the dynamic contact angle. In short, some of the phenomena described here—transient or inertial effects, and a dynamic contact angle—may be masked by artifact associated with the lack of control of the chemical purity of the liquids and, especially, the surfaces under consideration.

An interesting and related problem is that of the dynamics of rise of an annular meniscus surrounding a vertical cylinder (as a fiber, for example) that is suddenly touched to the free surface of a wetting liquid. Theory and experiments are described by Quéré and Di Meglio (1994).

## 8-3 RADIAL PENETRATION THROUGH A NARROW PLANAR SPACE

Suppose, as suggested in Fig. 8-3.1, that a drop of liquid is brought into contact with a small hole that leads into the narrow space separating two parallel plates or disks. If the liquid wets the surfaces, wicking will occur radially from the entry region. We want to develop a model for the rate of penetration of the liquid, analogous to the capillary wicking model. We will assume that the space through which wicking occurs is horizontal and neglect any effect of gravity on the dynamics of the flow.

Anticipating the use of a quasi-steady model, as in the case of capillary wicking, we first need an analog of the Poiseuille equation for radial flow between parallel surfaces subject to a pressure gradient. This is not difficult to develop.

The geometry that we will study is shown in Fig. 8-3.1. A liquid is supplied continuously at the entrance to the center of a pair of parallel planar surfaces. The flow then moves radially outward, and the circumferential region—the moving boundary—is exposed to atmospheric pressure. We will assume that the planes are rigid no-slip surfaces, that the flow is symmetric about the $z$ axis, and that there is no velocity component $u_\theta$. In addition, we assume that the liquid is Newtonian, isothermal, and incompressible.

The first thing we do is solve for the flow field under steady fully developed conditions. In effect, we take the flow to be leaving the system at a finite fixed radius $R$, and the region between the surfaces is completely filled with liquid. At

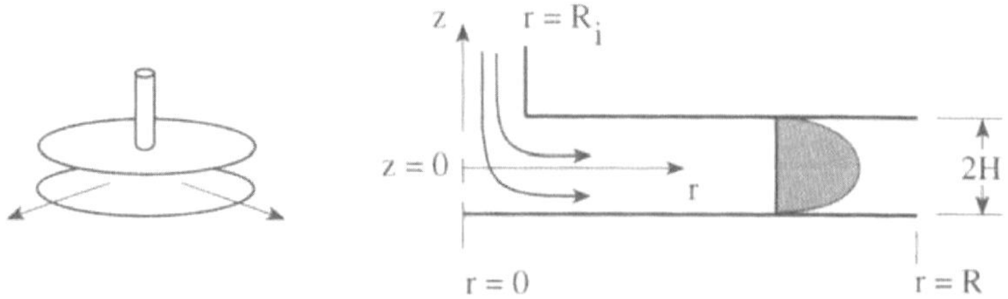


***FIGURE 8-3.1*** Radial flow between parallel planes.

this stage of development of a radial penetration model there is no free surface, and capillarity does not enter the model at all. Our goal is simply to develop an analog of the Poiseuille equation for this geometry.

Let's look first at the continuity equation. If we assume that the flow is fully developed from the radial position $R_i$ outward, this takes the form

$$0 = \frac{1}{r}\frac{\partial(ru_r)}{\partial r}. \tag{8-3.1}$$

The assumption that the flow is fully developed is equivalent to the assumption that there is no axial velocity component:

$$u_z = 0. \tag{8-3.2}$$

The Navier–Stokes equations, for low Reynolds number flows, simplify to the form

$$0 = -\frac{\partial p}{\partial r} + \mu\left\{\frac{\partial}{\partial r}\left[\frac{1}{r}\frac{\partial(r\,u_r)}{\partial r}\right] + \frac{\partial^2 u_r}{\partial z^2}\right\}. \tag{8-3.3}$$

Boundary conditions on the solid surfaces are simple. The no-slip condition is

$$u_r = 0 \quad \text{on } z = \pm H,\ R_i \le r \le R. \tag{8-3.4}$$

A boundary condition on pressure can be written if we assume that the fluid exiting the disks along the circumference $r = R$ is at atmospheric pressure, $p = 0$, and so our boundary condition on pressure at the outlet is expressed as

$$p = 0 \quad \text{for } -H \le z \le +H,\ r = R. \tag{8-3.5}$$

We must impose some condition on the flow into the planar region. We have two choices. One is a condition on the entry pressure:

$$p = P_{\text{entr}} \quad \text{at } r = R_i. \tag{8-3.6}$$

The flow rate through the system would then be determined from the solution to the dynamic equations. Alternatively, we could impose a boundary condition on the flow rate itself:

$$Q = \int_{-H}^{+H} 2\pi r u_r\, dz. \tag{8-3.7}$$

Then the pressure required to produce this flow would be determined from the solution. Either choice of boundary condition yields a well-defined boundary value problem, and the physics of the case at hand (what we actually impose on the system—pressure or flow) would determine the choice.

Before we proceed further, let's nondimensionalize Eq. 8-3.3 by defining

$$\frac{u_r}{U} = u^* \quad \text{and} \quad \frac{p}{\mu U/H} = p^*$$

and

$$\frac{r}{H} = r^* \quad \text{and} \quad \frac{z}{H} = z^*. \tag{8-3.8}$$

For the moment, $U$ is undefined except that it has the units of a velocity. With these definitions Eq. 8-3.3 takes the form

$$0 = -\frac{\partial p^*}{\partial r^*} + \frac{\partial^2 u^*}{\partial z^{*2}}. \tag{8-3.9}$$

We can now choose a sensible definition of a characteristic velocity $U$. The volume flow rate through the system is defined in Eq. 8-3.7. Whether we know $Q$ or not, there certainly is a well-defined $Q$ for this steady-state system, and we can use it to define a velocity scale by writing

$$Q = 4\pi R_i H U. \tag{8-3.10}$$

Note that since $4\pi R_i H$ is the cross-sectional area of the entrance to the flow region, $U$ is the average velocity at the entrance. This is a perfectly reasonable choice of a velocity scale to use for nondimensionalization. Keep in mind, however, that Eq. 8-3.10 is only a *definition* of $U$ in terms of $Q$. Since we don't know $Q$ yet, neither do we know $U$.

To solve Eq. 8-3.9 we begin by noting that the continuity equation (Eq. 8-3.1) has the solution (in dimensionless variables)

$$u^* = \frac{1}{r^*} C(z^*). \tag{8-3.11}$$

$C$ is not a function of $r^*$, but it could be a function of $z^*$. If we introduce this result into Eq. 8-3.3 we find

$$0 = -\frac{\partial p^*}{\partial r^*} + \frac{1}{r^*}\frac{d^2 C}{dz^{*2}}, \tag{8-3.12}$$

which we rearrange to the form

$$r^* \frac{\partial p^*}{\partial r^*} = \frac{d^2 C}{dz^{*2}} = A. \tag{8-3.13}$$

Because $C$ is not a function of $r^*$, its second derivative (denoted $A$) is not a function of $r^*$. Since there is no flow in the $z$ direction the only source of a dependence of $p^*$ on the $z$ coordinate is hydrostatic pressure. It follows then that the solution for $p^*$ is

$$p = -\rho g z + \text{fn}(r), \tag{8-3.14}$$

where fn($r$) is some unknown function only of $r$. If gravity is neglected the hydrostatic term vanishes, but regardless, it follows that $\partial p/\partial r$ is not a function of $z$, so the far left-hand term in Eq. 8-3.13 is not a function of $z^*$. We conclude, then, that since $A$ is not a function of either $r^*$ or $z^*$ it must be a constant. This permits us to integrate Eq. 8-3.13 immediately to yield

$$C = -\frac{A}{2}\left[1 - (z^*)^2\right]. \tag{8-3.15}$$

(In the course of the integration we find two constants which are determined by imposing the no-slip condition on $u^*$ at $z^* = \pm 1$.) The pressure distribution also follows from integration of Eq. 8-3.13, with the result

$$p^* = A \ln r^* + P^*. \tag{8-3.16}$$

For a boundary condition on pressure at the outer rim of the disk we set (see Eq. 8-3.5)

$$p^* = 0 \quad \text{at } r^* = R/H, \tag{8-3.17}$$

from which it follows that

$$P^* = -A \ln(R/H). \tag{8-3.18}$$

Note that the constant $A$ is still undetermined. It can be found by imposing conservation of mass on the solution. We write the volume flow rate in the form

$$Q = 2\int_0^H 2\pi\,(ru_r)\,dz = 2H^2U\int_0^1 2\pi C\,dz^*. \tag{8-3.19}$$

After introducing Eq. 8-3.15 for $C$ into the integral, and using Eq. 8-3.10 for $U$, we find

$$A = -\frac{3R_i}{H}. \tag{8-3.20}$$

The velocity field may be expressed in the form

$$u_r(r,z) = \left(\frac{3Q}{8\pi H}\right)\frac{1}{r}\left[1 - \left(\frac{z}{H}\right)^2\right]. \tag{8-3.21}$$

To find the pressure at the entrance to the disk region, i.e., $p$ at $r = R_i$, we use Eqs. 8-3.16, 18, and 20 and we find

$$p^* = \frac{pH}{\mu U} = -\frac{3R_i}{H}\ln\frac{r}{R}. \tag{8-3.22}$$

The overall pressure drop may be defined as

$$\Delta P = p(R_i) - p(R) = p(R_i). \tag{8-3.23}$$

From Eq. 8-3.22, after setting $r = R_i$, we may find the pressure drop as

$$\Delta P = -\frac{3\mu Q}{4\pi H^3}\ln\left(\frac{R_i}{R}\right). \tag{8-3.24}$$

We have achieved the desired result—a radial flow analog to Poiseuille's law for flow through a capillary. This yields the pressure drop–flow rate relationship for radial flow between parallel surfaces. We may now turn to the use of this result in development of a penetration rate model for this geometry. Figure 8-3.2 will aid the discussion.

In the *quasi-steady* model of this system we assume that Eq. 8-3.24 holds, with $R(t)$ substituted for $R$. We write the volume flow rate in terms of the rate of advancement of the interface:

$$Q = 4\pi H R(t)\,\frac{dR}{dt}. \tag{8-3.25}$$

We use the Young–Laplace equation for the capillary pressure. In this geometry, there are two radii of curvature. In the $rz$ plane the radius is

$$\frac{1}{R_{rz}} = -\frac{\cos\,\theta_d}{H}. \tag{8-3.26}$$

Note that the sign of the curvature is negative in this plane.

In the $r\theta$ plane the radius is

$$\frac{1}{R_{r\theta}} = \frac{1}{R}. \tag{8-3.27}$$

Thus we write the capillary pressure at the moving meniscus as

$$p_c = \sigma\left(-\frac{\cos\,\theta_d}{H} + \frac{1}{R}\right). \tag{8-3.28}$$

Then the driving pressure that goes into Eq. 8-3.24 is

$$\Delta P = \sigma\left(\frac{\cos\,\theta_d}{H} - \frac{1}{R}\right). \tag{8-3.29}$$

(We have reversed the boundary conditions on pressure from those of the derivation of Eq. 8-3.24. Now we take the pressure at the entrance to be $p = 0$, and the meniscus pressure is given by Eq. 8-3.28. In other words, the reservoir liquid at the entrance to the region is taken to be at zero pressure.) After combining Eqs. 8-3.24, 25 and 29 we find a differential equation for $R(t)$:

$$-\frac{3\mu\, R(t)\, dR/dt}{H^2}\ln\left(\frac{R_i}{R}\right) = \sigma\left(\frac{\cos\,\theta_d}{H} - \frac{1}{R}\right). \tag{8-3.30}$$

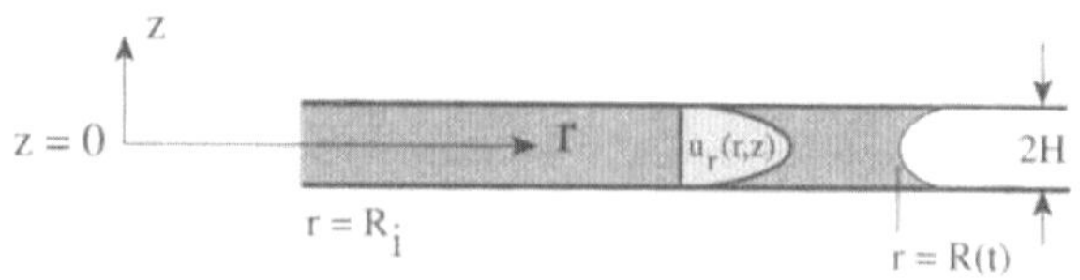


*FIGURE 8-3.2* Geometry of the radial penetration model.

This may be nondimensionalized to the form

$$\int_{X_i}^{X} \frac{\ln (X/X_i)\, X\, dX}{1 - [1/(X \cos \theta_d)]} = \frac{\sigma \cos \theta_d}{3\mu H} t \equiv T^*, \tag{8-3.31}$$

where

$$X = R(t)/H, \tag{8-3.32}$$

and $T^*$ is defined in Eq. 8-3.31. Numerical integration of the left-hand side yields $X(T^*)$. An analytical solution is possible with the approximation

$$\frac{1}{X \cos \theta_d} << 1. \tag{8-3.33}$$

The solution in that case is

$$\left(\frac{X}{X_i}\right)^2 \ln \left(\frac{X}{X_i}\right)^2 - \left[\left(\frac{X}{X_i}\right)^2 - 1\right] = \frac{4T^*}{X_i^2}. \tag{8-3.34}$$

Figure 8-3.3 shows the extent of penetration according to this expression.

Although Eq. 8-3.34 and Fig. 8-3.3 are limited by the approximation of Eq. 8-3.33, most of the penetration takes place after this inequality is satisfied, and so this model should be useful for predicting the extent of penetration. The primary modification that would be worth considering is the inclusion of a model for the dynamic contact angle, as per Eq. 8-2.9. There is no evidence available, however, to support the idea that Eq. 8-2.9, based on data for capillary penetration, would hold for penetration into a planar space.

## 8-4 PENETRATION FED BY A DROP OF FINITE VOLUME

In the models developed to this point, we have assumed that the liquid was continuously supplied to the entry region of the capillary or planar space. Under

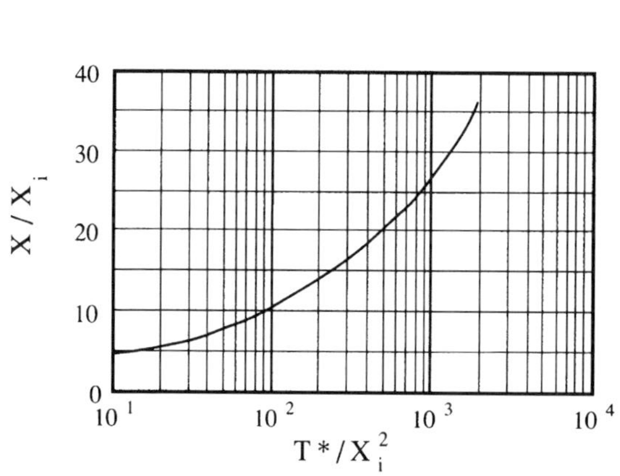


***FIGURE 8-3.3*** Radial penetration as a function of time, according to Eq. 8-3.34.

some circumstances of practical interest penetration is supported by a finite drop which, of course, disappears as penetration proceeds. Furthermore, the drop may be small enough at some stage that the positive pressure associated with its curvature aids the penetration. Figure 8-4.1 shows the geometry at the entrance to the "wick," whether it be a capillary or a planar (radial) space.

It is assumed that the drop is pinned to a circular opening of radius $r_c$. This could be the radius of a capillary, or the orifice for entry of the liquid into a parallel plate opening. We will need a relationship between the volume of the drop at any time, $V$, and its radius of curvature, $R_d$. We will assume that the drop is small enough that gravity does not distort it from a spherical shape. For a spherical geometry the following relationships are available:

$$\frac{R_d}{r_c} = \frac{1}{\pi}\left(\frac{r_c}{h}\right)^2 \phi + \frac{h}{3r_c} \tag{8-4.1a}$$

and

$$\frac{h}{r_c} = \left[\frac{3\phi}{\pi} + \sqrt{\left(\frac{3\phi}{\pi}\right)^2 + 1}\right]^{1/3} + \left[\frac{3\phi}{\pi} - \sqrt{\left(\frac{3\phi}{\pi}\right)^2 + 1}\right]^{1/3}, \tag{8-4.1b}$$

where a dimensionless drop volume is defined as

$$\phi \equiv \frac{V}{r_c^3}. \tag{8-4.2}$$

These equations follow strictly from geometrical considerations—they have nothing to do with capillary hydrodynamics or hydrostatics. Because of the algebraic complexity of these expressions it is useful to plot $R_d/r_c$ vs. $\phi$. This is shown in Fig. 8-4.2. For large values of $\phi$ a good approximation, one that follows exactly from the equations above, is

$$\frac{R_d}{r_c} = \left(\frac{3}{4\pi}\phi\right)^{1/3} = 0.620\phi^{1/3}. \tag{8-4.3}$$

We must now make two modifications to the dynamic equations for penetration that we considered earlier. First, the pressure drop is increased by the addition of

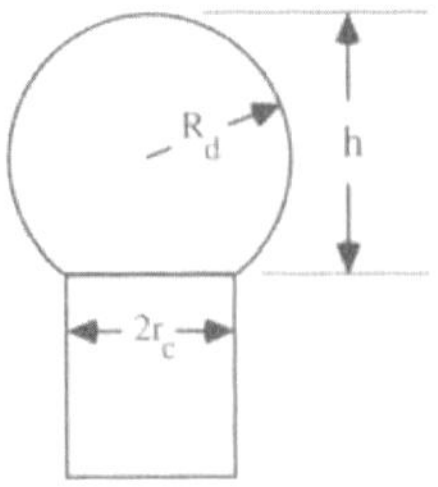


*FIGURE 8-4.1* Geometry of a drop on a circular opening.

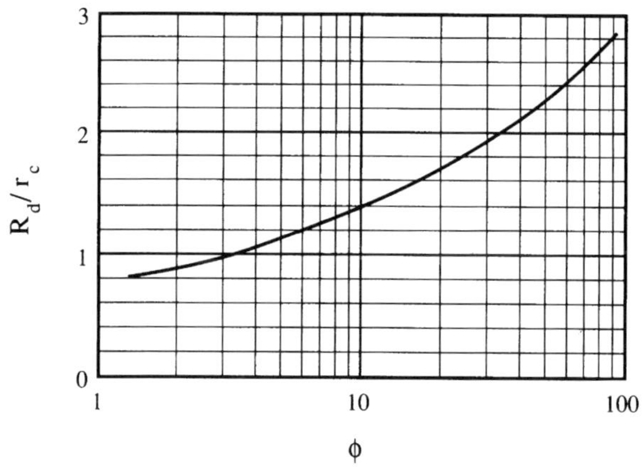


***FIGURE 8-4.2*** Radius of curvature of a spherical segment of a drop of volume $V$, pinned to an opening of radius $r_c$.

a term

$$\delta p = \frac{2\sigma}{R_d}. \tag{8-4.4}$$

In addition, the volume of the drop at any instant is related to the volume $V_p$ that has penetrated by

$$V = V_0 - V_p. \tag{8-4.5}$$

The volume that has penetrated may be written as either of the following for the two cases of interest to us:

$$V_p(t) = \pi R^2 x(t) \tag{8-4.6}$$

for *axial penetration through a capillary* of radius $R$, or

$$V_p(t) = 4\pi HR(t). \tag{8-4.7}$$

for *radial penetration through an axisymmetric planar region* of separation $2H$.

If we use the approximation of Eq. 8-4.3, for example, with Eq. 8-4.7, we have

$$R_d = 0.620\left[V_0 - 4\pi HR(t)\right]^{1/3}. \tag{8-4.8}$$

Combining this with Eq. 8-4.4 and 8-3.30, we obtain a modified dynamic equation for the extent of radial penetration $R(t)$ through a planar space:

$$-\frac{3\mu\, R(t)\, dR/dt}{H^2} \ln\left(\frac{R_i}{R}\right) = \sigma\left(\frac{\cos\theta_d}{H} - \frac{1}{R} + \frac{2}{0.620\left[V_0 - 4\pi HR\right]^{1/3}}\right). \tag{8-4.9}$$

Keep in mind that Eq. 8-4.3 is valid only for large drops ($\phi >> 1$ ), so Eq. 8-4.9 is not to be used in the later stages of penetration when $V_p$ approaches $V_0$. At longer times it would be necessary to use Eqs. 8-4.1a,b, and the resulting differential equation would be algebraically complex. It would still be possible, of course, to develop a numerical solution for the longer time period. For that matter, even at early times of penetration we cannot integrate Eq. 8-4.9 exactly, and a numerical method must be employed.

For the case of penetration of a finite drop into a *capillary* it is possible to obtain an analytical solution for short times and thereby to gain some insight into the importance of the pressure associated with a finite drop. A detailed discussion of this problem is presented by Marmur (1988). We present the simpler *gravity-free* analysis here. We begin with Eq. 8-2.4, drop the gravity term, and add the excess pressure (Eq. 8-4.4) due to the existence of a finite drop as a reservoir. The rate of penetration is then

$$\frac{dx}{dt} = \frac{R^2}{8\mu x}\left(\frac{2\sigma\cos\theta_d}{R} + \frac{2\sigma}{R_d}\right). \tag{8-4.10}$$

Keep in mind that in this geometry $R$ is the constant radius of the capillary—not the extent of penetration, which is denoted $x(t)$. Now we should use Eq. 8-4.6, 8-4.5, and 8-4.1a,b. The result is algebraically complex because of the forms of Eqs. 8-4.1a,b. However, if we look at short times, and invoke Eq. 8-4.3 as an approximation, and then use Eq. 8-4.6 in Eq. 8-4.5, we find

$$\frac{dx}{dt} = \frac{R^2}{8\mu x}\left(\frac{2\sigma\cos\theta_d}{R} + \frac{2\sigma}{0.620\left[V_0 - \pi R^2 x(t)\right]^{1/3}}\right). \tag{8-4.11}$$

Consistent with the short-time approximations already made, we take the extent of penetration small enough that we may write

$$\frac{1}{\left[V_0 - \pi R^2 x(t)\right]^{1/3}} \approx \frac{1 + \pi R^2 x(t)/3V_0}{V_0^{1/3}}. \tag{8-4.12}$$

This permits us to write Eq. 8-4.11 in the dimensionless form

$$X\frac{dX}{dT} = A + BX, \tag{8-4.13}$$

where $X$ and $T$ are defined in Section 8-2, Eqs. 8-2.6. The coefficients $A$ and $B$ are given by

$$A = 1 + \frac{1.612\Phi_0^{-1/3}}{\cos\theta_d},$$

$$B = \frac{0.422\Phi_0^{-4/3}}{\cos\theta_d} \tag{8-4.14}$$

and $\Phi_0$ is a dimensionless initial drop volume:

$$\Phi_0 \equiv \frac{V_0}{R^3}. \tag{8-4.15}$$

Equation 8-4.13 may be solved analytically, and the result is

$$\frac{X}{B} - \frac{A}{B^2} \ln\left(1 + \frac{B^2}{A}\frac{X}{B}\right) = T. \tag{8-4.16}$$

Hence, if we wish, we may plot this solution as curves of $X/B$ vs. $T$, with $A/B^2$ as a parameter. Alternatively, as in Fig. 8-4.3, we may compare the penetration rate of a drop of finite volume with that for an infinite reservoir of the liquid (based on the Washburn equation, Eq. 8-2.8 for the gravity-free case). In this case, $\Phi_0$ and $\theta_d$ are parameters. As expected, the extent of penetration ($X$) is greater, at a given time $T$, for drops of finite volume.

While an analytical solution is potentially useful, it is of limited use here because of the requirement implied by Eq. 8-4.12. To satisfy this approximation it is necessary that

$$\frac{\pi R^2 x}{V_0} << 1 \tag{8-4.17}$$

or

$$X << \frac{4\Phi_0}{\pi}. \tag{8-4.18}$$

For the case $\Phi_0 = 8$, for example, this means that the analytical solution is not accurate for $X$ larger than about $X = 0.8$. In view of the definition of $X$ (see Eqs. 8-2.6) we see that there is hardly any penetration at this point. It is only for large values of $\Phi_0$ that the analytical solution is useful to us. (A small circle appears on each line in Fig. 8-4.3 indicating the penetration limit for that value of $\Phi_0$.)

Again we must point out that the model presented here, as in the original work of Marmur (1988), fails to account for the presence of a dynamic contact angle. A numerical solution of the dynamic equation for this case, valid for

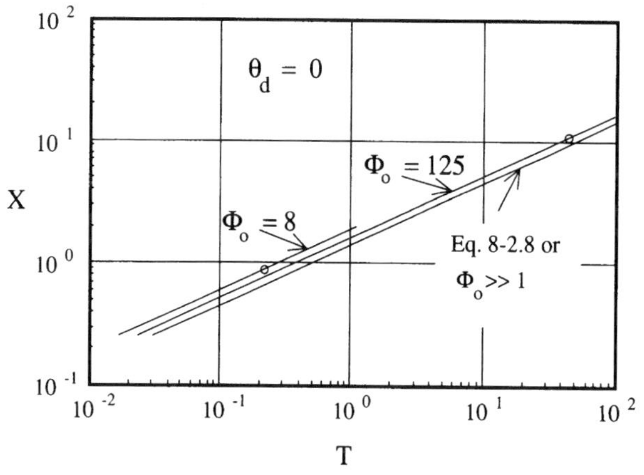


***FIGURE 8-4.3*** Effect of initial drop size on capillary penetration.

long-time penetration, would be possible, using a model such as Eq. 8-2.9 for the dynamic contact angle for axial flow through a capillary. In a similar manner it would be possible to carry out the dynamic model for penetration between parallel disks, but the first order of business would be the development of a correlation similar to Eq. 8-2.9, for the geometry of parallel planar penetration.

## 8-5 WICKING OF A DROP INTO A POROUS SURFACE

We turn now to an application of the simple models of capillary wicking to a complicated problem in practical fluid dynamics. Printing is a process in which ink must be transferred to a porous surface such as paper, followed by the penetration of the ink into the porous substrate. Capillarity plays a significant role in the dynamics of printing. The quality of the printed image depends on the rate of wicking of the ink into the substrate. Printing involves a sequence of complex phenomena that depend on the *structure* of the porous medium through which the liquid moves. By "structure" we mean factors such as porosity, the size and orientation of the fibers or particles that make up the paper, and possibly the surface chemistry of the paper itself. Based on the single capillary model presented earlier in this chapter, we can develop a simple dynamic model of a drop, initially sitting on a porous surface, wicking into the porous medium below the surface.

The volume flow rate of liquid across the wetted interface of the porous surface and the time required for the initial volume of the drop to disappear across that surface are important features of a model of this aspect of printing dynamics. In the development of a simple model we will regard the contact area of the drop to be independent of time—a circle of radius $R_d$. The interface between the drop and the paper will be regarded as one end of a bundle of capillaries, each individually of radius $R$. This surface is characterized by a fractional open area given by

$$\varepsilon = N_c \pi R^2 / \pi R_d^2, \tag{8-5.1}$$

where $N_c$ is the number of capillaries that open into the footprint of the drop. Figure 8-5.1 suggests this model of the porous substrate. Figure 8-5.2 (see color insert) is a sequence of photographic images of a drop as it wicks into a porous substrate.

The average velocity of flow into a single capillary follows from Eq. 8-2.7 as

$$u = \frac{dx}{dt} = \left(\frac{\sigma R \cos\theta_d}{8\mu t}\right)^{1/2}. \tag{8-5.2}$$

The volume flow rate $Q$ is just

$$\pi R_d^2 u \varepsilon = Q = -dV/dt. \tag{8-5.3}$$

After combining Eqs. 8-5.2 and 3 we have a differential equation for the volume of the drop at any time . The solution is easily found and written in the form

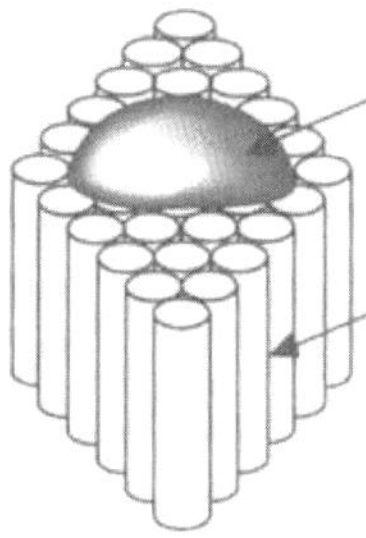


***FIGURE 8-5.1*** The capillary bundle model of a porous substrate. The drop has radius $R_d$, and the capillaries have radii $R$.

$$V(t) = V_0 - \pi R_d^2 \varepsilon \left( \frac{\sigma R \cos \theta_d}{2\mu} t \right)^{1/2}, \quad (8\text{-}5.4)$$

where $V_0$ is the initial volume of the drop.

If all of the drop were wicked into the porous medium, then the time for the wicking process to be complete would be obtained upon setting $V(t) = 0$ in Eq. 8-5.4. We do this and find, after some algebraic rearrangement,

$$\frac{\sigma t_w}{\mu L} = \left( \frac{2}{RL} \right) \left( \frac{V_0}{\pi R_d^2 \ \varepsilon \left( \cos \ \theta_d \right)^{1/2}} \right)^2, \quad (8\text{-}5.5)$$

where $t_w$ is the "wicking time." Note that we have introduced a length scale $L$ into the nondimensionalization, but we do not define $L$ at this point. Although $L$ appears in Eq. 8-5.5, it divides out of both sides of the equation.

According to this model, the wicking time should depend on the initial drop size, through the terms $V_0$ and $R_d$. The wick properties enter through the parameters $R$ and $\varepsilon$ that characterize the porous medium and through the contact angle $\theta_d$, none of which is known *a priori*. The model predicts that if the wicking time is nondimensionalized as

$$T_w = \frac{\sigma t_w}{\mu L} \quad (8\text{-}5.6)$$

we should find $T_w$ to vary as $V_0^2/R_d^4$ for a given porous medium. Since $V_0$ and $R_d$ are likely to be related to the same length scale $L$, we expect that $T_w$ varies as $L$, where $L$ is some characteristic length scale such as $V_0^{1/3}$.

Data on the wicking times of drops of varying viscosity and surface tension are available (Chien, 1994) to provide a test of Eq. 8-5.5. Since the parameters that characterize the wicking surface are not known, we can only test the proposition to a limited extent. Figure 8-5.3 shows these data, which indicate that the model is in approximate agreement with the observations.

In Eq. 8-5.5, $L$ is taken as $(3V_0/4\pi)^{1/3}$ in calculating the dimensionless time $T_w$. The initial drop radius $R_d$ depends on the drop volume $V_0$. In a separate set

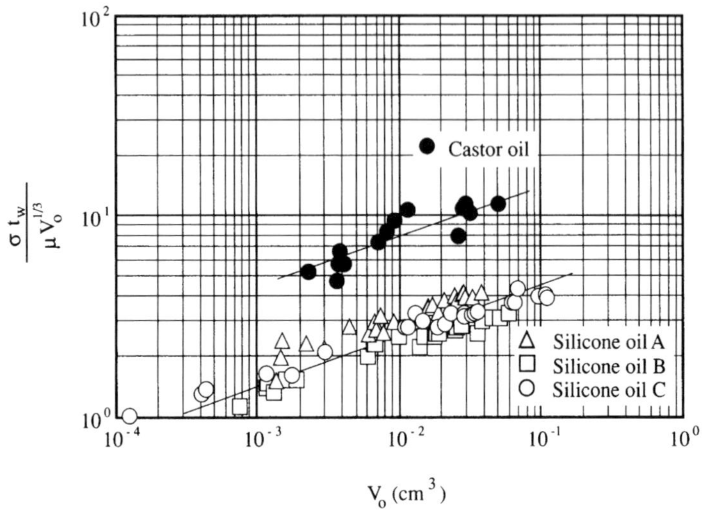


*FIGURE 8-5.3* Data on wicking time plotted as a test of Eq. 8-5.8. The medium is a porous polyethlyene sheet. Unpublished data of Chien and Middleman (1994).

of experiments (Mandrusov, 1990) the relationship was found, within 10%, to be given by

$$R_d = L = (3V_0/4\pi)^{1/3}. \tag{8-5.7}$$

As a consequence we may write Eq. 8-5.5 in the form

$$\frac{\sigma t_w}{\mu V_0^{1/3}} = \left(\frac{1.37}{R\varepsilon^2 \cos\theta_d}\right) V_0^{1/3}. \tag{8-5.8}$$

The parenthetical term contains the unknown properties of the porous medium, as well as the unknown wetting angle of the liquid on this medium. The data shown in Fig. 8-5.3 were obtained using a porous polyethylene sheet as the model for the wick or paper. The dependence of the wicking time on surface tension, viscosity, and drop size is approximately accounted for by the theory when plotted in this format. When, as in this case, a single porous medium is studied (so that $\varepsilon$ and $R$ are fixed), the simple model outlined here predicts that the data should fall on separate curves for different liquids, corresponding to different wetting angles $\theta_d$. Considering the range of viscosities (8.2 to 34 poise) and surface tensions (19 to 36 dyn/cm), the data may be regarded as clustering about a single line with the predicted slope (from Eq. 8-5.8) of one-third.

This model is subject to a number of approximations. Among them is the assumption that the contact radius remains constant during the wicking process, with the value $R_d$. This assumption may be relaxed upon introducing a model for capillary wicking *in the plane* of the porous surface. Thus, as in Fig. 8-5.4, we include capillaries that are radially oriented beneath the surface. A simple approach is to use Eq. 8-5.2 for $dr/dt$ and upon integration find

$$R_p = R_d + (\sigma \cos \theta_d Rt/2\mu)^{1/2}. \tag{8-5.9}$$

In a loose sense wicking now occurs through two regions: the initial contact footprint of radius $R_d$, and a growing annular ring of inner radius $R_d$ and outer radius $R_p(t)$ given by Eq. 8-5.9. At any radial position $r$ in the range $[R_d, R_p(t)]$ the upper surface is assumed to have been wet for a time (obtained by solving for $t$ from Eq. 8-5.9 ) given by

$$t' = 2\mu(r - R_d)^2/\sigma \cos \theta_d R. \tag{8-5.10}$$

Within the annular region wicking occurs in the direction *normal* to the surface at a velocity given by Eq. 8-5.2, except that account must be taken of the fact that at each position $r$ there is a different "age" of wetting. Hence we write

$$u = \frac{dx}{dt} = \left(\frac{\sigma \cos \theta_d R}{8\mu(t - t')}\right)^{1/2}. \tag{8-5.11}$$

The term $(t - t')$ accounts for this age of wetting. When $t'$ from Eq. 8-5.10 above is substituted into Eq. 8-5.11 we have an expression for the velocity of penetration at any radial position in the annular ring.

Now if we return to the equation for $dV/dt$ given earlier we may sum the wicking terms into the two regions. Equation 8-5.4 is then replaced by the following expression:

$$-[V - V_0] = \pi R_d^2 \varepsilon \left(\frac{\sigma \cos \theta_d R}{2\mu} t\right)^{1/2}$$

$$+ \int_0^t \left\{ 2\pi\varepsilon \left(\frac{\sigma \cos \theta_d R}{8\mu}\right)^{1/2} \int_{R_d}^{R_d + \left(\frac{\sigma \cos \theta_d R}{2\mu} t\right)^{1/2}} \left[t - \frac{2\mu}{\sigma \cos \theta_d R}(r - R_d)^2\right]^{-1/2} r\, dr \right\} dt. \tag{8-5.12}$$

***FIGURE 8-5.4*** Model of a porous medium with transverse and radial capillaries.

Despite the formidable-looking mathematical expression, the integrations may be performed analytically, and the drop volume may be written in the following dimensionless format:

$$V^* = V^*_0 - \frac{1}{4}\pi\varepsilon\beta^{1/2}T\left(\frac{\beta}{3}T^{1/2} + \frac{\pi}{2}\beta^{1/2} + 2T^{-1/2}\right). \tag{8-5.13}$$

In the above we have defined a dimensionless time $T$ as

$$T = 4\sigma \cos \theta_d \; t/\mu R. \tag{8-5.14}$$

and a dimensionless parameter $\beta$ as

$$\beta = R^2/2R_d^2. \tag{8-5.15}$$

A dimensionless drop volume has been defined as

$$V^* = V(t)/R_d^3. \tag{8-5.16}$$

The dimensionless wicking time again follows upon setting $V^* = 0$ in Eq. 8-5.13, with the result

$$(\beta T_w\Delta)^{1/2}\left[1 + \frac{\pi}{2}(\beta T_w\Delta)^{1/2} + \frac{2}{3}\beta T_w\Delta\right] = \frac{V_0}{\pi\varepsilon R_d^3}, \tag{8-5.17}$$

where $T_w$ is defined as in Eq.8-5.6, and

$$\Delta = L/R. \tag{8-5.18}$$

Figure 8-5.5 shows the result of Eq. 8-5.17, plotted in terms of the product $\beta\Delta T_w$, which occurs as a group in Eq. 8-5.17. We see that

$$\beta\Delta T_w \equiv \frac{R\sigma \cos \theta_d \, t_w}{2R_d^2 \mu}. \tag{8-5.19}$$

We may verify that for small values of the group $\beta\Delta T_w$, Eq. 8-5.19 is identical to Eq. 8-5.5.

If we have enough information regarding properties of the wicking medium, paper in the case of most printing processes, then we may estimate wicking times from Fig. 8-5.5 or 8-5.3. Clearly, one important research task would be to develop an experimental protocol for extracting the wicking parameters ($\varepsilon$, $\theta_d$, $R$) from observations of the wicking of a drop into various substrates. Of course, this discussion begs the question of whether the paper can in fact be characterized so simply by two parameters, $\varepsilon$ and $R$, and by a contact angle $\theta_d$ characteristic of the wetting of the liquid on the substrate.

In addition to the wicking time, the final footprint or spot radius of the drop is of importance. It is not difficult to write a simple model for spreading of the drop as it wicks. We begin by adopting Eq. 8-5.9 for spreading in the radial direction, in the plane of the porous surface, but rewritten slightly into the form

$$R_p(t) = R_d + (\sigma \cos \theta_d \, R't/2\mu)^{1/2} \tag{8-5.20}$$

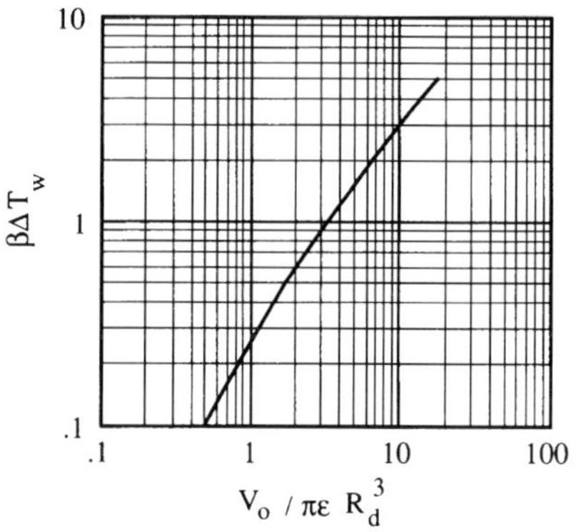


*FIGURE 8-5.5* Wicking time vs. drop size, based on Eq. 8-5.17.

to reflect the possibility that in the surface the capillary length scale $R'$ might be different from that in the bulk of the porous medium. In this equation, the footprint radius of the drop is denoted $R_p(t)$.

If we use now Eqs. 8-5.7 and 8 we find

$$\frac{R_p}{R_d} = 1 + \frac{4}{3\varepsilon}\left(\frac{R'}{R}\right)^{1/2}. \tag{8-5.21}$$

The ratio $R_p/R_d$ is sometimes called the "spread factor" (Bares, 1988). This result can be tested as well. All of the wicking experiments shown in Fig. 8-5.3 yielded "spot radii" $R_p$. This was measured from the wet image of the drop in the plane of the upper side of the porous surface, after wicking of the drop had ceased. In all cases, the porous sheet was thick enough that the drop did not wick through to the adjacent parallel surface. This corresponds to the so-called phase I wicking of Gillespie (1958, 1959). Since only a single porous material was used, we expect (from Eq. 8-5.21) that a single value of $R_p/R_d$ would be found, depending only on the medium parameters $\varepsilon$ and $R'/R$. Figure 8-5.6 shows data for $R_p/R_d$ for all the liquids studied on a single porous medium. For the several silicone oils studied, as well as for the castor oil, the data indicate a value of about $R_p/R_d = 2$ over a wide range of drop volumes. These results imply, from Eq. 8-5.21 and the fact that $\varepsilon < 1$, that $R'/R$ is much smaller than unity, and so the surface is indeed characterized by a very different capillary length scale than the interior of the medium.

It is interesting to note that studies by Bares (1988) using a Hewlett–Packard DeskJet printhead and commercial papers indicate that $R_p/R_d = 2$ for ink drop diameters in the range of 50 to 75 μm. Ink is much less viscous than the model liquids studied by us, and the commercial ink-jet drop sizes are well over an order of magnitude smaller. This suggests that our use of the larger, more viscous drops does provide a model that can be extrapolated into the commercial range of ink-jet printing. Bares also notes that by a judicious choice of paper it is possible to increase the ratio of $R_p/R_d$ by a factor of two.

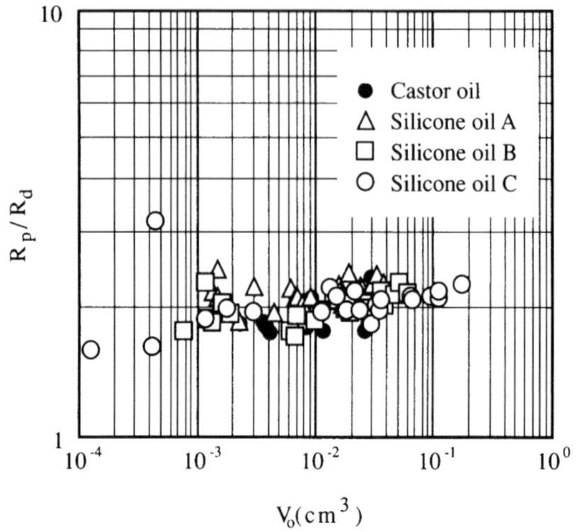


***FIGURE 8-5.6*** Data for the "spread factor." Unpublished data of Chien and Middleman (1994).

## Problems

P8-1 In Chapter 7 we presented a model, due to Tanner, that indicates that the dynamic contact angle depends upon the speed of the moving contact line according to Eq. 7-2.20. Show that for small dynamic contact angles, Eq. 8-2.9 is of the form, for the case $\theta_c = 0$,

$$\theta_d = A\,\text{Ca}^{1/4}. \tag{P8-1.1}$$

P8-2 It is difficult to measure a contact angle accurately, from a direct measurement of the slope of the meniscus at the capillary wall. Using Fig. 8-2.1, derive a simple relationship between the contact angle and the "depth" of the meniscus.

P8-3 Consider the case of water penetrating a horizontal capillary of radius 1 mm. What is the value of the parameter $\Omega$ for this case? Assume that the water completely wets the capillary. What is the value of the capillary number when the liquid has penetrated a distance equal to 10 capillary diameters? Give the same for 100 diameters. How far does the water penetrate before the velocity field is essentially quasi-steady?

P8-4 For the system described by Fig. 8-3.2, are there any constraints on the choice of $R_i$ relative to $H$? Use Eq. 8-3.30 as a guide to answering this question.

P8-5 According to the physical ideas that lead to Eq. 8-5.17 as a model for drop penetration, what would be the shape of the penetrating interface of the liquid as time proceeds?

P8-6 Fisher and Lark (1979) presented the data shown in Fig. P8-6.1, for the penetration of cyclohexane through Pyrex glass capillaries with radii less

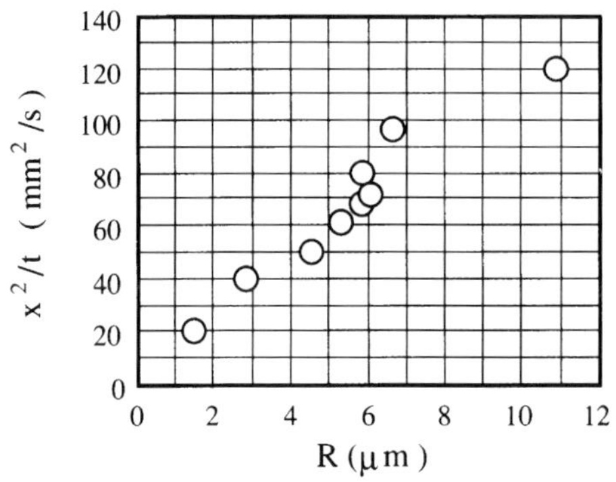


***FIGURE P8-6.1*** Data of Fisher and Lark (1979).

than 10 μm. Measured values of the time required for the liquid to traverse an axial distance $x$ are presented for various capillaries. At the temperature of the experiments the physical properties of the cyclohexane are

$$\mu = 0.00102 \text{ Pa-s}, \qquad \sigma = 0.0255 \text{ N/m}, \qquad \theta_c = 8°.$$

Test the ability of Eq. 8-2.8 to describe these data.

P8-7 Denesuk *et al.* (1993) presented data such as those shown in Fig. P8-7.1, for the wicking of a drop of a silicone liquid (polydimethylsiloxane) into a glass surface penetrated by parallel noninterconnected capillaries of

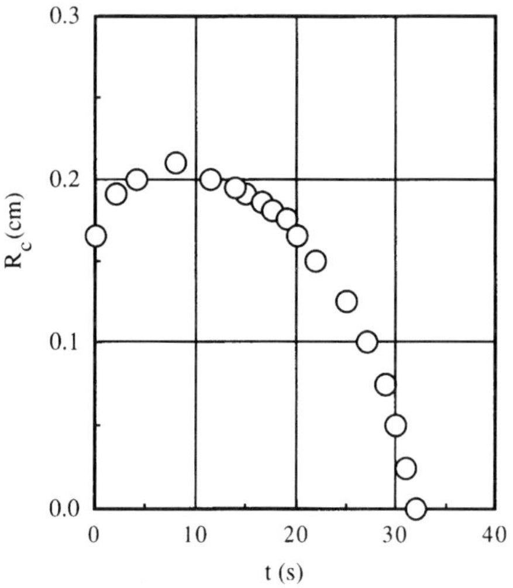


***FIGURE P8-7.1*** Representation of the data of Denesuk *et al.* (1993) for the contact radius of a drop wicking into a porous surface.

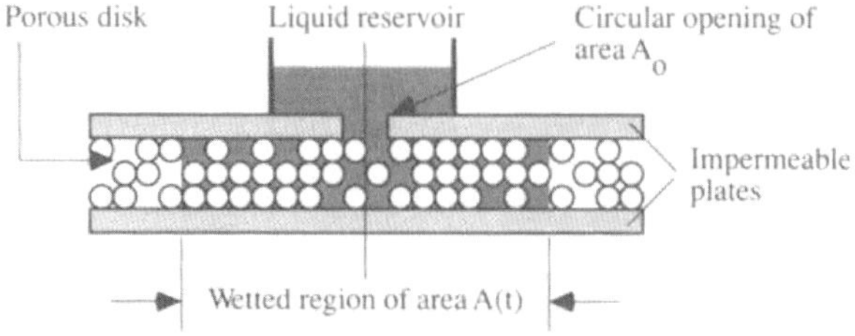


***FIGURE P8-8.1*** Radial spreading of a liquid in a porous disk.

radius $R = 5$ μm (see Fig. 8-5.1). The liquid completely wets this glass, so we can assume that $\theta = 0$.

Predict the wicking time from Fig. 8-5.5 and compare it to the data. Explain the result of this comparison, in view of the physics of their experiment vs. the assumed physics of the model presented in Section 8-5.

Their porous medium has $\varepsilon = 0.64$, and the liquid has a viscosity $\mu = 9.9$ poise and a surface tension $\sigma = 20$ dyn/cm. The initial drop mass was 0.01 g and the liquid density was 0.97 g/cm$^3$.

P8-8 Develop a mathematical model for the radial spreading of liquid in a disk-shaped porous medium, as defined by the sketch in Fig. P8-8.1. Determine any unknown parameters of the medium by fitting your model to the data of Borhan and Rungta (1993) shown in Fig. P8-8.2. Estimate the parameter values independently of the data, and compare them to your value(s) obtained from fitting the data with your model.

The following parameter values will be helpful: The liquid is a 50% ethylene glycol solution; $\sigma = 53.9$ dyn/cm; $\mu = 3.6$ cP. The porous disk

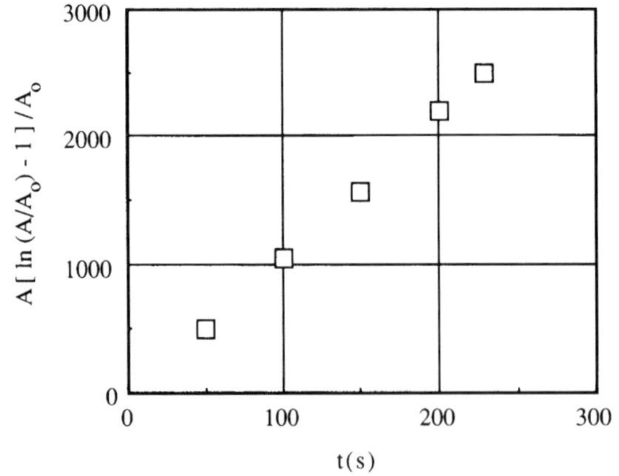


***FIGURE P8-8.2*** Data on radial spreading, from Borhan and Rungta (1993). $A_0$ is the initial footprint area of the spreading drop, and $A$ is the instantaneous value.

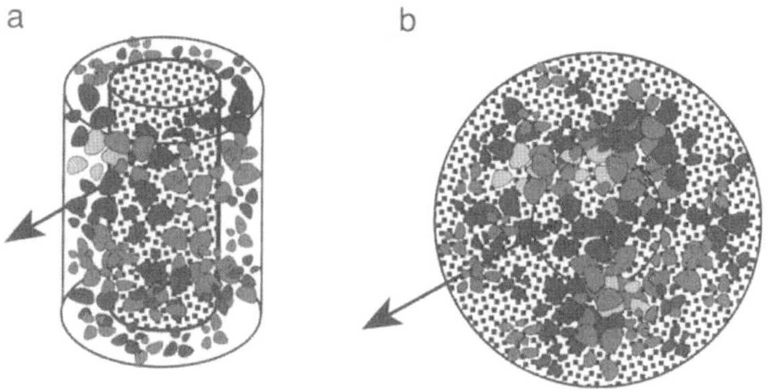


***FIGURE P8-9*** Sketch for a model of wicking through a porous cylinder (a) or a porous sphere (b).

is made up of glass beads with diameters in the range 590 to 840 μm. The beads are between parallel plates with an open spacing of approximately 23 mm.

P8-9 Equation 8-5.9 does not account for the radial geometry implied by the discussion. Equation 8-3.34 is a wicking model for a radial geometry but bounded between parallel planes. Develop a model for radial wicking through a porous medium, where the liquid is supplied from an axial reservoir, as in Fig. P8-9a. Do the same for wicking through a spherically symmetric porous medium, as in Fig. P8-9b. Do any simple scaling rules arise, of the form $R(t) = At^n$?

## *References*


Bares, S. J. (1988). Printing on plain paper with a thermal inkjet printer. *Hewlett-Packard J.* **39**(6), 39.

Batten, G. L., Jr. (1984). Liquid imbibition in capillaries and packed beds. *J. Coll. Int. Sci.* **102**, 513.

Borhan, A., and K. K. Rungta. (1993). An experimental study of the radial penetration of liquids in thin porous substrates. *J. Coll. Int. Sci.* **158**, 403.

Chien, S. (1994). Unpublished data, Univ. of California, San Diego.

Denesuk, M., G. L. Smith, B. J. J. Zelinski, N. J. Kreidl, and D. R. Uhlmann. (1993). Capillary penetration of liquid droplets into porous materials. *J. Coll. Int. Sci.* **158**, 114.

Fermigier, M., and P. Jenffer. (1991). An experimental investigation of the dynamic contact angle in liquid-liquid systems. *J. Coll. Int. Sci.* **146**, 226.

Fisher, L. R., and P. Lark. (1979). An experimental study of the Washburn equation for liquid flow in very fine capillaries. *J. Coll. Int. Sci.* **69**, 486.

Gillespie, T. (1958). The spreading of low vapor pressure liquids on paper. *J. Coll. Sci.* **13**, 32.

Gillespie, T. (1959). The capillary rise of a liquid in a vertical strip of filter paper. *J. Coll. Sci.* **14**, 123.

Good, R. J., and N.-J. Lin. (1976). Rate of penetration of a fluid into a porous body. *J. Coll. Int. Sci.* **54**, 52.

Hoffman, R. L. (1975). A study of the advancing interface. *J. Coll. Int. Sci.* **50**, 228.

Jeje, A. A. (1979). Rates of spontaneous movement of water in capillary tubes. *J. Coll. Int. Sci.* **69**, 420.

Joos, P., P. van Remoortere, and M. Bracke. (1990). The kinetics of wetting in a capillary. *J. Coll. Int. Sci.* **136**, 189.

Kissa, E. (1981). Capillary sorption in fibrous assemblies. *J. Coll. Int. Sci.* **83**, 265.

Letelier S., M. F., H. J. Leutheusser, and C. Rosas Z. (1979). Refined mathematical analysis of the capillary penetration problem. *J. Coll. Int. Sci.* **72**, 465.

Mandrusov, E. (1990). Unpublished data, Univ. of California, San Diego.

Marmur, A. (1988). Penetration of a small drop into a capillary. *J. Coll. Int. Sci.* **122**, 209.

Oliver, J. F., and R. C. Forsyth. (1989). A dynamic liquid sorption apparatus for studying interactions of microscopic drops *in situ* on porous surfaces. *Coll. Surf.* **43**, 295.

Quéré, D., and J.-M. Di Meglio. (1994). The meniscus on a fibre. *Adv. Coll. Int. Sci.* **48**, 141.

Rillaerts, E., and P. Joos. (1980). The dynamic contact angle. *Chem. Eng. Sci.* **35**, 883.

Szekely, J., A. W. Neumann, and Y. K. Chuang. (1971). The rate of capillary penetration and the applicability of the Washburn equation. *J. Coll. Int. Sci.* **35**, 273.

van Remoortere, P., and P. Joos. (1991). The kinetics of wetting. The motion of a three phase contact line in a capillary. *J. Coll. Int. Sci.* **141**, 348.

Washburn, E. W. (1921). The dynamics of capillary flow. *Phys. Rev.* **17**, 273.

# 9 Ultrathin Films: Disjoining Pressure

## 9-1 INTRODUCTION

In previous chapters we have assumed that the dynamics of the flows under consideration were determined by the bulk properties of the liquid(s) of interest. In particular, the dynamic equations included terms that represented the contributions of viscous effects, and surface tension entered generally through a boundary condition. Surface tension gave rise to a force or pressure associated with the curvature of the interface.

When the film thickness approaches molecular dimensions it would appear that additional forces, associated with long-range electrostatic interactions between the interfaces that bound the film, must be taken into account in order to explain observed behavior. We will describe some of these issues briefly in this chapter. These ideas are described more fully in several excellent references cited at the end of this chapter (de Gennes, 1985; Leger and Joanny, 1992; Teletzke *et al.*, 1987, 1988).

Figure 9-1.1 will aid our definition of terms relevant to the forces acting on and within ultrathin films.

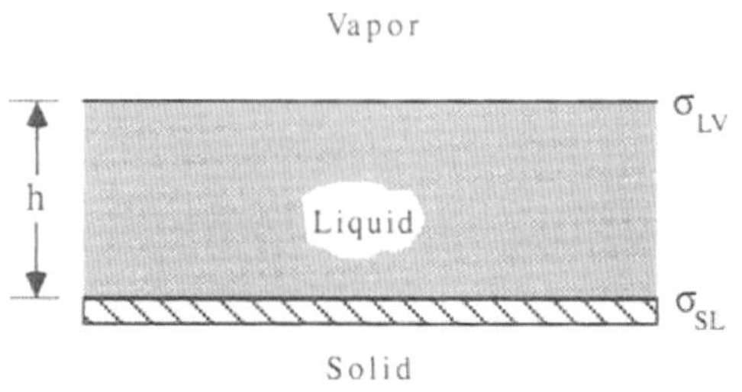


***FIGURE 9-1.1*** A liquid film separates a solid from a vapor.

When a *planar* liquid film intervenes between a solid and a vapor there is a free energy per unit of interfacial surface area, the magnitude of which is a sum of contributions from the solid-liquid surface tension $\sigma_{SL}$ and the liquid–vapor surface tension $\sigma_{LV}$. We may write this in the form

$$F = \sigma_{SL} + \sigma_{LV}. \quad (9\text{-}1.1)^1$$

Since the energies involved are those of the interfaces, the free energy $F$ does not depend upon the film thickness itself. This is a macroscopic view. It treats the boundaries of the liquid film as mathematical interfaces. In fact, these boundaries separate physical inter*phases* within which the liquid molecules exhibit variations in density and orientation as one moves from each interface into the liquid phase. If the film thickness $h$ is small enough, these interphases overlap in the sense that there can be molecular (steric) interactions between the molecules in these interphases. In addition, long-range intermolecular forces between molecules in the interphases exist. In such thin films, then, $F$ can be a function of the film thickness, and we can define a contribution to the free energy due to these molecular interactions by adding a term to Eq. 9-1.1:

$$F = \sigma_{SL} + \sigma_{LV} + \Phi(h). \quad (9\text{-}1.2)$$

The form taken by the function $\Phi$ depends upon the nature of the liquid, for example, whether the liquid is polar or nonpolar, and upon the surface as well. One of the commonest representations of the function $\Phi$ is

$$\Phi(h) = \frac{A}{12\pi h^2}, \quad (9\text{-}1.3)$$

in which $\Phi$ is said to arise from nonretarded London–van der Waals interactions. The constant $A$, or rather its negative, is the so-called Hamaker constant, and it depends upon the properties of the three phases shown in Fig. 9-1.1. This would be a typical model for a nonvolatile and nonpolar oil film on a metal surface. Water films are more complex because the possibility of overlapping of electrical double layers at the surfaces must be accounted for in thin aqueous films. In

[1]Note that since the interface is planar, and not curved, there is no pressure term (arising through the Young–Laplace equation) to contribute to the free energy.

addition, hydrogen bonding may occur between the water molecules and the solid surface. A detailed discussion of the various types of molecular interactions of interest is given in Teletzke *et al.* (1988).

## 9-2 EFFECT OF THE DISJOINING PRESSURE ON EQUILIBRIUM FILMS

In hydrodynamic theories the energy density function associated with the interfaces can be thought of as corresponding to (i.e., giving rise to) a pressure within the liquid phase. This pressure is called the "disjoining pressure," denoted $\Pi_d$ and calculated from

$$\Pi_d(h) \equiv -\frac{d\Phi(h)}{dh}. \qquad (9\text{-}2.1)$$

The disjoining pressure so defined is the free energy of a film of thickness $h$ relative to that of the bulk liquid on that surface, that is, relative to that of a very thick film such that $\Pi_d \to 0$. If $\Phi(h)$ is given by Eq. 9-1.3 we find the disjoining pressure to be[2]

$$\Pi_d(h) = \frac{A}{6\pi h^3}. \qquad (9\text{-}2.2)$$

For films that satisfy such an expression, and for $A > 0$, $\Pi_{d(h)}$ increases as the film thickness grows small. This corresponds (from Eq. 9-2.1) to an *increase* in the free energy, and so very thin films (i.e., of nearly atomic dimensions) of this liquid are energetically disfavored. Then we might expect that van der Waals forces could support a finite film thickness under conditions that classical continuum physics would not permit. Consider, for example, a horizontal planar solid surface connected to a reservoir of liquid, as depicted in Fig. 9-2.1.

Suppose that a small hydrostatic head $H$ is imposed. Then a small drop of liquid will form on the solid surface. We assume that the solid is "ideal," in the sense that it is perfectly smooth and chemically homogeneous. There are three interfaces across which an energy (per unit of contact area) exists. We denote these energies per area by $\sigma_{SV}$ for the bare solid–vapor interface, $\sigma_{SL}$ for the liquid–solid interface, and $\sigma_{LV}$ for the liquid-vapor interface. These $\sigma$s are just the surface tensions between each pair of phases. If the drop were in fact a flat (and, on a molecular scale, thick) film covering the entire surface, we would define a quantity $S$ as the free energy difference (per covered surface area) between the dry surface in contact with the vapor and the wetted surface:

[2]Reader beware! One finds commonly in the literature Eq. 9-2.2, but with a minus sign. Hence discussions of positive and negative disjoining pressures are meaningless and confusing unless the reader understands which sign convention is used. This is similar to the situation in fluid mechanics where Eq. 1-1.20 can be found with a negative sign, corresponding to a different convention on positive and negative pressure and stress.

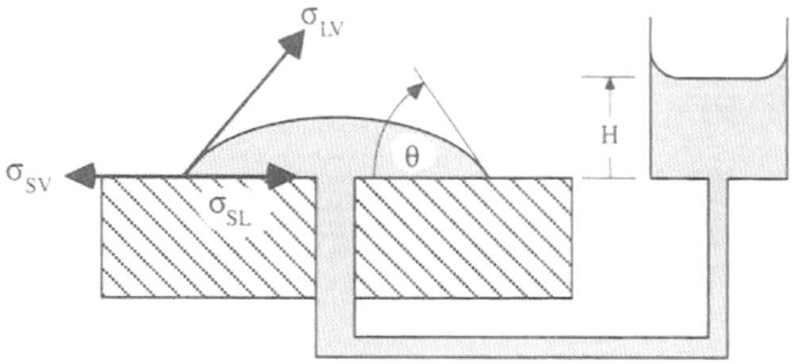


*FIGURE 9-2.1* Interfacial forces on a drop on a surface.

$$S \equiv \sigma_{SV} - (\sigma_{SL} + \sigma_{LV}). \qquad (9\text{-}2.3)$$

If $S > 0$ the wetted surface is energetically favored because it has a lower energy than the dry surface. Hence a liquid drop would spread spontaneously, without the need for a hydrostatic driving pressure, and the extent of spreading would be limited only by the total volume of the drop.[3] We would refer to this as "total wetting." As defined above, $S$ is called the "spreading parameter" for this liquid–solid–vapor combination. Note that if $S$ is negative then spreading would not be spontaneous, and a positive hydrostatic pressure would be required to partially wet the solid surface. The drop would adopt a shape that would satisfy a static force balance, subject to a drop volume constraint, and some equilibrium static contact angle $\theta$ would be observed. The case $S < 0$ is often referred to as "partial wetting." If the liquid does not wet the surface at all we would observe a drop that would be spherical in shape (if the drop is small enough that gravity does not distort it) and the equilibrium contact angle would be $\theta = 180°$. (Keep in mind that $\theta$ is the angle measured from *within* the liquid.)

In the definition of $S$ (Eq. 9-2.3) the term $\sigma_{LV}$ for the liquid–vapor interface is what we normally refer to as the surface tension of the liquid. (It is really the interfacial tension of the liquid with respect to its own vapor. Since the vapor phase will ordinarily exert no measurable influence on the interface, the term "surface tension," which implies that $\sigma_{LV}$ is a property only of the liquid, is a common usage.) Since $\sigma_{LV}$ is measurable by a number of techniques, some of which have been described in earlier chapters of this book, it is useful to relate the spreading parameter $S$ to $\sigma_{LV}$. This is achieved through a static force balance, which assumes that each surface tension corresponds to a force per unit of contact line length, acting tangentially to the contact line. In mechanical equilibrium then, the horizontal (i.e., in the plane of the solid) components of these forces balance and we find

$$\sigma_{SV} - \sigma_{SL} = \sigma_{LV} \cos \theta$$

[3]In principle a drop of finite volume could spread to the point that its thickness was infinitesimal. When the disjoining pressure is accounted for, as it must be when the film gets thin enough, the drop will stop at a finite, albeit small, thickness.

or

$$S = \sigma_{LV}\left(-1 + \cos\theta\right). \tag{9-2.4}$$

The contact angle is nonzero for the case $S < 0$, as expected. The first of this pair of equations, without $S$, is often referred to as Young's law.

Now suppose that a *drop* (rather than a film) of a wetting liquid is placed upon the solid. The free energy of the drop, relative to that of the dry solid, follows from Eqs. 9-1.2 and 9-2.3:

$$(F - \sigma_{SV})\mathsf{A} = \left[-S + \Phi(h)\right]\mathsf{A} = \left[-S + \Phi(h)\right]\left(\frac{V_d}{h}\right). \tag{9-2.5}$$

Note that in this equation we use $\mathsf{A}$ for the interfacial area, whereas in earlier equations we used the same letter (but a different typeface) $A$ for the Hamaker constant. We are assuming here that the drop, of constant volume $V_d$, spreads and at equilibrium is a thin "pancake" of uniform thickness $h$ except very near the three-phase contact line. The equilibrium thickness will be such as to minimize the free energy relative to the dry solid. Hence the constraint on $h$ is that the derivative of the left-hand side of Eq. 9-2.5, with respect to $h$, must vanish. When we perform the differentiation, and introduce the definition of the disjoining pressure, we find

$$S = h\,\Pi(h) + \Phi(h). \tag{9-2.6}$$

If we use Eq. 9-2.2 for the disjoining pressure we may find the equilibrium film thickness to be

$$h_{\text{equi}} = \sqrt{A/4\pi S}. \tag{9-2.7}$$

For the totally wetting case ($S > 0$) the drop of finite volume spreads spontaneously until a film thickness equal to $h_{\text{equi}}$ is achieved. Hence the disjoining pressure prevents a totally wetting liquid from totally wetting a surface.

If we examine typical values for $A$ and $S$, say $A = 10^{-20}$ J and $S = 0.001$ N/m, we find

$$h_{\text{equi}} = \sqrt{\frac{10^{-20}}{4\pi \times 0.001}} = 8.92 \times 10^{-10}\ \text{m} \approx 9\ \text{Å}. \tag{9-2.8}$$

Although it is difficult to do so, it is possible to obtain good estimates of the Hamaker constant for some liquids. Hence measurements of the equilibrium film thickness would permit one to estimate the spreading parameter for a particular liquid on a surface.

A similar analysis may be carried out for a film of axial length $L$ which forms an annular layer on a cylindrical solid surface of radius $b$ (such as a fiber). We modify Eq. 9-2.5 by including the contribution of the pressure $p_\sigma$ that arises from the curvature of the liquid–vapor interface, through the Young–Laplace equation. Thus we have

$$\begin{aligned}(F - \sigma_{SV})\mathsf{A} &= [-S + \Phi(h)]\,\mathsf{A} + p_\sigma V \\ &= [-S + \Phi(h)]\,\mathsf{A} + \frac{\sigma_{LV}}{b+h}\mathsf{A}h.\end{aligned} \tag{9-2.9}$$

On the assumption that the film thickness $h$ is very small compared to the fiber radius $b$, we write the interfacial area $\mathsf{A}$ and the volume $V$ as

$$\mathsf{A} = 2\pi bL \quad \text{and} \quad V = \mathsf{A}h. \tag{9-2.10}$$

With the constraint of constant volume we find the film thickness that minimizes the free energy to be

$$h_{\text{equi}} = \left[\frac{b^2 S}{2\sigma_{LV}}\left(1 - \sqrt{1 - \frac{A\sigma_{LV}}{\pi S^2 b^2}}\right)\right]^{1/2}. \tag{9-2.11}$$

Curvature of the interface reduces the equilibrium film thickness, relative to that for a planar film.

Equation 9-2.11 gives the equilibrium film thickness *when* the film spontaneously spreads into an annular sheath. *Whether* this happens depends upon the difference in free energies between the liquid-coated fiber and the dry fiber upon which a drop of the liquid has been placed. Figure 9-2.2 shows this case. By comparing the free energies of the drop and the sheath we can find a criterion for the spreading of a drop onto a cylinder. This is of interest in the consideration of the wetting of fibers, and Quéré *et al.* (1988) discuss this problem in some detail.

The free energy per unit of volume for a drop is

$$f_d = \frac{4\pi R^2 \sigma_{LV}}{\frac{4}{3}\pi R^3} = \frac{3\sigma_{LV}}{R}, \tag{9-2.12}$$

while for the annular sheath we find (dividing Eq. 9-2.9 by the volume and assuming that $b \gg h$)

$$f_s = \frac{[-S + \Phi(h)]}{h} + \frac{\sigma_{LV}}{b}. \tag{9-2.13}$$

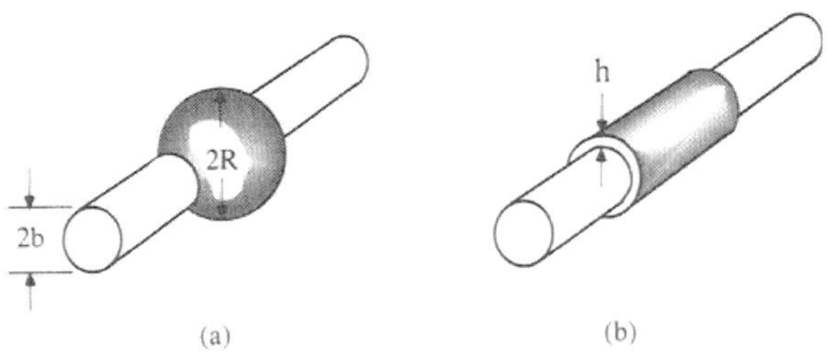


***FIGURE 9-2.2*** Schematic for comparison of the free energy of a drop and an annular sheath on a cylinder.

The sheath will be the preferred state if

$$f_s - f_d < 0 \tag{9-2.14}$$

and the critical condition for spreading is

$$f_s = f_d \tag{9-2.15}$$

or

$$\frac{\left[-S + \Phi(h)\right]}{h_c} + \frac{\sigma_{LV}}{b} = \frac{3\sigma_{LV}}{R}, \tag{9-2.16}$$

where $h_c$ is the sheath thickness at the critical condition. This film will be stable only if

$$\frac{\partial(f_s - f_d)}{\partial h} = 0. \tag{9-2.17}$$

This leads to the result that $h_c$ must also satisfy (see Eq. 9-2.6)

$$S = h_c\,\Pi(h_c) + \Phi(h_c). \tag{9-2.18}$$

If we adopt the nonretarded London–van der Waals model (Eq. 9-1.3) we may eliminate $S$ from Eqs. 9-2.16 and 18, and with the assumption that $R >> h_c$ we find

$$h_c = \left(\frac{Ab}{6\pi\sigma_{LV}}\right)^{1/3}. \tag{9-2.19}$$

This is the sheath thickness that occurs right at the critical condition for wetting. Note that the result does not depend on the size of the drop, $R$, so long as $R$ is large enough that $R >> h_c$ holds. Since volume must be conserved in the spreading of the drop, we conclude that the sheath length will increase or decrease according to the volume of the drop, but the thickness will satisfy Eq. 9-2.19, when the spreading parameter satisfies the relationship

$$S_c = \frac{3}{2}\sigma_{LV}\frac{h_c}{b} = \frac{3}{2}\left(\frac{A\,\sigma_{LV}^2}{6\pi b^2}\right)^{1/3}. \tag{9-2.20}$$

Equation 9-2.20 gives the threshold value of the spreading parameter for wetting of a fiber. Since the analysis assumes that the reservoir of liquid in contact with the fiber, the drop, has a large radius of curvature, we may take Eq. 9-2.20 as a criterion for wetting of a fiber in contact with a reservoir of zero curvature, as suggested in Fig. 9-2.3. In the absence of any additional forces, the liquid from the reservoir would coat the fiber with a film of thickness $h_c$, along the entire length of the fiber, and regardless of its orientation.

In a gravitational field, the thickness of an annular sheath on a vertical fiber would be position dependent. The analysis for this case is not difficult to carry out.

The free energy of a portion of the film with a (differential) vertical length $dz$

is given by

$$dF = \left(2\pi\left[b + h(z)\right]\sigma + 2\pi b\Phi(h) + \rho g z\pi\left\{\left[b + h(z)\right]^2 - b^2\right\}\right) dz. \qquad (9\text{-}2.21)$$

(We will now denote $\sigma_{LV}$ by $\sigma$ without subscripts.) If we assume that $h \ll b$ everywhere this becomes

$$dF = 2\pi\left\{\left[b + h(z)\right]\sigma + b\Phi(h) + \rho g z h\left(b + \frac{h}{2}\right)\right\} dz. \qquad (9\text{-}2.22)$$

The equilibrium film occurs when $dF/dz = 0$, from which we find that the disjoining pressure is balanced by terms corresponding to the hydrostatic pressure and the pressure arising from the curvature of the sheath:

$$\Pi(h) = \frac{\sigma}{b} + \rho g z. \qquad (9\text{-}2.23)$$

For the case of the nonretarded van der Waals potential this expression permits us to find the $h(z)$ profile as

$$h(z) = \left(\frac{A}{6\pi\rho g}\right)^{1/3} \left(\frac{\sigma}{\rho g b} + z\right)^{-1/3}. \qquad (9\text{-}2.24)$$

This result predicts that the sheath thickness gets very small as the height above the reservoir increases, but that $h(z)$ vanishes asymptotically in the limit of $z \rightarrow \infty$. In reality Eq. 9-2.24 fails as atomic dimensions are approached, and the wetting behavior becomes extremely complex. These issues are described in some detail by Quéré *et al.* (1988), as well as by Joanny and de Gennes (1986). We will not pursue them further here.

There is another technological setting in which the thickness of microscopic films plays an important role. In the protection of the surface of the magnetic medium layer on a data storage disk used in a hard disk drive, a thin layer of lubricant is applied. This overcoat film has a thickness that is typically of the order of 30 Å (Mate, 1992). Figure 9-2.4 shows some of the important features of such a system. In the common case of *particulate* magnetic media, the overcoat layer partially fills the porous surface and partially resides as a nonuniform film across the asperities of the media. It is this nonuniformity that is of concern, because of the possibility that in some regions the film thickness can be too small to provide adequate protection from impact with the recording head and slider.

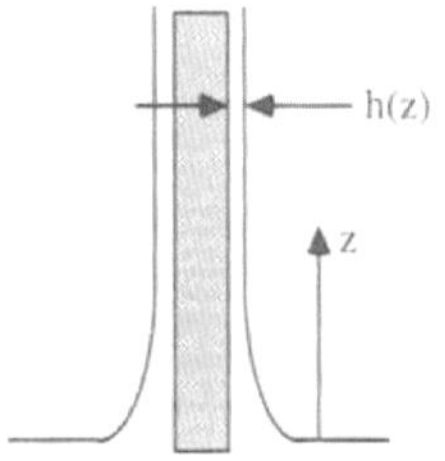


***FIGURE 9-2.3*** Liquid film on a vertical fiber.

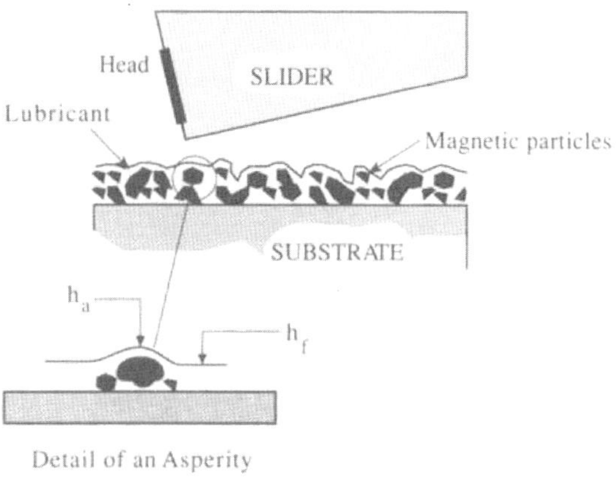


***FIGURE 9-2.4*** The slider–disk interface on a hard disk with particulate media.

We will idealize the asperity as a small sphere on a planar surface and imagine that the planar surface has a coating of lubricant of uniform thickness $h_f$. Our goal is to determine the equilibrium thickness of lubricant on the asperity that arises from the balance of surface tension and disjoining pressures. The static equilibrium condition is the equality of the pressures at the apex of the asperity with the disjoining pressure in the planar film:

$$\Pi_a - \frac{2\sigma}{R_a} = \Pi_f. \tag{9-2.25}$$

Using Eq. 9-2.2 for the disjoining pressure, we find

$$\frac{A}{6\pi h_a^3} - \frac{2\sigma}{R_a} = \frac{A}{6\pi h_f^3}. \tag{9-2.26}$$

Then the lubricant thickness at the apex of the asperity is

$$h_a = \left(\frac{1}{h_f^3} + \frac{12\pi\,\sigma}{AR_a}\right)^{-1/3}. \tag{9-2.27}$$

Figure 9-2.5 shows that the lubricant layer becomes quite thin at the top of asperities in the case of small particulate media. To protect (lubricate) the smallest asperities arising from the smallest particles of medium, it would be necessary to use a thicker lubricant layer than might otherwise be called for. However, for submicron asperities, the lubricant thickness becomes nearly independent of the thickness $h_f$ of the lubricant film in the neighborhood of the asperity and in fact is predicted by this model to depend only upon the asperity size and the physical properties $A$ and $\sigma$:

$$h_a \approx \left(\frac{AR_a}{12\pi\,\sigma}\right)^{1/3}. \tag{9-2.28}$$

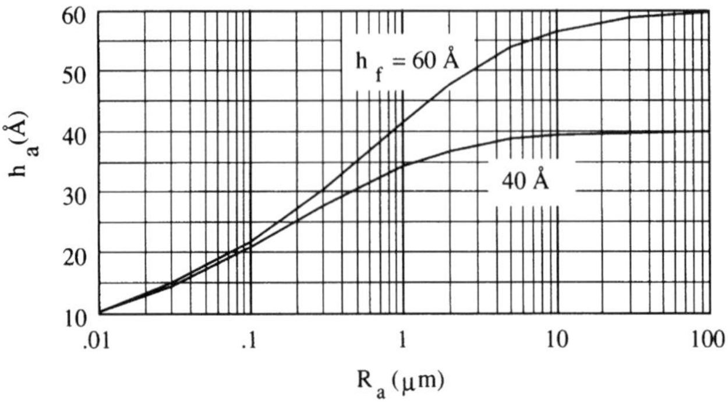


***FIGURE 9-2.5*** Lubricant thickness on asperities, as a function of particle size.

Hence the further addition of lubricant would not effectively lubricate the smallest particles, if the physical properties are such as to correspond to the case illustrated here. A Hamaker constant corresponding to $A = 10^{-19}$ J is argued by the studies of Mate and Novotny (1991) (but see Problem P9-8 at the end of this chapter), and Bhushan and Blackman (1991) present evidence for asperity radii in the range $0.1 < R_a < 1$ mm.

## 9-3 DEWETTING

Earlier the comment was made that the effect of a disjoining pressure is to thicken a film. If this is so, then any tendency toward local thinning of a film would give rise to forces that would restore the film to its original thickness. This sounds like a statement that disjoining pressure stabilizes a film, and this is true. We must be careful with the language, however. For some liquid–solid systems the disjoining pressure function has the form suggested in Fig. 9-3.1b. Recall that $\Pi$ is a chemical potential or *free energy* associated with intermolecular surface forces. In the case $d\Pi/dh < 0$ (Eq.9-2.2 with $A > 0$) any decrease in film thickness leads to an increase in free energy. This is not thermodynamically favored. Hence any mechanical event that leads to a local thinning of the film gives rise to a force that returns the system (the film thickness) to its original value. This is a stable film. In case (b) of Fig. 9-3.1 just the opposite occurs: a local thinning of the film reduces the free energy and the system moves toward a thinner film.

Let's consider now the fate of a small axisymmetric disturbance to the surface of a thin film. Figure 9-3.2 shows the geometry. The evolution of the film thickness will be determined from a solution to Eq. 7-2.11, but with the addition of a term for disjoining pressure. Then we must solve a dynamic equation of the form

$$\frac{\partial h}{\partial t} + \frac{1}{r}\frac{\partial}{\partial r}\left[\frac{\sigma}{3\mu} rh^3 \frac{\partial}{\partial r}\left(\frac{\partial^2 h}{\partial r^2} + \frac{1}{r}\frac{\partial h}{\partial r} + \frac{\Pi}{\sigma}\right)\right] = 0. \tag{9-3.1}$$

We will take Eq. 9-2.2 for $\Pi(h)$. We assume that the film is so thin that gravity has no effect on it. When we define a set of dimensionless variables as

$$\tau \equiv \frac{\sigma}{3\mu h_0} t, \qquad h^* \equiv \frac{h}{h_0}, \qquad r^* \equiv \frac{r}{h_0}. \tag{9-3.2}$$

the evolution equation takes the form

$$\frac{\partial h^*}{\partial \tau} + \frac{1}{r^*}\frac{\partial}{\partial r^*}\left[r^* h^{*3} \frac{\partial}{\partial r^*}\left(\frac{\partial^2 h^*}{\partial r^{*2}} + \frac{1}{r^*}\frac{\partial h^*}{\partial r^*} + \frac{\mathsf{Di}}{h^{*3}}\right)\right] = 0, \tag{9-3.3}$$

where a new dimensionless group containing the Hamaker coefficient appears:

$$\mathsf{Di} = \frac{A}{6\pi\sigma h_0^2}. \tag{9-3.4}$$

We will not attempt to solve Eq. 9-3.3. Instead, we want to examine the fate of a slight depression in the film. In order to do this without a great deal of algebraic mess we have chosen the depression shown in Fig. 9-3.2, because that shape will permit us to make an approximation that considerably simplifies the algebra that ensues. The approximation is based on the idea that if the depression in the film is centered about a circumferential perimeter whose radius is large compared to the radial width of the depression, then we may neglect the effects of curvature in the angular direction. In effect, we rewrite Eq. 9-3.3 in Cartesian coordinates, with the result

$$\frac{\partial h^*}{\partial \tau} + \frac{\partial}{\partial r^*}\left[h^{*3} \frac{\partial}{\partial r^*}\left(\frac{\partial^2 h^*}{\partial r^{*2}} + \frac{\mathsf{Di}}{h^{*3}}\right)\right] = 0. \tag{9-3.5}$$

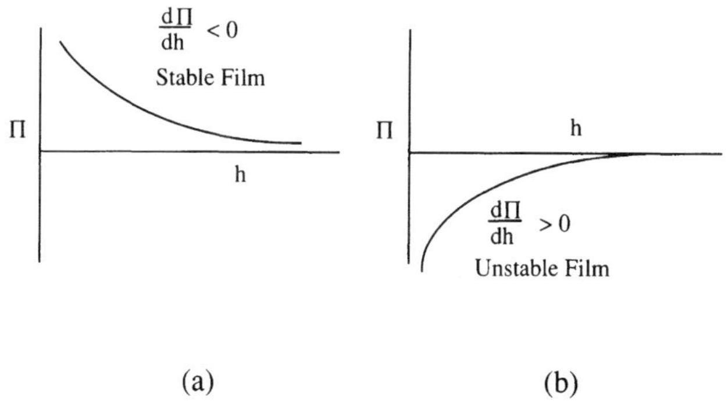


***FIGURE 9-3.1*** Disjoining pressure curves for a stable film (a) and for an unstable film (b).

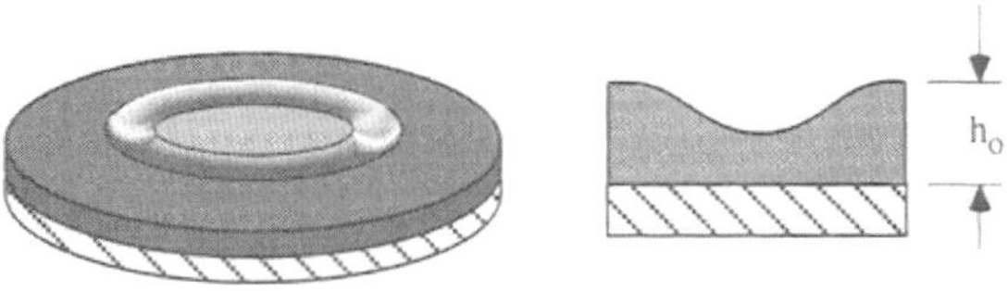


*FIGURE 9-3.2* An annular depression in a thin liquid film.

This is still a nonlinear partial differential equation, and no analytical solution is possible. However, we may analyze the behavior of small disturbances to the film profile by a simple perturbation method, following the approach of Kheshgi and Scriven (1991) to a very similar problem. (They include gravity in their analysis, and their nondimensionalization is therefore different from ours.)

We carry out a linear stability analysis of the film profile by expanding the height of the film, measured from the solid surface, in a series of functions such that

$$h^* = 1 + \varepsilon h_1^* + O(\varepsilon^2). \tag{9-3.6}$$

This is substituted into Eq. 9-3.5, and only linear terms in $\varepsilon$ are retained. The result is a *linear* partial differential equation of the form

$$\frac{\partial h_1^*}{\partial \tau} = \mathsf{Di}\,\frac{\partial^2 h_1^*}{\partial r^{*2}} - \frac{\partial^4 h_1^*}{\partial r^{*4}}. \tag{9-3.7}$$

Now we take the disturbance function to have the form

$$h^* = 1 + \varepsilon e^{\beta\tau} \sin \alpha r^*. \tag{9-3.8}$$

Here $\beta$ is a growth rate of the surface disturbance. We want to determine conditions under which $\beta$ is positive, to define parametric regions of instability, i.e., spontaneous film thinning subsequent to a small disturbance. We expect that $\beta$ will depend upon the wave number of the disturbance, $\alpha$, and upon the dimensionless number Di. To proceed, we simply introduce Eq. 9-3.8 into Eq. 9-3.7, and the result is an algebraic equation for $\beta$:

$$\beta = -\mathsf{Di}\,\alpha^2 - \alpha^4. \tag{9-3.9}$$

Figure 9-3.3 shows $\beta(\alpha)$ with Di as a parameter. We see that a small-amplitude disturbance is stable ($\beta < 0$) so long as the disjoining parameter is positive, i.e., $A > 0$. When Di (or $A$) is negative there is a wave number region of instability, defined by a critical wave number that satisfies

$$\alpha_{\text{crit}} = \sqrt{-\mathsf{Di}}. \tag{9-3.10}$$

For negative values of Di a disturbance is an unstable one if the wave number is less than $\alpha_{\text{crit}}$ defined above.

The mathematical description of an axisymmetric depression is much more complex than that of the two-dimensional (Cartesian) depression, even for the

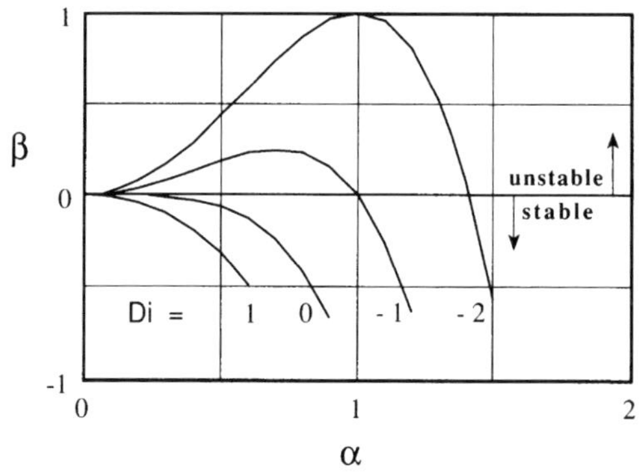


*FIGURE 9-3.3* Growth rate parameter β as a function of the wave number α of a small disturbance.

linearized stability analysis (see Problem P9-4). However, the axisymmetric case is important in the consideration of the rupture of a thin surface film and the subsequent dewetting of the underlying substrate. The mechanism of rupture is very complex, and although there are experimental and theoretical studies of film rupture (e.g., Buevich and Lipkina, 1978; and Foister, 1987), an adequate model of the dynamics of rupture of a liquid film on a rigid surface, based on hydrodynamic principles, is not yet available. We will discuss film rupture again in Chapter 10, in the context of its role in drop and bubble coalescence, and offer a more complete bibliography for further study.

## *Problems*

P9-1 Derive Eq. 9-2.11.

P9-2 Derive Eqs. 9-2.19 and 20.

P9-3 Derive Eq. 9-3.7.

P9-4 Instead of the film disturbance geometry shown in Fig. 9-3.2, suppose the disturbance is a "dimple" centered at $r = 0$. We can no longer use the Cartesian approximation (Eq. 9-3.5) and must revert to Eq. 9-3.3. What equation replaces Eq. 9-3.7? Can we use Eq. 9-3.8 for this case?

P9-5 Show that Eq. 9-3.9 leads to a maximum growth rate parameter of

$$\beta_{max} = \frac{Di^2}{4}. \qquad \text{(P9-5.1)}$$

What is the wave number of the most rapidly growing disturbance?

P9-6 A small spherical particle sits on a planar surface, as indicated in Fig. P9-6.1. Find the radius of curvature $R_m$ of the meniscus, and the contact radius $R_c$, in terms of physical properties $A$ and $\sigma$ and the (assumed uniform) film thickness $h$ far from the point of contact. Show that there is a

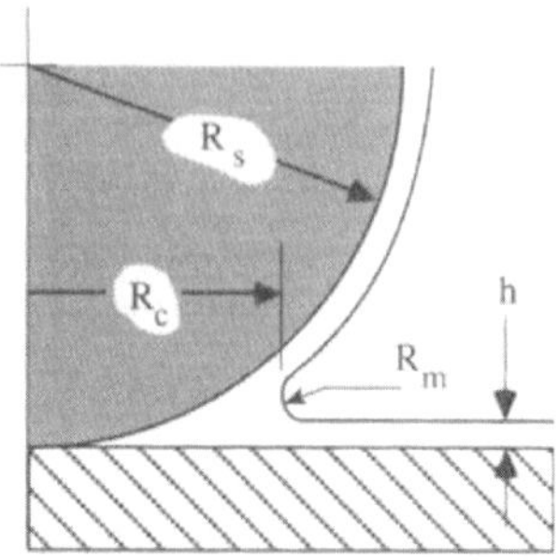


***FIGURE P9-6.1*** A liquid-coated particle on a planar surface.

force holding the sphere to the plane, arising from the (negative) capillary pressure in the meniscus, and that this force is given approximately by

$$F_m = 4\pi R_s \sigma \tag{P9-6.1}$$

For solid particles of densities of the order of 2000 to 5000 kg/m$^3$, what is the ratio of particle weight to meniscus force?

P9-7 The data shown in Fig. P9-7.1 were obtained by Aronson *et al.* (1978) for aqueous films on silica. Force Eq. 9-2.2 to fit these data, and give the resulting value of the Hamaker constant. Comment on the degree to which Eq. 9-2.2 describes these data.

P9-8 The data shown in Fig. P9-8.1 were obtained by Mate and Novotny (1991) for films of a perfluoropolyether polymer of molecular weight 15,000 on smooth silicon dioxide surfaces. Force Eq. 9-2.2 to fit these data, and give the resulting value of the Hamaker constant. Comment on the degree to which Eq. 9-2.2 describes these data.

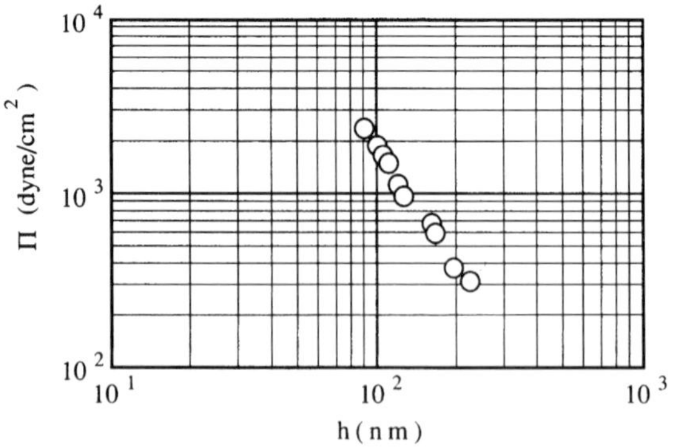


***FIGURE P9-7.1*** Disjoining pressure data of Aronson *et al.* (1978).

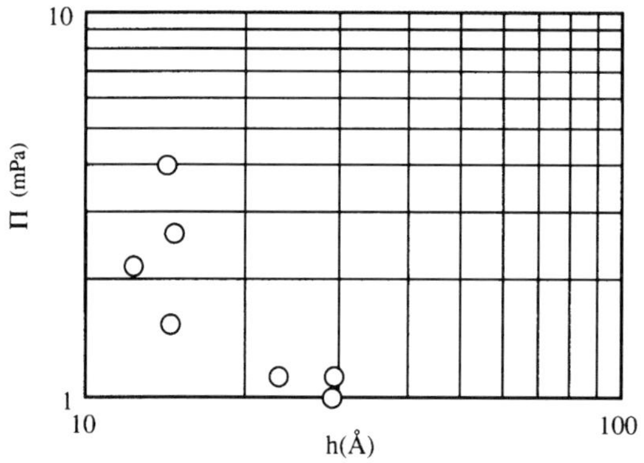


***FIGURE P9-8.1*** Disjoining pressure data of Mate and Novotny (1991).

## *References*


Aronson, M. P., M. F. Petko, and H. M. Princen. (1978). On the stability of aqueous films between oil and silica. *J. Coll. Int. Sci.* **65**, 296.

Bhushan, B., and G. S. Blackman. (1991). Atomic force microscopy of magnetic rigid disks and sliders and its applications to tribology. *ASME Trans. J. Tribol.* **113**, 452.

Brochard, F. (1986). Spreading of liquid drops on thin cylinders: The 'Manchon/droplet' transition. *J. Chem. Phys.* **84**, 4664.

Buevich, Yu. A., and E. Kh. Lipkina. (1978). Disruption of thin liquid films. *Koll. Zh.* **40**, 201.

Foister, R. T. (1987). Three-phase boundary expansion in thin liquid films. *J. Coll. Int. Sci.* **116**, 109.

de Gennes, P. G. (1985). Wetting, statics and dynamics. *Rev. Mod. Phys.* **57**, 827.

Joanny, J. F. and P.-G. de Gennes. (1986). Upward creep of a wetting fluid. A scaling analysis. *J. Phys.* **47**, 121.

Kheshgi, H. S., and L. E. Scriven. (1991). Dewetting. Nucleation and growth of dry regions. *Chem. Eng. Sci.* **46**, 519.

Leger, L., and J. F. Joanny. (1992). Liquid spreading. *Rep. Prog. Phys.* **55**, 431.

Mate, C. M. (1992). Application of disjoining and capillary pressure to liquid lubricant films in magnetic recording. *J. Appl. Phys.* **72**, 3084.

Mate, C. M., and V. J. Novotny. (1991). Molecular conformation and disjoining pressure of polymeric liquid films. *J. Chem. Phys.* **94**, 8420.

Quéré, D., and J.-M. Di Meglio. (1994). The meniscus on a fibre. *Adv. Coll. Int. Sci.* **48**, 141.

Quéré, D., J.-M. Di Meglio and F. Brochard-Wyart. (1988). Wetting of fibers. Theory and experiments. *Rev. Phys. Appl.* **23**, 1023.

Teletzke, G. F., H. T. Davis, and L. E. Scriven. (1987). How liquids spread on solids. *Chem. Eng. Comm.* **55**, 41.

Teletzke, G. F., H. T. Davis, and L. E. Scriven. (1988). Wetting hydrodynamics. *Rev. Phys. Appl.* **23**, 989.

# 10 The Interaction of a Deformable Drop with a Surface

In *Chapter 6* we examined the impact of a liquid drop with a solid surface, when the drop falls freely to the surface. In this chapter we examine two related phenomena. One is the approach toward impact of a drop when a second viscous liquid intervenes between the approaching drop and a rigid planar surface. The drop decelerates as it approaches the surface, and viscous stresses have the capacity to deform the drop, against the restraining forces at the interface (arising from interfacial tension) that act to keep the drop spherical. The second phenomenon arises in the case of a drop, or even a *solid* spherical particle, *plunging across* a liquid interface. Here the primary issue is the dynamics of the "target" interface during and subsequent to impact, which, as we will see, can be quite complex.

The problem of a drop approaching a solid surface is central to the issue of wetting of a surface by the liquid drop. In addition, the coalescence of a drop with another drop, or of two bubbles in a liquid, is determined in part by the dynamics of approach of the drops or bubbles. Under some conditions it is possible for a drop to approach to within a very small distance of another surface and then to remain in a stable steady state separated from the surface by an ultrathin film. Under other conditions this intervening film may rupture spontaneously and permit wetting of the surface by the drop. This is a very interesting and challenging area of fluid dynamics, and suffice it to say that the formulation of models that reflect the essential physics is a difficult task. After these models are formulated, it is usually necessary to seek solutions for the dynamics of the interface by numerical methods. However, some analytical methods provide simple but realistic models of at least some of the essential features of the physics, and we shall examine some of the successes of analytical approaches, while describing as well some numerical simulations.

## *10-1 THE APPROACH OF A SOLID SPHERE TO A RIGID PLANE*

We will begin with a conceptually simpler problem than the ones described above. We consider first the slow approach of a *rigid solid* sphere toward a rigid planar surface. Hence there are no interfacial dynamics to consider, and only the flow squeezed out by the approaching surfaces is at issue. An analytical solution[1] to this problem was given by Brenner (1961), who wrote the approach velocity of the sphere as

$$U = U_s / \lambda, \tag{10-1.1}$$

where $U_s$ is the Stokes' law velocity:

$$U_s = \frac{F}{6\pi\mu R} \tag{10-1.2}$$

and $F$ is the force acting on the sphere of radius $R$. Stokes' law is valid only for small Reynolds numbers, say

$$\mathrm{Re} = \frac{2U_s R\rho}{\mu} \leq 1, \tag{10-1.3}$$

where $\rho$ is the fluid-phase density. Brenner found the factor $\lambda$ by which the Stokes velocity must be reduced in the form of an infinite series:

$$\lambda = \frac{4}{3} \sinh\alpha \sum_{n=1}^{\infty} \frac{n(n+1)}{(2n-1)(2n+3)} \times \left[ \frac{2\sinh(2n+1)\alpha + (2n+1)\sinh 2\alpha}{4\sinh^2\left(n+\frac{1}{2}\right)\alpha - (2n+1)^2 \sinh^2\alpha} - 1 \right], \tag{10-1.4}$$

[1]Brenner writes the solution for the stream function in a bipolar coordinate system, in the form of an infinite series of Legendre polynomials. Only the final result for the function $\lambda$ is displayed here.

where

$$\alpha = \cosh^{-1}\frac{H}{R} \tag{10-1.5}$$

and $H$ is the distance of the center of the sphere from the planar surface.

Two simple asymptotic results are available for the calculation of $\lambda$ when the sphere is either very close, or very far from the surface. They are

$$\lambda = \left(\frac{H}{R} - 1\right)^{-1} \quad \text{for} \quad \frac{H}{R} - 1 << 1 \tag{10-1.6}$$

and

$$\lambda = 1 + \frac{9R}{8H} \quad \text{for} \quad \frac{H}{R} >> 1. \tag{10-1.7}$$

Figure 10-1.1 shows the exact solution, and the two asymptotic results. Experimental verification of Brenner's result is given by MacKay *et al.* (1963).

As $H/R$ approaches unity the force resisting the motion of the sphere is dominated by the viscous flow being squeezed out from the approaching surfaces. The sphere slows down and approaches the lower plane asymptotically. Figure 10-1.2 shows an approximate solution for the approach of the sphere (see Problem P10-1) that reflects this behavior. The time scale is dimensionless and is defined by

$$T \equiv \frac{F}{6\pi\mu R^2} t. \tag{10-1.8}$$

The force $F$ is taken as a constant, as it would be if the sphere were falling under the action of gravity.

Although the case of the approach of a solid sphere to a rigid planar surface is of considerable interest, we will not deal with it further here. Instead we turn immediately to the case of the approach of a *deformable* drop.

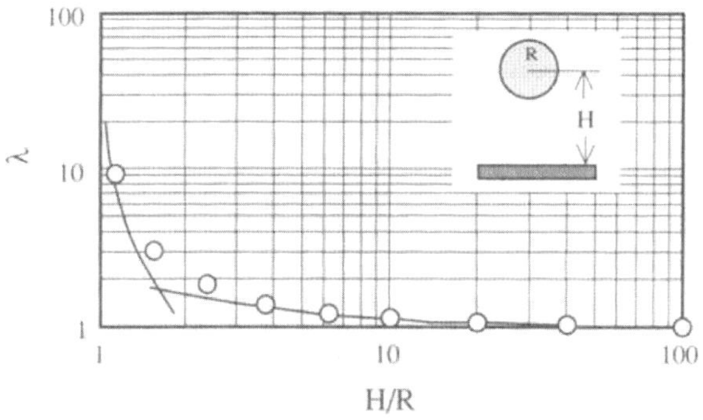


***FIGURE 10-1.1*** The correction factor $\lambda$ of Eq. 10-1.4 (O) and the asymptotes from Eqs. 10-1.6 and 7.

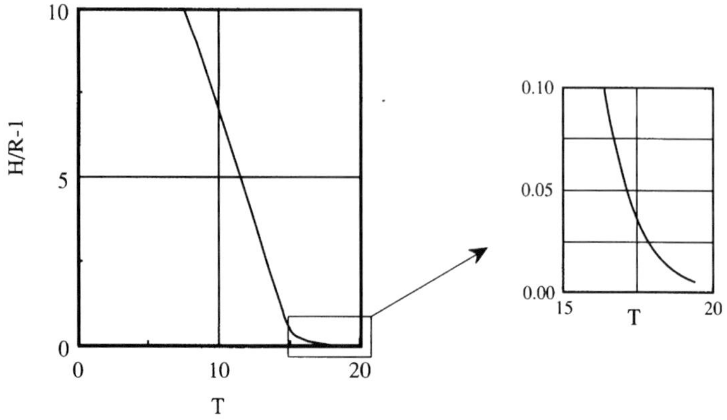


*FIGURE 10-1.2* Approach of a rigid sphere to a rigid planar surface.

## 10-2 SLOW APPROACH OF A VISCOUS DROP TO A SOLID SURFACE

The next problem we examine arises in the case of the slow approach of a viscous deformable drop to a rigid planar surface, the motion taking place through a viscous liquid. As the leading surface of the drop approaches the planar surface it is necessary that the intervening liquid flows (largely radially) out of the way of the drop. Eventually the drop is so close to the surface that the depletion of the separating liquid from the space between the drop and the plane is retarded by viscous resistance to this radial flow. In consequence, the drop slows near the axis of symmetry, and the lower surface of the drop flattens toward a disk-like shape. The drop then attains a near-equilibrium shape, with respect to the balance of forces of gravity and surface tension, and it continues to squeeze out the intervening liquid so long as gravity provides a positive force of attraction. Hartland (1967) presents photographs of falling drops at this stage (see Fig. 10-2.1 for a schematic of the drop shape) and he finds that there is a period of final approach when the drop shape is essentially invarient with time.

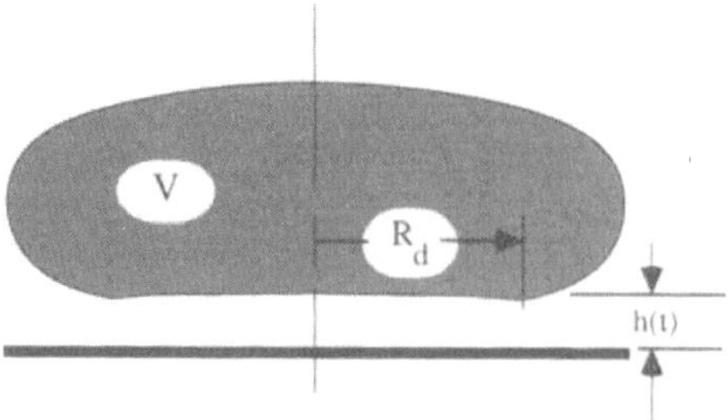


*FIGURE 10-2.1* Idealization of the approach of a drop to a planar surface.

Thus, with reference to Fig. 10-2.1, we may attempt to model the flow in the separating region as a squeezing flow between a plane and a disk of radius $R_d$. In order to estimate the value of $R_d$ at this stage, it is assumed that the drop is identical to a sessile drop with a contact angle of 180°. Knowledge of the drop volume $V$ and the parameter $\Delta\rho g/\sigma$ is sufficient to find the radius $R_d$ by using tables of equilibrium drop shapes (Bashforth and Adams, 1883). This is described in Hartland (1967).

The flow analysis is based on the model of viscous-dominated flow in the narrow parallel region between the lower rigid plane, $y = 0$, and the upper liquid–liquid interface[2] at $h(t)$. A no-slip condition is imposed on the lower rigid surface. The upper surface is that of a fluid, and hence it has some potential mobility. At the one extreme, if the drop is of a very high viscosity, the drop interface will be nearly immobile,[3] and an appropriate boundary condition would be that of no slip at $h(t)$. On the other hand, if the drop is of a very low viscosity fluid, relative to that of the surrounding liquid, a suitable boundary condition would reflect a negligible shear stress at the interface:

$$-\mu \frac{\partial u_r}{\partial y} = 0 \quad \text{at} \quad y = h(t). \tag{10-2.1}$$

In the parallel region the radial component of the Navier–Stokes equation, neglecting inertial terms, would be

$$-\frac{\partial p}{\partial r} + \mu \frac{\partial^2 u_r}{\partial y^2} = 0\,. \tag{10-2.2}$$

If we invoke the usual lubrication approximation that $p$ is independent of the $y$ coordinate, then we may integrate Eq. 10-2.2 twice, impose the boundary conditions, and find the velocity field between the drop and the plane to be

$$u_r = \frac{1}{2\mu}\frac{dp}{dr}\left(y^2 - \frac{2yh}{n}\right). \tag{10-2.3}$$

If the no-slip boundary condition is appropriate at $y = h(t)$, then we set $n = 2$; otherwise $n = 1$ corresponds to Eq. 10-2.1 as the boundary condition at the interface. In either case, we have assumed no slip at the lower rigid planar surface, $y = 0$.

Across any radial perimeter at $r$ the volumetric flow rate of the liquid squeezed out from between the drop and the fixed planar surface is simply

[2]In the previous section the approach of the rigid sphere was in terms of $H$, the distance of the center of the sphere from the lower surface. For the deformable drop the center is not easily defined, so the approach is written in terms of the thickness $h$ of the separating liquid film, at the axis of the drop.

[3]The presence of surfactants can also effectively immobilize an otherwise mobile interface, in which case the no-slip condition would be the more suitable boundary condition.

$$\int_0^{h(t)} 2\pi r u_r \, dy = \pi r^2 U(t), \tag{10-2.4}$$

where $U = -\,dh/dt$ is the velocity of approach of the drop to the plane. Using Eq. 10-2.3, we find the radial pressure gradient to be

$$\frac{dp}{dr} = -\frac{3n^2}{2}\frac{\mu U r}{h^3}. \tag{10-2.5}$$

We integrate this with the boundary condition that

$$p = 0 \quad \text{at} \quad r = R_\mathrm{d} \tag{10-2.6}$$

and find the radial pressure distribution:

$$p(r,t) = \frac{3n^2}{4}\frac{\mu U R_\mathrm{d}^2}{h^3}\left(1 - \frac{r^2}{R_\mathrm{d}^2}\right). \tag{10-2.7}$$

Note that since we are using a planar lubrication approximation, the interfacial tension of the drop boundary does not enter explicitly into the model.[4] The pressure distribution in the draining film retards the downward motion of the drop by exerting an upward force given by

$$F = \int_0^{R_\mathrm{d}} 2\pi r p(r,t) \, dr = \frac{3\pi n^2}{8}\frac{\mu U R_\mathrm{d}^4}{h^3}. \tag{10-2.8}$$

For slow motion we may neglect the deceleration of the drop and simply take $F$ as the net weight of the drop in the surrounding liquid (i.e., with buoyancy accounted for):

$$F = V\left(\rho_\mathrm{drop} - \rho_\mathrm{liquid}\right) g = V \, \Delta\rho \, g, \tag{10-2.9}$$

where $V$ is the drop volume.

Equation 10-2.8 is just a differential equation for $h(t)$, since $U = -\,dh/dt$, and we may write this as

$$U = -\frac{dh}{dt} = \frac{8Fh^3}{3\pi n^2 \mu R_\mathrm{d}^4}. \tag{10-2.10}$$

While it is tempting to solve this differential equation (as Hartland, 1967, does) we do not have an *a priori* initial condition, since the drop changes from a spherical shape to that of the flattened shape shown in Fig. 10-2.1 over a distance not defined by this model, and the model is not valid until the drop is so near the wall that the lubrication approximation to the flow field (Eq. 10-2.3) is accurate. In other words, we do not know where to start the clock on this model.

[4]But surface tension enters implicitly, since the drop shape (particularly $R_\mathrm{d}$ ) will depend on surface tension and gravity through a Bond number.

The best we can say is that we would expect that the model is realistic for the later stages of approach, where $h << R_d$.

On the other hand, if our primary interest is in the final stages of approach of the drop to the surface, the initial conditions may play a minor role. To see one aspect of this, we solve Eq. 10-2.10 for the film thickness $h(t)$ and plot the results in Fig. 10-2.2. The solution for long times may be written in the form

$$\frac{h(t)}{R} \approx \frac{3n}{8}\left(\frac{R_d}{R}\right)^2\left(\frac{\mu}{R\,\Delta\rho\;gt}\right)^{1/2}, \qquad (10\text{-}2.11)$$

where the length scale $R$ is the radius of a sphere corresponding to the volume $V$, and $T$ is a dimensionless time defined by

$$T \equiv \frac{R\,\Delta\rho\;gt}{\mu}. \qquad (10\text{-}2.12)$$

Figure 10-2.2 shows (curves A) the approach of a spherical drop to the planar surface, assuming that it remains spherical. Far from the surface the drop falls under its terminal velocity. For a spherical drop the terminal velocity in the Stokes flow regime is given by

$$U_\infty = \frac{2R^2 g\;\Delta\rho}{3\mu}\left(\frac{1+\kappa}{2+3\kappa}\right), \qquad (10\text{-}2.13)$$

where

$$\kappa \equiv \frac{\mu_{\text{drop}}}{\mu_{\text{liquid}}} \qquad (10\text{-}2.14)$$

is the ratio of viscosities of the fluid interior and exterior to the drop. For a gas bubble, for example, or a drop of much lower viscosity than the surrounding liquid, $\kappa = 0$. This corresponds to the case $n = 1$, where we neglect the shear stress at the drop interface. A rigid sphere, or a drop of extremely high viscosity relative to the surrounding liquid, would correspond to the case $\kappa = \infty$ and $n = 2$.

In preparing Fig. 10-2.2 it was assumed that the drop remained spherical along the path described by curve A. This is not a realistic assumption as the drop gets within a few radii of the surface and indeed may be a poor assumption over the entire path of the drop, since the drag forces could distort the shape of the drop. Similarly, curves B, which follow from a lubrication analysis, would not be a good model until $h/R_d$ becomes small. For the selected value[5] of $R_d/R = 0.88$, this means that curves B would not be reliable above $h/R = 0.1$. Hence there is some curve that would join the heavy portions of curves A and B, but we cannot obtain that curve from any simple analytical theory.

It would be necessary to resort to a numerical analysis in order to produce a more realistic picture of the approach of this drop to a surface. Such an analysis

[5]This value corresponds to Hartland's (1967) $V = 0.25$ ml drop of golden syrup falling through sextol phthalate. This drop has a spherical radius of $R = 0.3912$ cm, and Hartland calculates $R_d = 0.345$ cm for this drop.

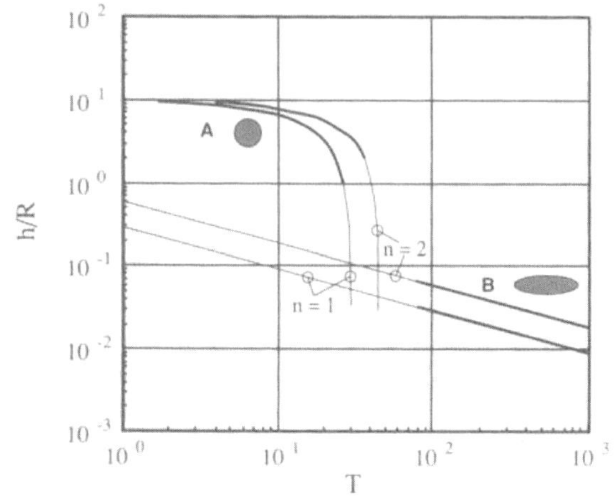


***FIGURE 10-2.2*** Approach of a drop to a planar surface. Curves A are for a spherical drop falling at its terminal velocity, far from the lower surface. Curves B are for the flattened drop, according to Eq. 10-2.11, with $R_d/R = 0.88$. Curves for $n = 1$ or 2 correspond to the no-shear or no-slip condition at the drop surface, respectively.

would include the interfacial tension, and hence it would yield the shape of the drop as it moves toward the planar surface. Pozrikidis (1990) has developed the numerical tools for analyzing this class of problem, and an example of his numerical solutions for the evolution of the drop shape is presented in Fig. 10-2.3. In part (a) of this figure we see the shape of a drop as it approaches the lower rigid boundary. The viscosity of both fluids is the same ($\kappa = 1$), and the surface tension corresponds to a Bond number of $\mathsf{Bo} = 10$, where

$$\mathsf{Bo} \equiv \frac{|\Delta\rho|\; g\; R_{\text{drop}}^2}{\sigma} \tag{10-2.15}$$

and the *magnitude* of the density difference of the two fluid phases appears. In the cases shown, $R_{\text{drop}}$ is the radius of the initially spherical drop. For the choice of Bond number selected, $\mathsf{Bo} = 10$, the drop shape develops a "dimple" as it nears the lower boundary. However, this feature of the drop shape is dependent on the Bond number. In part (b) of Fig. 10-2.3 we see the shapes of three different drops, including the drop of part (a) of the figure, as the lower boundary is approached. It is apparent that the Bond number has a strong effect on the shape of the drop and that an observable dimple appears under the drop only for large Bond numbers.

## *10-3 THE DIMPLED DROP*

The lubrication analysis is oversimplified in several aspects but particularly in its treatment of the drop boundary. As Fig. 10-2.1 suggests, the lower drop

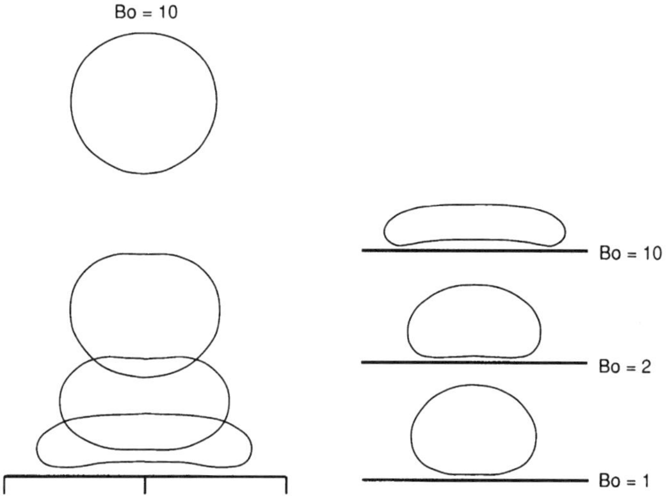


***FIGURE 10-2.3*** Evolution of the shape of an initially spherical drop as it approaches a planar rigid surface. The drop and the surrounding fluid have the same viscosity. (a) Bo = 10. (b) Bond numbers as shown on the figure. Adapted from originals kindly supplied by Dr. C. Pozrikidis.

boundary is regarded as a planar disk in deriving Eq. 10-2.10. Since there is a radial pressure distribution in the thin fluid film region separating the lower boundary of the drop from the rigid planar surface, and since the drop–liquid interface has mobility, we expect that the interface could deform. Of course, surface tension would affect the extent of deformation. As a consequence of the radial pressure distribution, the lower drop interface would appear as suggested in Fig. 10-3.1. The high pressure region near the axis of the drop would retard the approach of that section of the interface. Toward the periphery of the drop, as the pressure approaches that of the liquid surrounding the drop, an annular region could develop in which the film thickness is relatively small. Thus a centrally located "dimple" in the drop would be formed. The extent to which this occurs would depend upon the relative balance of viscous, capillary, and gravitational forces.

An analysis of this evolution of the drop shape to the form suggested in Fig. 10-3.1 is clearly a complex problem. There is a large literature describing a variety of methods of modeling this flow. In addition, quite a bit of experimental work has been done, some of which has guided the theoretical approaches.

The classical theory of the drainage of liquid from beneath a dimpled film is usually cited as that of Frankel and Mysels (1962), although they clearly cite the earlier work of Derjaguin and Kussakov (1939).[6] Their work rests on the assumption that a minimum in $h$ occurs, as indicated in Fig. 10-3.1—it does not

[6]See also the approach of Buevich and Lipkina (1975), who use a combination of numerical analysis and simple aproximations.

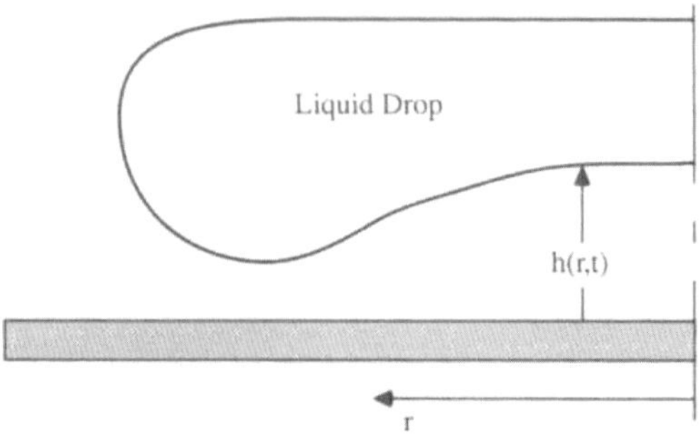


***FIGURE 10-3.1*** The interface of a dimpled drop. The vertical scale is exaggerated.

predict the conditions for the appearance of this so-called barrier ring. An approximate analytical model that leads to a criterion for the appearance of the dimple is presented by Tsekov and Ruckenstein (1994).

The radius at which this minimum occurs is dependent on the dynamics of approach of the drop to the surface, but a quasi-steady state occurs in the case of small drops, and the barrier radius may be estimated from a simple force balance. Suppose, as in Fig. 10-3.2, that a small drop approaches the lower plane very slowly. The drop shape is nearly spherical except in the thin film region. If the drop is nearly spherical, as would be the case for a very small drop, then its net weight is given by

$$F = \frac{4\pi R^3 \,\Delta\rho\, g}{3}, \tag{10-3.1}$$

where $\Delta\rho$ is the difference in densities of the drop and the surrounding liquid. This weight is supported by the pressure in the liquid film beneath the drop. However, for a nearly spherical drop near static equilibrium, this pressure is balanced by the pressure within the drop, which would be given approximately by

$$P_\sigma = \frac{2\sigma}{R}. \tag{10-3.2}$$

If the dimple is shallow, and if the radius of the dimple is nearly to the outer edge of the drop (as suggested in the sketch of Fig. 10-3.2), then the upward force acting on the dimpled surface, of radius $R_{\min}$, balances the weight of the drop and is given by

$$\frac{2\sigma}{R}\pi R_{\min}^2 = \frac{4\pi R^3 \,\Delta\rho\, g}{3} \tag{10-3.3}$$

or

$$R_{\min} = R^2\left(\frac{2\,\Delta\rho\, g}{3\sigma}\right)^{1/2}. \tag{10-3.4}$$

An early experimental study presented by Platikanov (1964) suggested that

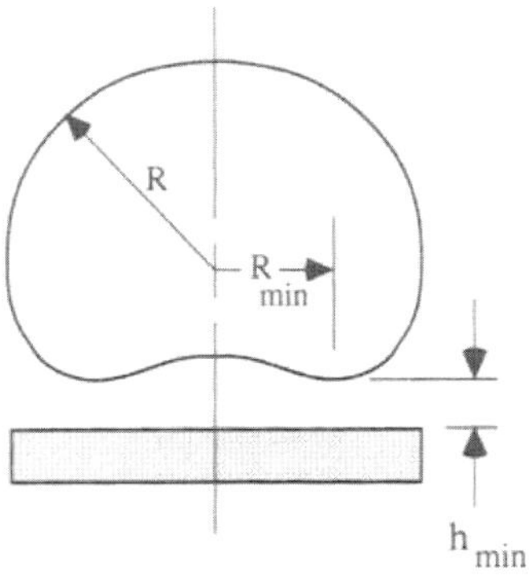


*FIGURE 10-3.2* A small dimpled drop.

the theory of Frankel and Mysels described well several of the features of the approach of a dimpled drop to a planar rigid surface. Figure 10-3.3 shows a schematic of the experimental system. Instead of a drop, a bubble of gas is formed at the tip of a capillary surrounded by liquid and forced against a glass plate. Optical methods were used by Platikanov to measure the profile of the liquid film separating the "bubble" from the lower solid surface. A set of data on film profiles is shown in Fig. 10-3.4.

In Platikanov's experiments the force that squeezes the liquid film from beneath the bubble arises from the pressure within the bubble, which is maintained at a constant value. Except near the contact region the bubble is a portion of a sphere of radius $R_b$. Hence to a good approximation the bubble pressure is

$$P = \frac{2\sigma}{R_b} \tag{10-3.5}$$

and the force $F$ that induces the motion of the liquid film from between the lower interface of the bubble and the solid surface is

$$F = \frac{2\sigma}{R_b} \pi R_d^2. \tag{10-3.6}$$

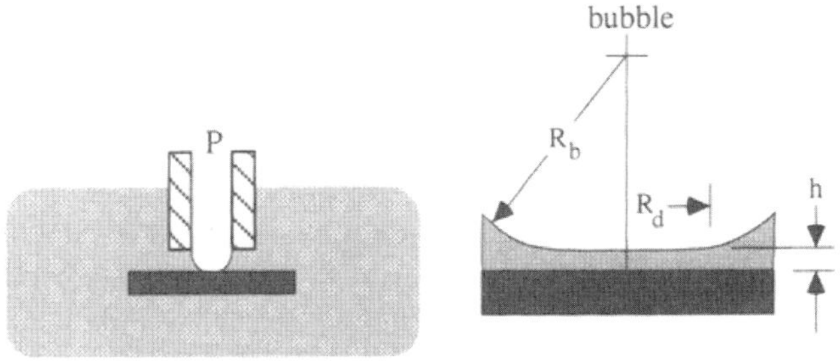


*FIGURE 10-3.3* Platikanov's method for studying the film beneath a bubble.

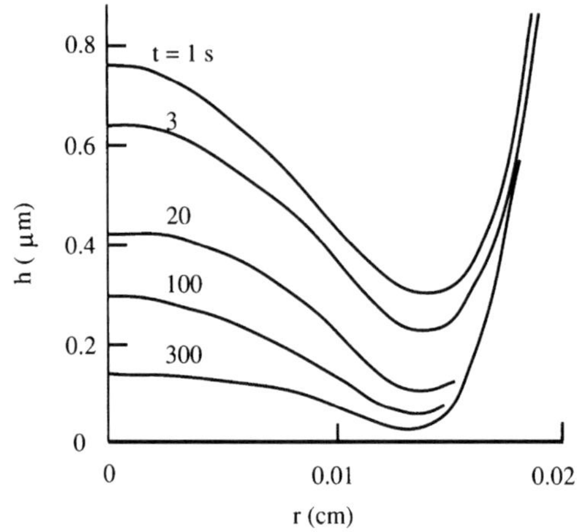


***FIGURE 10-3.4*** Platikanov's data on a film profile beneath a bubble. The liquid is 0.1 N KCl on glass. Profiles are at times after formation of the film, shown on each profile curve in seconds. Reprinted with permission from Platikanov, 1964. Copyright 1964 American Chemical Society.

If the liquid film were flat, as in Fig. 10-2.1, then Eq. 10-2.10 would lead us to the following expression for the film thickness:

$$h(t) = \left[ \alpha \,(t - t_0) + \frac{1}{h_0^2} \right]^{-1/2}, \tag{10-3.7}$$

where $t_0$ is the time at which the film thickness is measured as $h_0$, and the coefficient $\alpha$ follows from integration of Eq. 10-2.10 and is

$$\alpha = \frac{\sigma}{0.09375 n^2 \mu R_d^2 R_b}. \tag{10-3.8}$$

For long observation times we may approximate Eq. 10-3.7 as

$$h(t) = (\alpha t)^{-1/2}. \tag{10-3.9}$$

Equation 10-3.7 was derived for the case of a film of uniform thickness—not for a film that shows a dimple under the drop or bubble. The behavior in the neighborhood of the maximum in film thickness (the dimple region) is much more difficult to model by any quasi-analytical method. Numerical methods have been used by Lin and Slattery (1982) to integrate the film evolution equations (see Section 10-4) and they compare their results favorably to the data of Platikanov.

The model presented by Frankel and Mysels gives Eq. 10-3.7 (with a slightly different numerical coefficient), but for the film thickness at the closest approach of the bubble surface to the lower plane, *not for the film thickness at the axis of the drop*. For the film thickness at the center (the dimple thickness) Frankel and Mysels give a result that at large times can be written as

$$h_{\text{d}} = \left(\frac{\sigma R_{\text{b}} t}{0.0096 n^2 \mu R_{\text{d}}^6}\right)^{-1/4}. \tag{10-3.10}$$

A test of the Frankel–Mysels theory against a sample of Platikanov's data is shown in Fig. 10-3.5. The film thickness at the rim is well predicted by Eq. 10-3.7. This seems odd, since that theory assumes a disk-like drop interface, not a dimpled interface. The film thickness at the center of the dimple is well described by Eq. 10-3.10 until long times. It would appear that the Frankel–Mysels theory is a useful alternative to a numerical solution for the evolution of the interface. However, one must keep in mind that the theory requires knowledge of two length scales characteristic of the drop or bubble shape. For nearly spherical (small) drops one might identify $R_{\text{b}}$ with the radius $R$ of a drop of equivalent volume. Under some circumstances $R_{\text{d}}$ might be approximated with $R_{\text{min}}$ given by Eq. 10-3.4. The parameter that controls the shape of the drop is the Bond number

$$\text{Bo} \equiv \frac{R^2 g \, \Delta\rho}{\sigma}. \tag{10-3.11}$$

The work of Pozrikidis (1990), described at the end of Section 10-2, presents numerical calculations of drop shapes for drops approaching a solid planar surface, beginning with the spherical drop far from the surface. In addition to the Bond number, the viscosity ratio of the two liquids plays a role. His simulations show that Eq. 10-3.4 is not a good approximation, to an extent that depends upon the Bond number.

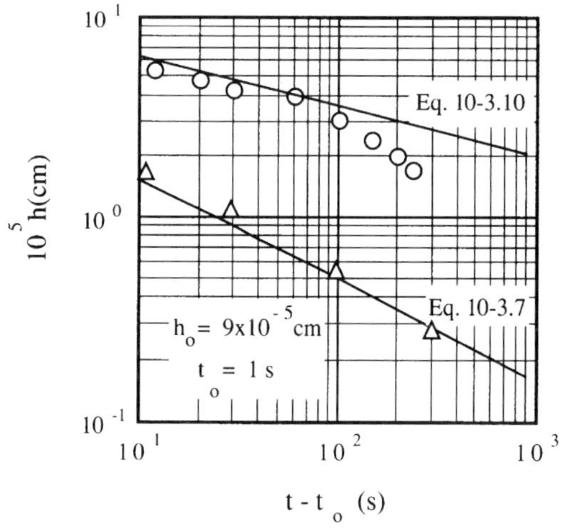


***FIGURE 10-3.5*** Test of Eqs. 10-3.7 and 10 against Platikanov's data for a gas bubble displacing 0.1N KCl.

## 10-4 FILM EVOLUTION EQUATION FOR THE DIMPLED DROP

In the previous section we mentioned the fact that Lin and Slattery (1982) integrated the film evolution equations for the dimpled drop, but we did not display those equations. In this section we present a derivation of the equations, following the method of Lin and Slattery. Our primary goal is to clarify the approximations that have been made in the course of deriving these equations and discuss some of the limitations of the approach.

We begin with the axisymmetric form of the continuity equation:

$$\frac{1}{r}\frac{\partial r u_r}{\partial r} + \frac{\partial u_z}{\partial z} = 0. \tag{10-4.1}$$

If we invoke the usual lubrication approximations by assuming that the film thickness satisfies

$$\left(\frac{h_0}{R^0_{\text{min}}}\right)^2 << 1, \tag{10-4.2}$$

then the radial momentum equation becomes

$$0 = -\frac{\partial p}{\partial r} + \left(\frac{R^0_{\text{min}}}{h_0}\right)^2 \frac{\partial^2 u_r}{\partial z^2}. \tag{10-4.3}$$

The length scales $h_0$ and $R^0_{\text{min}}$ are defined as follows. $h_0$ is the film thickness at the axis $r = 0$, at some initial time $t = 0$, and $R^0_{\text{min}}$ is the radius of the rim at that time.

These equations are written in terms of dimensionless variables defined as

$$r^* = \frac{r}{R^0_{\text{min}}}, \quad z^* = \frac{z}{h_0}, \quad t^* = \frac{\mu t}{\rho R^{0\,2}_{\text{min}}},$$

$$h^* = \frac{h(r,t)}{h_0}, \quad u_r^* = \frac{\rho R^0_{\text{min}} u_r}{\mu}, \quad u_z^* = \frac{\rho R^{0\,2}_{\text{min}} u_z}{h_0 \mu},$$

$$p^* = \frac{\rho R^{0\,2}_{\text{min}} p}{\mu^2}, \tag{10-4.4}$$

but we have dropped the * notation in writing Eqs. 10-4.1 and 3, and subsequently any unstarred variables $v$ are understood to be those defined as $v^*$ in Eqs. 10-4.4.

As we usually do in lubrication analysis, we regard $\partial p/\partial r$ to be independent of the $z$ coordinate, and we simply integrate Eq. 10-4.3 twice with respect to $z$ to find (cf.: Eq. 10-2.3)

$$u_r = \left(\frac{h_0}{R^0_{\text{min}}}\right)^2 \frac{\partial p}{\partial r}\left(\frac{z^2}{2} - \frac{hz}{n}\right). \tag{10-4.5}$$

This is then substituted into Eq. 10-4.1, the indicated differentiation with respect to $r$ is performed, and then the resulting equation is integrated once with respect to $z$ to yield a relationship for $u_z$ at $h(r,t)$. When this result for $u_z$, and Eq. 10-4.5 for $u_r$ are introduced into the kinematic boundary condition

$$u_z = \frac{\partial h}{\partial t} + u_r \frac{\partial h}{\partial r} \quad \text{at } z = h(r,t), \tag{10-4.6}$$

the velocity components are eliminated in favor of the film thickness, but we still have a pressure term appearing in this equation.

The pressure is now related to the film thickness through the Young–Laplace equation, which we write in the form

$$p = \frac{2\rho R\sigma}{\mu^2}\left(\frac{\partial^2 h}{\partial r^2} + \frac{1}{r}\frac{\partial h}{\partial r}\right) = 2\mathsf{S}\left(\frac{\partial^2 h}{\partial r^2} + \frac{1}{r}\frac{\partial h}{\partial r}\right). \tag{10-4.7}$$

The dimensionless group $\mathsf{S}$ was defined earlier in Chapter 3 (see Section 3-3, for example). In writing this equation we are neglecting viscous normal stresses relative to the capillary pressure. This is equivalent to the assumption that $\mathsf{S} >> 1$.

When this expression is used in place of the pressure terms we obtain the film evolution equation. It is

$$-\frac{\partial h}{\partial t'} = \frac{h^3}{3}\left(\frac{\partial^4 h}{\partial r^4} + \frac{2}{r}\frac{\partial^3 h}{\partial r^3} - \frac{1}{r^2}\frac{\partial^2 h}{\partial r^2} + \frac{1}{r^3}\frac{\partial h}{\partial r}\right)$$
$$+ h^2 \frac{\partial h}{\partial r}\left(\frac{\partial^3 h}{\partial r^3} + \frac{1}{r}\frac{\partial^2 h}{\partial r^2} - \frac{1}{r^2}\frac{\partial h}{\partial r}\right), \tag{10-4.8}$$

where another dimensionless time $t'$ is defined so as to remove all parameters from the evolution equation:

$$t' = \frac{3-n}{2n}\left(\frac{h_0}{R^0_{\text{min}}}\right)^3 \frac{\sigma}{\mu R^0_{\text{min}}}\, t. \tag{10-4.9}$$

In this equation only, $t$ is the real time—not the dimensionless time.

The integration (numerically) of Eq. 10-4.8 is complicated by the need for an initial configuration $h(r,0)$ of the film thickness. The calculation cannot start at some time when the drop or bubble is far from the surface, since this would correspond to a film that would violate the lubrication approximation. Hence the computation must begin with a film thickness profile with a shape similar to that shown in Fig. 10-3.2. One way to do this is to select an experimental film profile at some point in the evolution of the draining of the film, use that as a starting condition on the solution of the film evolution equation, and then examine the profile shapes at subsequent times. Another and less restrictive approach is to use the observation that there is a time in the evolution of the film such that the thinning rate at the center ($h = 0$) is the same as that at the rim ($R^0_{\text{min}}$). Lin and Slattery (1982) assume that there is a time at which the thinning rate is independent of radial position $r$, and they develop an equation (solved numerically) for the shape of the film at that time. This becomes the initial configuration $h(r,0)$ used for solution of Eq. 10-4.8.

Lin and Slattery discuss the details of the procedure they used for such an integration of Eq. 10-4.8, and they compare their results to some experimental profiles given by Platikinov (1964) and described above. Agreement of their computed shapes with the data of Platikinov is fairly good until long times. The predicted thinning rates of the dimple and the rim are in good agreement with the equations of Frankel and Mysels (1962), Eqs. 10-3.7 and 10. The main disadvantage of the approach of Lin and Slattery is that it is not capable of yielding the evolution of the drop (or bubble) shape until the drop is so close to the barrier plane that the lubrication equations hold. Hence the most promising approach is that of Pozrikidis (1992), which is not restricted to such an approximation and so has the potential to yield the full evolution of the drop shape as it approaches the surface.

### 10.4.1 Inertial Effects in Film Thinning

The analyses presented above assume that viscous effects dominate the hydrodynamics, and the Reynolds lubrication equations are used to provide the base flow in the draining film. Whether this is a good assumption in low-viscosity liquids is not intuitively obvious. A Reynolds number characteristic of the balance between viscous and inertial forces would be based on the film thickness, and so it may seem reasonable to assume that when the draining film is extremely thin, the Reynolds number will eventually be small enough that inertial forces may be neglected. However, this statement ignores the role that is played by the forces that lead to the initial creation of the separating film. One could imagine, for example, a liquid droplet or a bubble immersed in a second, low-viscosity liquid, with that second liquid confined between large parallel planes. At some instant of time, if the confining planes were driven toward one another at very high speed (obviously under the action of a very large force), the ensuing flow in the draining film could very well be subject to significant inertial effects. Hence it seems useful to describe a simple model of film drainage in the absence of viscous effects—the evolution equation that would describe the drainage of an inviscid liquid film. Figure 10-4.1 will aid in the derivation.

If we assume that the interfaces are free of shear, consistent with the inviscid assumption, then the radial velocity is independent of the axial coordinate, and the velocity profile is flat, as shown in Fig. 10-4.1. The conservation of mass statement then takes the form (from Eq. 7-1.6)

$$\frac{\partial h}{\partial t} = -\frac{1}{r}\frac{\partial}{\partial r}(hur), \tag{10-4.10}$$

where $u(r,t)$ is the radial component of velocity. The radial component of the momentum equation, in the absence of viscosity, is simply

$$\frac{\partial u}{\partial t} + u\,\frac{\partial u}{\partial r} = -\frac{1}{\rho}\frac{\partial p}{\partial r}. \qquad (10\text{-}4.11)$$

As before, the pressure term arises from the curvature (we neglect disjoining pressure, as we have in the previous cases as well) and

$$-\frac{\partial p}{\partial r} = \sigma\,\frac{\partial}{\partial r}\left(\frac{\partial^2 h}{\partial r^2} + \frac{1}{r}\frac{\partial h}{\partial r}\right). \qquad (10\text{-}4.12)$$

subject to the approximation that the curvature is small $[(\partial h/\partial r)^2 << 1]$.

This lets us write the momentum equation in the form

$$\frac{\partial u}{\partial t} + u\,\frac{\partial u}{\partial r} = \frac{\sigma}{\rho}\left(\frac{\partial^3 h}{\partial r^3} + \frac{1}{r}\frac{\partial^2 h}{\partial r^2} - \frac{1}{r^2}\frac{\partial h}{\partial r}\right). \qquad (10\text{-}4.13)$$

In the viscous-dominated case we used the lubrication analysis to provide the velocity in the draining film, in terms of the film thickness (Eq. 10-4.5). This lets us eliminate velocity and write a single partial differential equation for $h(r,t)$—the film evolution equation. In the inviscid case we cannot do this, and so we have to solve Eqs. 10-4.10 and 13 together.

Before we can discuss the solution of this set of equations we must propose some set of initial and boundary conditions. One possible set of conditions would correspond to the case that at some distance $h_0 = h(0,0)$ the bubble is an undeformed sphere of radius $R$, with a velocity $U$. Then initial conditions on $h$ and $u$ would be

$$h = h_0 + R - R\left[1 - \left(\frac{r}{R}\right)^2\right]^{1/2} \qquad (10\text{-}4.14)$$

and

$$u = \frac{rU}{2h} \quad \text{at } t = 0. \qquad (10\text{-}4.15)$$

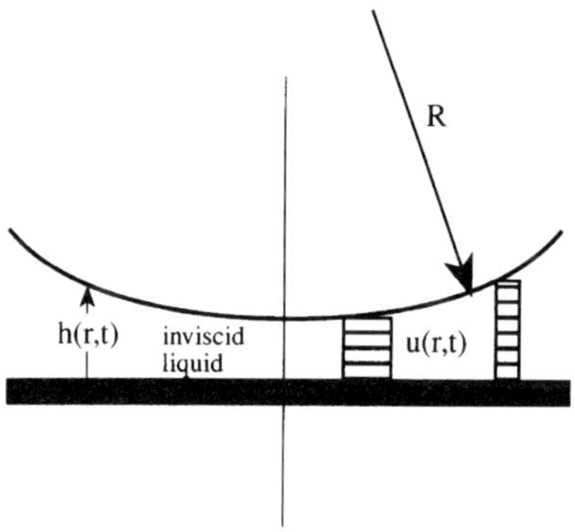


***FIGURE 10-4.1*** Thinning of an inviscid film. With the assumption of no shear stresses in the film or on the boundaries, the velocity profile is flat.

Symmetry of the solutions for $h$ and $u$ at the axis $r = 0$ would require that

$$\frac{\partial h}{\partial r} = 0 \quad \text{and} \quad \frac{\partial u}{\partial r} = 0 \quad \text{at } r = 0. \tag{10-4.16}$$

Symmetry of the pressure would lead to

$$\frac{\partial p}{\partial r} = 0 \quad \text{or} \quad \frac{\partial}{\partial r}\left(\frac{\partial^2 h}{\partial r^2} + \frac{1}{r}\frac{\partial h}{\partial r}\right) = 0 \quad \text{at } r = 0. \tag{10-4.17}$$

The third boundary condition on $h(r,t)$ would be imposed far from the axis, at some position $r = \alpha R$, where $\alpha > 1$. The pressure far from the axis would be taken to be equal to that of the undisturbed liquid. This would be the same as the pressure in equilibrium with the spherical bubble, so we would write this boundary condition in the form

$$\frac{\partial^2 h}{\partial r^2} + \frac{1}{r}\frac{\partial h}{\partial r} = \frac{2}{R} \quad \text{at } r = \alpha R. \tag{10-4.18}$$

Also, far from the axis of symmetry, the velocity would satisfy

$$-\frac{\partial h}{\partial t} = U. \tag{10-4.19}$$

It is convenient to nondimensionalize these equations with the following definitions:

$$h^* \equiv \frac{h}{R\,\mathrm{We}}, \qquad u^* \equiv \frac{u\,\mathrm{We}^{1/2}}{U}$$
$$t^* \equiv \frac{Ut}{R\,\mathrm{We}}, \qquad r^* \equiv \frac{r\,\mathrm{We}^{-1/2}}{R}, \tag{10-4.20}$$

where a Weber number appears, defined as

$$\mathrm{We} \equiv \frac{\rho U^2 R}{\sigma}. \tag{10-4.21}$$

As a consequence of this choice of nondimensionalization we solve for $h^*$ and $u^*$ from

$$\frac{\partial h^*}{\partial t^*} = -\frac{1}{r^*}\frac{\partial}{\partial r^*}(h^* u^* r^*) \tag{10-4.22}$$

and

$$\frac{\partial u^*}{\partial t^*} + u^*\frac{\partial u^*}{\partial r^*} = \frac{\partial^3 h^*}{\partial r^{*3}} + \frac{1}{r^*}\frac{\partial^2 h^*}{\partial r^{*2}} - \frac{1}{r^{*2}}\frac{\partial h^*}{\partial r^*}. \tag{10-4.23}$$

The initial conditions become

$$h^* = h_0^* + \frac{1}{\mathrm{We}} - \frac{1}{\mathrm{We}}\left(1 - \mathrm{We}\, r^{*2}\right)^{1/2} \quad \text{at} \quad t^* = 0$$

$$u^* = \frac{r^*}{2h^*} \tag{10-4.24}$$

where $h_0^* = h_0/R\ \mathrm{We}$. The symmetry conditions (Eqs. 10-4.16 and 17) are unchanged in form, except that $h$, $r$ and $u$ are everywhere replaced by $h^*$, $r^*$ and $u^*$.

Far from the axis of the film, the boundary condition Eq. 10-4.18 becomes

$$\frac{\partial^2 h^*}{\partial r^{*2}} + \frac{1}{r^*}\frac{\partial h^*}{\partial r^*} = 2 \quad \text{at} \quad r^* = \frac{\alpha}{\mathrm{We}^{1/2}} \tag{10-4.25}$$

and Eq. 10-4.19 is simply

$$-\frac{\partial h^*}{\partial t^*} = 1 \quad \text{at} \quad r^* = \frac{\alpha}{\mathrm{We}^{1/2}}. \tag{10-4.26}$$

To review, we have a pair of partial differential equations to solve (Eqs. 10-4.22 and 23) subject to a pair of initial conditions (Eqs. 10-4.24), symmetry conditions (Eqs. 10-4.16 and 17, but in terms of the starred variables), and boundary conditions far from the axis (Eqs. 10-4.25 and 26).

Before proceeding further, let's examine the nondimensionalization defined in Eqs. 10-4.20. What was the purpose in introducing a Weber number into the definitions of these dimensionless variables? Why not simply use $R$ and $U$ to make all length variables ($h$, $r$) and the velocity $u(r,t)$ dimensionless, and $R/U$ to make the time variable dimensionless? The specific choice indicated does remove the Weber number from the differential equations (Eqs. 10-4.22 and 23) but the Weber number remains in the initial and boundary conditions. Hence the solution still depends upon the Weber number.

The analysis presented here follows closely that presented by Chesters and Hofman (1982). One difference is that they write the initial shape of the film boundary using an approximation that is consistent with writing the radius of curvature in the form used in Eq. 10-4.12. This approximation is only good in the region near the film axis, where $(r/R)^2 << 1$. If we adopt that approximation we may write the initial condition on $h(r)$ from Eq. 10-4.24 in the form

$$h^* = h_0^* + \frac{r^{*2}}{2}. \tag{10-4.27}$$

and we see that the Weber number disappears from this initial condition. However, the Weber number remains in the boundary conditions given in Eqs. 10-4.25 and 26. Chesters and Hofman ignore this point, but note that their solutions are not influenced by the value of $r^*$ at which the boundary conditions of

Eqs. 10-4.25 and 26 are applied, so long as this value of $r^*$ is of order 10. In addition, $h_0^*$ is a parameter in the first of Eqs. 10-4.24, but Chesters and Hofman find that the solutions are not affected by the specific value chosen so long as it is significantly larger than unity. However, this requires that the Weber number satisfy a constraint on its magnitude, a point made by Chesters and Hofman, who indicate that a restriction on their model is We < 0.01. In summary, with the nondimensionalization shown, the evolution equations are free of parameters, and a single solution suffices to define the dynamics of the film drainage.

Chesters and Hofman present the film profiles as a function of time and find that a dimple appears, just as in the viscous case. The radius of the rim satisfies

$$r^* = 0.4 \tag{10-4.28}$$

as the film becomes very thin. In terms of real variables, this gives

$$R_{\min} = 0.4R\,\mathrm{We}^{1/2}. \tag{10-4.29}$$

Of course, this is a very different result from that given earlier in Eq. 10-3.4 for a viscous film.

Another result of significance in the inviscid case is that the velocity of the rim, that is, $\partial h/\partial t$ at $r = R_{\min}$, satisfies

$$-\frac{\partial h^*_{\min}}{\partial t^*} = 0.1 \tag{10-4.30}$$

or

$$-\frac{\partial h_{\min}}{\partial t} = 0.1U. \tag{10-4.31}$$

Since $h_{\min}$ goes to zero at a finite time, under the (questionable) assumption that viscous effects *never* intervene, this model predicts that film rupture occurs at a finite time following the event that "pushes" the drop or bubble toward the planar surface. In fact, the simulation of Chesters and Hofman shows that the rim touches the plane at a time corresponding to

$$t^* = 1, \tag{10-4.32}$$

where the time is measured from that time at which the drop would first touch the plane if there were no resistance to its motion at all.

Keep in mind that in the model of Chesters and Hofman the drop is pushed toward the surface at a constant speed, regardless of the force required to maintain this speed. In the viscous-dominated models presented earlier, the drop moves toward the surface under a constant force (such as gravity), or there is a constant internal pressure within a bubble, as in the experiments of Platikanov. Hence the models are different in several aspects, apart from the issue of the role of viscous effects.

## *10-5 DROPS AND BUBBLES CROSSING LIQUID INTERFACES*

When a gas bubble rises slowly through a viscous liquid it will distort the planar interface that it approaches. Figure 10-5.1 suggests the geometry just before the bubble bursts. The bubble continues to rise under the action of buoyancy (gravity) and the liquid film separating the two gas phases thins until it perforates. Surface tension causes the almost instantaneous retraction of the liquid membrane separating the two gas phases. At this point, a combination of capillary and hydrostatic pressures forces the liquid interface that forms the lower surface of the bubble to rise, and inertia carries the liquid layer that is just beneath the interface to rise above the original planar surface and to form into a short cylindrical column. Figure 10-5.2 shows a sequence of numerically simulated free-surface shapes that follow the bursting of the bubble, calculated by Boulton-Stone and Blake (1993) using a boundary integral numerical method. These authors compare their simulations, where possible, to experimental data of Kientzler *et al.* (1954) on bubble bursting, with favorable results.

Bubble bursting creates several consequences of technological interest. There is insufficient space to describe these in detail, but we mention two of particular interest. Bubble bursting can give rise to the creation of a spray of tiny droplets (Blanchard and Syzdek, 1972). Should the liquid have in it any surface-active material, the resulting aerosol will be enriched with respect to this material, often by more than an order of magnitude in comparison to the bulk concentration. A less visible consequence of bubble bursting is the high rate of dissipation of energy as the cavity collapses and deforms. Boulton-Stone and Blake (1993) speculate that this may be the cause of disruption of animal cells in aerated bioreactors.

Somewhat related phenomena occur when a liquid drop plunges across the interface that separates a liquid from a gas. Of particular interest to several investigators is whether or not a bubble will be entrained in the liquid subsequent to the "splash" of the drop. Figure 10-5.3 shows a sequence of events in the evolution of the interface following the impact of a small liquid drop with the surface of the liquid reservoir. We see that a large cavity of gas is formed by the impact of the drop and that as this cavity collapses a small gas bubble is entrained below the cavity. Pumphrey and Elmore (1990) have studied bubble

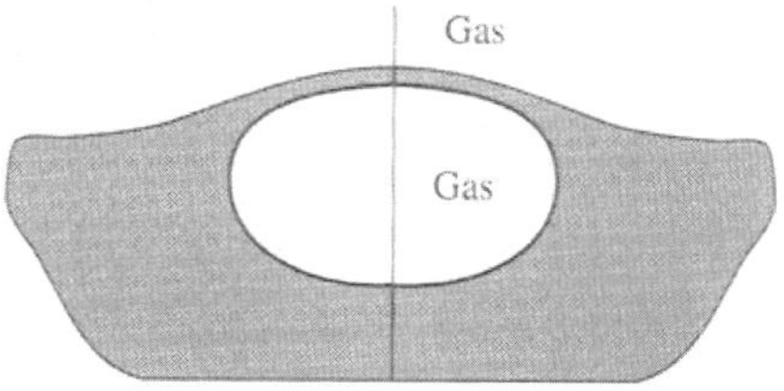


***FIGURE 10-5.1*** A rising gas bubble prior to bursting at the surface.

entrainment, and they distinguish two classes of bubble entrainment. The impact of large drops at high speed gives rise to what Pumphrey and Elmore call "irregular" entrainment, so named because every event does not create a bubble. Apparently, small variations in the dynamics of the cavity collapse can control whether or not a bubble will be formed.

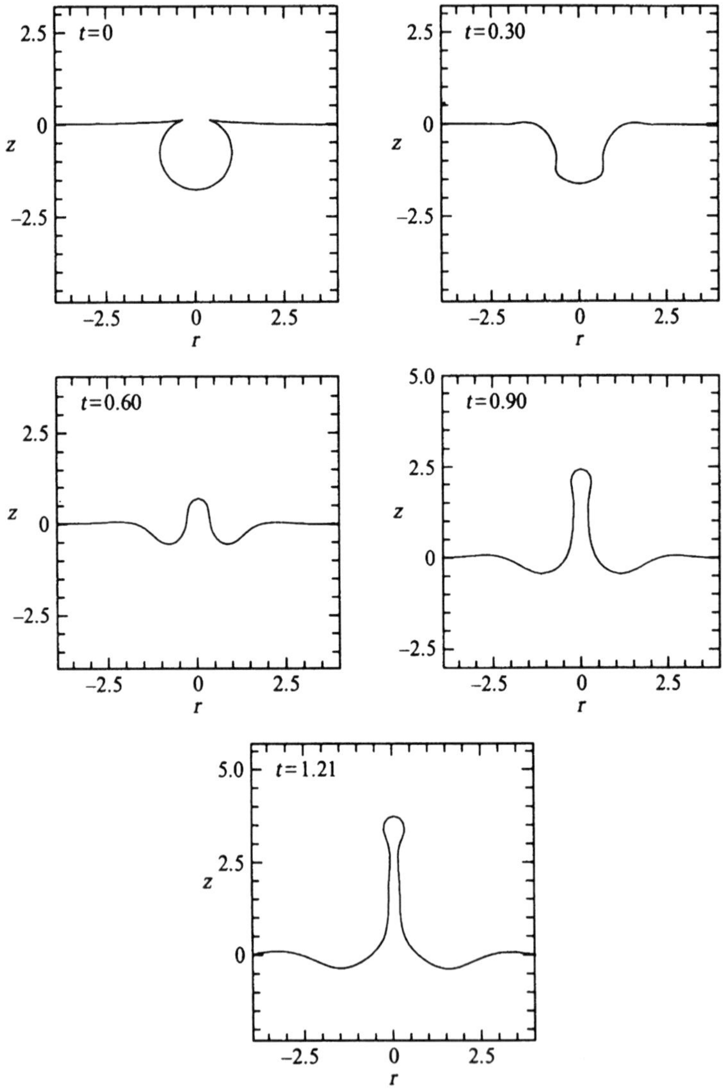


***FIGURE 10-5.2*** Sequence of free surface shapes subsequent to bursting at the surface. Reprinted with the permission of Cambridge University Press.

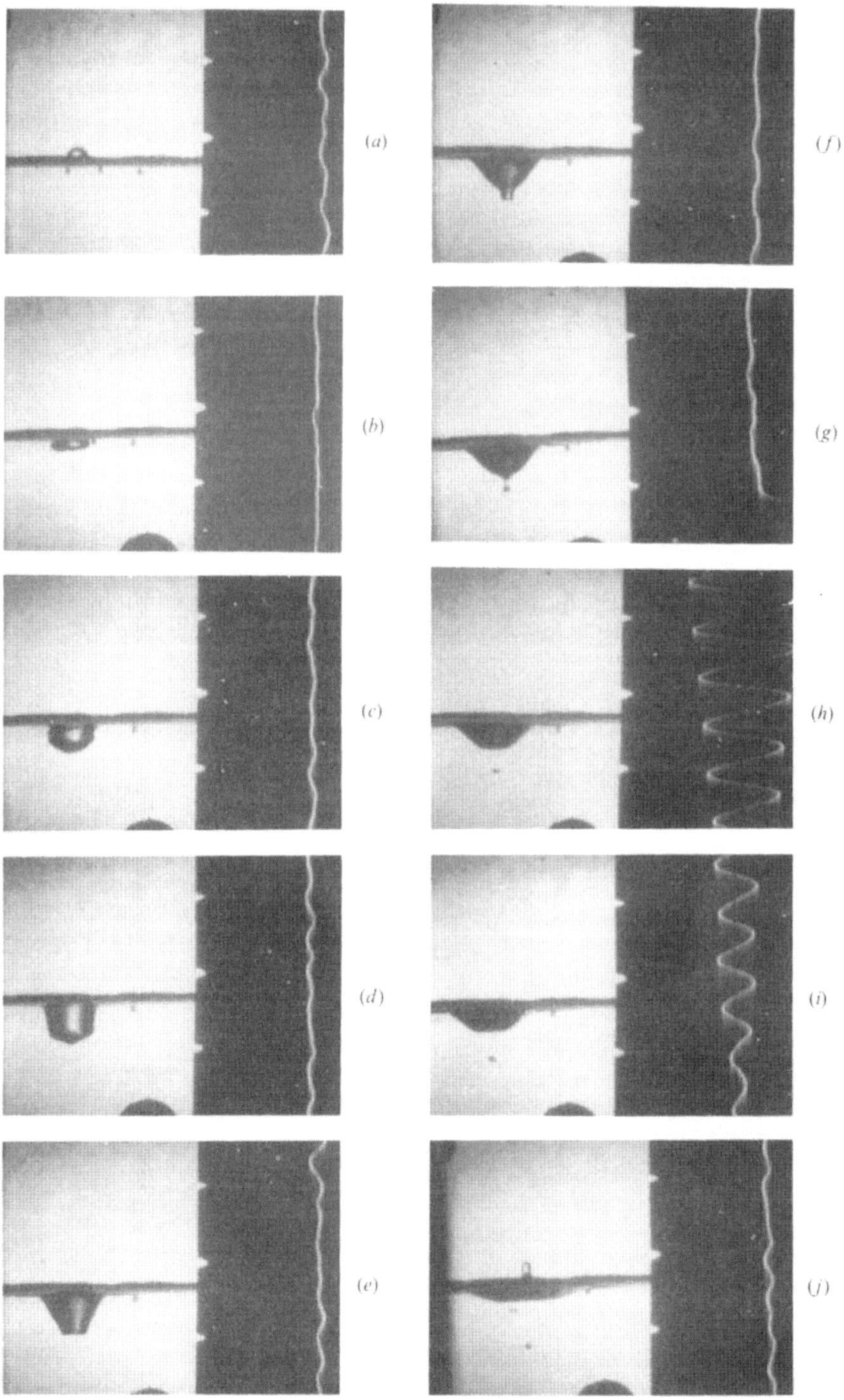


***FIGURE 10-5.3*** Sequence of events following the impact of a drop. Reprinted with the permission of Cambridge University Press.

On the other hand, "regular" entrainment, the formation of a bubble with *every* drop impact, occurs with relatively low-speed and small-diameter drops, at least within a well-defined range of the speed–diameter parameter space. Figure 10-5.4 shows the region (shaded) in which regular entrainment occurs.

The dynamics of bubble entrainment are clearly those of a high-speed free-boundary flow. It seems unlikely that any simple analytical model could reflect any of the physical features that lead to bubble entrainment, and this is largely the case. Indeed, the numerical simulation of these events is a task that pushes the current boundaries of computational fluid dynamics. Some of the successes of numerical analysis will be illustrated shortly.

First, let's look at what the application of some simple physical ideas can lead to. It seems reasonable to suppose that a very small and/or slowly moving drop will not disturb the original surface enough to lead to bubble entrainment. Hence an impacting drop must, first of all, carry enough kinetic energy to distort the free surface sufficiently that there is the potential for bubble formation. We might expect that this potential is related in some (complex) manner to the maximum depth of the cavity formed by the impact of the drop, just prior to the subsequent upwelling of the cavity under the action of gravitational and capillary forces. With reference to Fig. 10-5.5, let's calculate the energies creating this cavity, which we model as a hemisphere.

In the simplest case we consider an inviscid liquid and assume that the kinetic energy of the drop, just before impact, is completely converted to potential energy at the point in time when the maximum cavity is formed. The kinetic energy

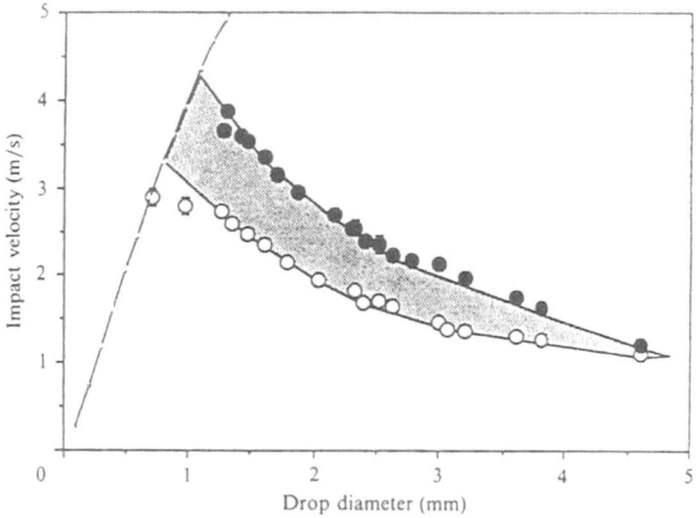


***FIGURE 10-5.4*** The boundaries of "regular" entrainment. Reprinted with the permission of Cambridge University Press.

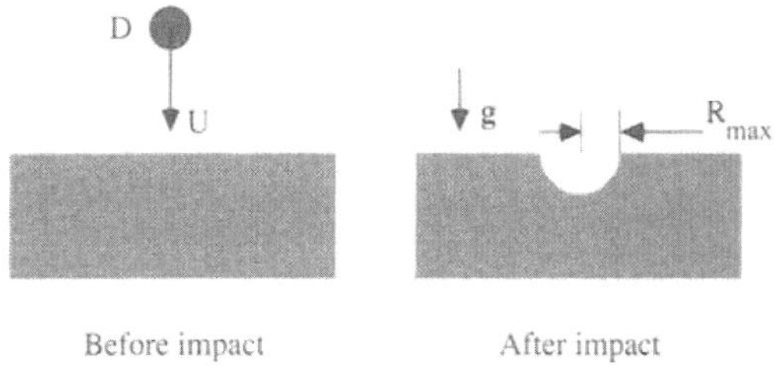


*FIGURE 10-5.5* Cavity formed upon drop impact.

is just

$$E_k = \frac{1}{2} mU^2 = \frac{\pi}{12} \rho D^3 U^2. \tag{10-5.1}$$

For the potential energy we regard the cavity as a hemisphere and calculate the potential energy from the mass of displaced liquid and the center of gravity of the cavity:

$$E_p = \frac{2\pi}{3} \rho R_{max}^3 \times \frac{3}{8} R_{max}. \tag{10-5.2}$$

Upon equating these two energies we may solve for the cavity radius and find

$$R_{max} = \frac{D^{3/4} U^{1/2}}{(3g)^{1/4}}. \tag{10-5.3}$$

A comparison of experimental data for water drops falling onto a pool of water is presented by Pumphrey and Elmore (1990) and, as Fig. 10-5.6 indicates, this simple model (shown as the line on the figure) is remarkably good. In nondimensional terms we may write this expression as

$$\frac{R_{max}}{D} = \frac{D^{-1/4} U^{1/2}}{(3g)^{1/4}} = a\, \mathrm{Fr}^{1/4}, \tag{10-5.4}$$

where a Froude number is defined here as

$$\mathrm{Fr} = \frac{U^2}{gD}. \tag{10-5.5}$$

Of course, this simple model of cavity formation does not address the issue of bubble entrapment. That requires the postulate of some connection between the size of the cavity and the creation of a bubble as the cavity retracts. Both Pumphrey and Elmore (1990) and Oguz and Prosperetti (1990) offer conjectures regarding the mechanism(s) of bubble entrapment which lean on a knowledge of the details of the kinematics that follow impact of the drop. These details include the time at which the outflow generated by the impact is reversed by retraction of the cavity and the formation and speed of small capillary waves on the surface of the cavity. Oguz and Prosperetti make some crude estimates of these features and

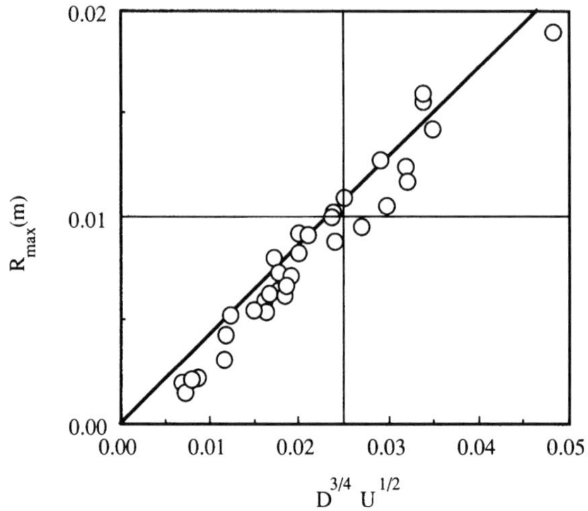


*FIGURE 10-5.6* Test of a simple model (Eq. 10-5.3) of the maximum size of the cavity formed upon drop impact. All parameters in SI units. Adapted from an original figure kindly supplied by Dr. H. C. Pumphrey.

produce predictions that the upper and lower boundaries of Fig. 10-5.4 correspond to relationships between the Weber and Froude numbers of the form

$$\mathrm{We} = b\,\mathrm{Fr}^n, \tag{10-5.6}$$

where the Weber number is defined as

$$\mathrm{We} = \frac{\rho U^2 D}{\sigma}.$$

When the data of Fig. 10-5.4 are replotted as in Fig. 10-5.7 we see that Eq. 10-5.6 fits the data quite well. However, it must be emphasized that only a single liquid was studied, and hence there is no variation in surface tension or density in these data. Furthermore, simple dimensional analysis would lead to Eq. 10-5.6 for an inviscid liquid, so the "success" demonstrated in Fig. 10-5.7 is not, of itself, a confirmation of the physical concepts that lead to the correlation. However, Oguz and Prosperetti actually predict the exponents $n$ of the two curves on the basis of their physical arguments, with excellent results. This suggests, but does not prove, that the assumptions they make regarding the mechanisms that are controlling bubble entrapment are consistent with the real physics of this complex process.

Oguz and Prosperetti (1990) present a series of numerical simulations of drop impacts, in a parameter range that corresponds to cases both inside and outside the bubble entrapment regime. They offer detailed discussions of their numerically generated "observations" of the interfacial dynamics, and the reader should study that paper carefully. While these authors point out the close correspon-

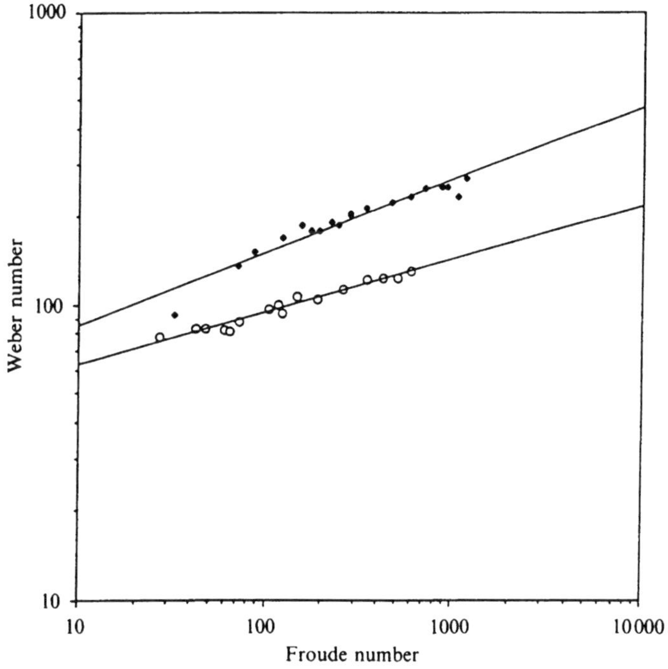


***FIGURE 10-5.7*** Test of Eq. 10-5.6. Reprinted with the permission of Cambridge University Press.

dence of their numerical "pictures" to their photographic data, they also note the difficulties inherent in implementing numerical methods for the simulation of free surface flows. We will not summarize their detailed studies here. Figure 10-5.8 shows an example of a simulated case of bubble entrapment.

In the laboratory of the real world the splash of a drop on a surface of the same liquid is extremely complex. Figure 10-5.9 shows a high-speed flash photograph of a water interface shortly after a small water drop has crossed the surface. The air cavity below the interface is clearly seen. Of more interest is the detailed structure of the film of water above the surface. Circumferential waves have formed on the open "bubble," and the bubble breaks into a set of small droplets around the rim. Smaller drops have been ejected and can be seen at a considerable height above the interface. A full numerical simulation of this sequence of events appears to be beyond current capabilities.

## *10-6 COALESCENCE OF DROPS AND BUBBLES*

If two drops or bubbles approach each other closely enough, the thin film that separates them may rupture, and the drops or bubbles will coalesce into a single drop or bubble. Clearly, an important feature affecting coalescence is the rate at

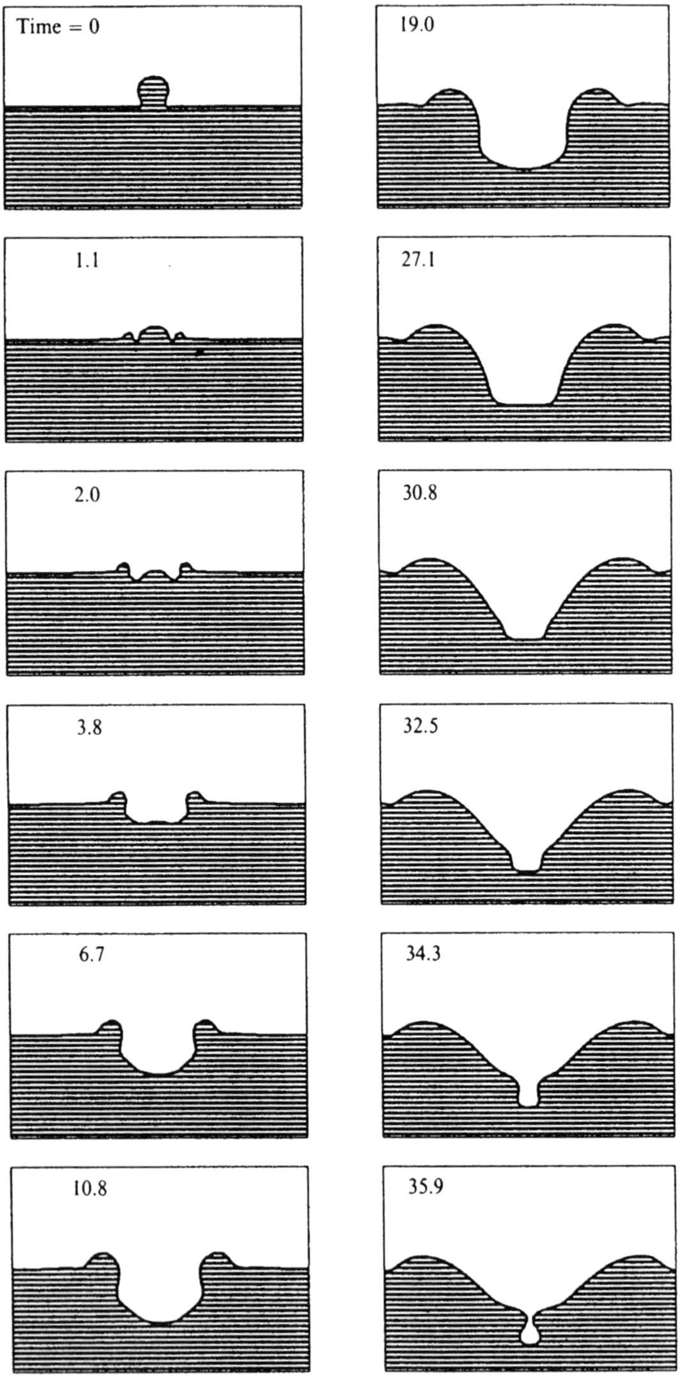


***FIGURE 10-5.8*** Computed surface shapes for a drop impacting a planar liquid surface. $D = 2$ mm, $U = 2.4$ m/s. The times indicated are nondimensional values $\tau = 2Ut/D$. Reprinted with the permission of Cambridge University Press.

***FIGURE 10-5.9*** Splash of a water drop on a water surface. (Photo by Mario Errico.)

which the intervening film thins. However, the models of this chapter suggest that the film thinning is an asymptotic process—the film thickness never goes to zero. Hence the critical conditions for coalescence must involve phenomena beyond the scope of the previous sections.

There are several physical models of film rupture that one can imagine but insufficient observations upon which to argue the validity of one model over another. One problem that inhibits the development of a clear picture of coalescence arises from the fact that rupture occurs when the intervening liquid film has a thickness in the range of 100 to 1000 Å, and the rupture "event" is itself extremely fast. Hence it is not easy to make detailed observations of rupture, such as the film thickness at rupture, or the radius of this thin film during the final stages that lead to rupture.

Another problem arises from the fact that there is likely to be a difference in mechanisms depending upon whether the liquid is a pure single component or a solution. Especially if one of the components of a solution has any surface activity, concentration gradients may be established in an ultrathin film that lead to surface tension gradients which in turn give rise to forces within the film that alter the flow. Since it is nearly impossible to guarantee purity in a laboratory study of coalescence, data are often confounded by experimental artifacts associated with the presence of surface active agents. Finally, in films of the thickness cited, disjoining pressure (see Chapter 9) can be expected to play a role in the final stages of thinning, especially at the stage of film rupture. One can find discussions of all of these ideas in the literature cited at the end of this chapter, especially in the papers of Chen (1984), Chen and Slattery (1982), Chesters (1975), Chesters and Hofman (1982), Drogaris and Weiland (1983), Gumerman and Homsy (1975), Jain and Ivanov (1980), Jones and Wilson (1978), Lee and Hodgson (1968), Marrucci (1969), Oolman and Blanch (1986), Radoev *et al.* (1983), and Vrij (1966).

We will not review this large literature here. To get a picture of some of the complexities of the issues that surround measuring and analyzing drop and bubble coalescence, a reading of the papers of Barber and Hartland (1976), Flumerfelt *et al.* (1981), and Yang and Maa (1984) is suggested. Some data from the latter are presented in Fig. 10-6.1. Pairs of nitrogen bubbles were formed in water containing minute concentrations of soluble surfactants. The time for coalescence of the bubbles was measured and found to be a strong function of the concentration of surfactant, as well as the type of surfactant. Clearly the presence and the amount of surfactant can alter the coalescence time by several orders of magnitude, and high concentrations of surfactant can prevent coalescence altogether, presumably by the action of disjoining pressure and other interfacial phenomena.

Apparently, coalescence involves two stages. The first is the thinning of the liquid layer separating the drops or bubbles. This is followed by the rupturing of this film, and this stage is usually treated as a problem in hydrodynamic stability. Stability theories have not been entirely satisfactory, especially when they

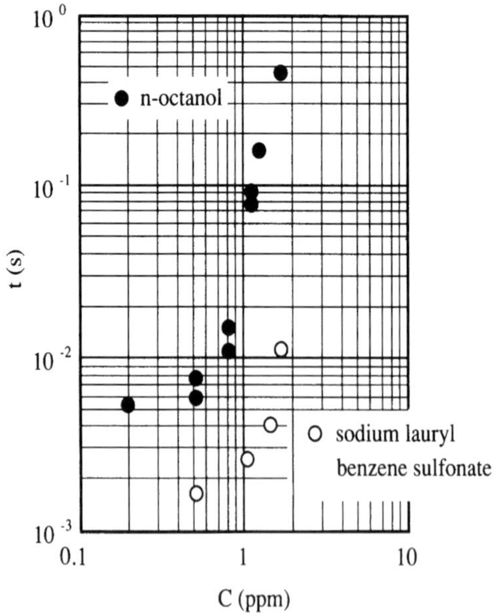


***FIGURE 10-6.1*** Coalescence times of nitrogen bubbles in aqueous solutions of surfactants (Data from Yang and Maa, 1984).

rest on oversimplifications of the surface physics. An example of a hydrodynamic stability theory for film rupture is provided in the work of Jain and Ivanov (1980). An analytical model can be developed under the assumptions that the disjoining pressure far exceeds that due to surface tension and when concentration gradients of surfactants in the surface film are neglected. With these assumptions we find that the model of Jain and Ivanov may be put in the form of a relationship between the critical film thickness $h_{crit}$ for rupture, and the radius $R$ of the film separating two drops or bubbles, or separating a single drop or bubble from a rigid planar surface. In a compact nondimensional format that relationship is found to be

$$\gamma^{-2} = 61.6 \left(\frac{k_B T}{|A|}\right) \exp\left(0.077 \mathsf{S_A} \gamma^{-4}\right), \tag{10-6.1}$$

where $\gamma$ is a dimensionless critical film thickness:

$$\gamma \equiv \frac{h_{crit}}{R}. \tag{10-6.2}$$

A dimensionless group involving the Hamaker constant appears, defined as

$$\mathsf{S_A} \equiv \frac{|A|}{6\pi\sigma R^2}, \tag{10-6.3}$$

where $\sigma$ is the interfacial tension. Note that we use the absolute value of the Hamaker constant here, since only $A < 0$ corresponds to the unstable case. Although differing in the choice of length scale, $\mathsf{S_A}$ is essentially the group defined as $\mathsf{Di}$ at the end of Chapter 9. The Boltzmann constant $k_B$ enters the theory of Jain and Ivanov through the assertion that instability occurs under the

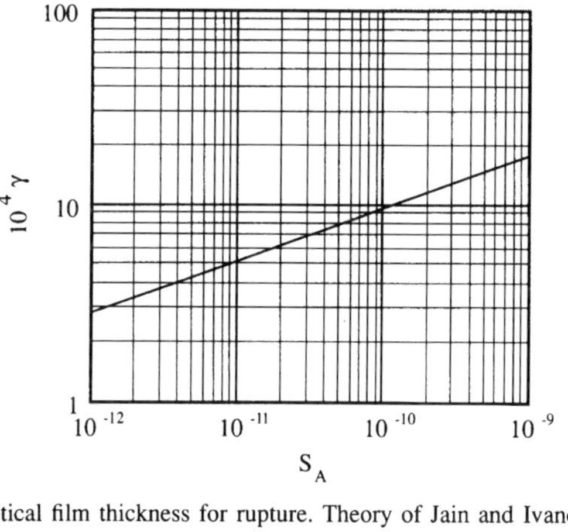


***FIGURE 10-6.2*** Critical film thickness for rupture. Theory of Jain and Ivanov (1980), from Eq. 10-6.1 with $k_B T/[A] = 1$.

action of thermal fluctuations in the fluid near the interface. The solution of Eq. 10-6.1 is given in Fig. 10-6.2, for $k_B T/|A| = 1$. The solutions are only weakly dependent on $k_B T/|A|$.

Of special interest is the dependence of the critical rupture thickness on the film radius $R$. Over the range of $S_A$ values plotted here, $\gamma$ is well represented by a power function of the form

$$\gamma = 0.59\, S_A^{0.275} \tag{10-6.4}$$

and hence

$$h_{crit} \sim R^{0.45}. \tag{10-6.5}$$

We find that this relationship is not consistent with available data on film rupture. Figure 10-6.3 shows some data of Radoev *et al.* (1983). The data show a much weaker dependence of $h_{crit}$ on $R$ than predicted by the theory of Jain and Ivanov. Radoev *et al.* place the blame for the failure of these "simple" hydrodynamic theories on the use of a velocity of thinning that comes from the lubrication model of the hydrodynamics. They present data suggesting that lubrication theory considerably underestimates the velocity of thinning and does so to a degree that depends significantly on the radius of the thinning film.

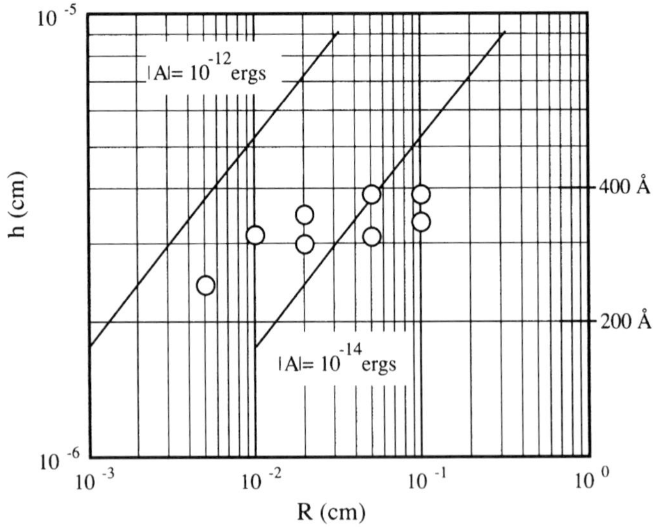


***FIGURE 10-6.3*** Critical film thickness for rupture. Theory of Jain and Ivanov (1980) and data of Radoev *et al.* (1983). $\sigma = 50$ dyn/cm, $T = 298$ K.

## *Problems*

P10-1 Derive an expression for $\lambda$, and compare it to that given in Eq. 10-1.6, for the case of close approach of a solid sphere to a rigid planar surface. Use lubrication theory. The algebra is simplified if the separation of the two surfaces $h(r)$ is approximated, in this "near-approach" region, by

$$h(r) = H - \left(R^2 - r^2\right)^{1/2} \approx H - R + \frac{r^2}{2R}. \qquad \text{(P10 -1.1)}$$

P10-2 Develop an approximate analytical model for the fall of a rigid sphere toward a rigid plane, by using the asymptotic results given in Eqs. 10-1.6 and 7 for $\lambda$. Pick the point $H/R = 1.5$ to switch from one expression to the other.

P10-3 Improve on the model developed in Problem P10-2 by curve-fitting the transition region between Eqs. 10-1.6 and 7. Select a simple power law expression of the form $\lambda = A(H/R)^{-n}$, for the region $1.25 < H/R < 4$. Give the analytical expressions for $H/R$ vs. $T$ over the three regions of the model, and compare the result to that from Problem 10-2.

P10-4 Hartland (1967) calculates $R_d$ of Fig. 10-2.2 from the Bashforth and Adams tables for equilibrium drop shapes. We might expect that $R_{min}$, as calculated from Eq. 10-3.4, would be close to $R_d$. Using the information given in Table 2 of Hartland's paper, compare $R_{min}$ to $R_d$ ($r_f$ in Hartland's nomenclature).

P10-5 Hartland (1967) shows (his Fig. 5) data on the interface shape of a small drop approaching a solid planar surface. Define $R_{min}$ for this shape, and compare the observation to the prediction of Eq. 10-3.4.

P10-6 Pozrikidis (1990) shows (his Fig. 18) a numerical simulation of a viscous drop moving toward a solid planar surface, for the case $\Gamma = 0.1$, where

$$\Gamma \equiv \frac{\sigma}{|\Delta\rho|\, g R_{drop}^2} \qquad \text{(P10-6.1)}$$

is an inverse Bond number. How does his barrier radius compare to that predicted from Eq. 10-3.4? Pozrikidis discusses the extent of agreement.[7] Does his argument apply as well to the previous problem?

P10-7 Derive Eq. 10-2.3.

P10-8 Return to the energy balance that leads to Eq. 10-5.3, and include the surface energy of the hemisphere. What form does Eq. 10-5.4 now take? Is surface energy negligible for the data shown in Fig. 10-5.6?

[7]There is a minor typographical error in his discussion on p. 357, where he gives, in his notation, $\sigma_{rim} = (2\Gamma/3)^{1/2}$. It should be written as Eq. 10-3.4, *viz* $\sigma_{rim} = (2/3\Gamma)^{1/2}$.

P10-9 Study the paper of Chesters and Hofman (1982). Their numerical study predicts the appearance of a dimple at a dimensionless radius (in their notation) $r^* = 0.4$. To what extent does their prediction differ from that given in Eq. 10-3.4? Discuss the differences in terms of the assumptions that characterize the approach of Chesters and Hofman.

P10-10 Study the paper of Jain and Ivanov (1980) and derive Eq. 10-6.1. Plot $\gamma$ vs. $S_A$ for $k_BT/|A| = 0.3$ and 3.0.

P10-11 Study the paper of Gumerman and Homsy (1975). These authors begin with a useful critique of much of the earlier literature on the stability of thin liquid films. They then proceed to present a specific model of film rupture or, more precisely, a model of the events leading to but preceding film rupture. In what essential ways, if at all, does the later work of Jain and Ivanov differ from this work? Replot the curves of Fig. 1 of Gumerman and Homsy, compare the results to those of our Fig. 10-6.2, and comment on the comparison.

P10-12 Manev *et al.* (1984) present data on the drainage of thin liquid films. An example of such data is shown in Fig. P10-12.1, for an aqueous liquid film surrounded on both sides by air at 25°C. They measured the time $t_d$ for the film to drain from a thickness of 200 nm to 50 nm. The planar film radius $R$ was measured, as well. The liquid contains a surfactant (sodium dodecyl sulfate) that reduces the surface tension to a very low value.

Derive a hydrodynamic model for drainage, beginning with Eq. 10-2.18. For consistency with the model used by Manev *et al.*, take $n = 2$ and $F = \pi R^2 \Delta P$, where $\Delta P$ is a sum of the capillary and disjoining pressures:

$$\Delta P = P_c - \Pi_d. \tag{P10-12.1}$$

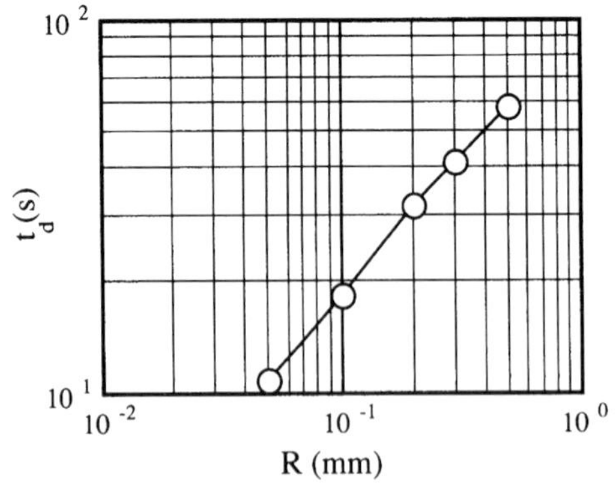


***FIGURE P10-12.1*** Data of Manev *et al.* (1984) for the drainage time of a thin liquid film from an initial thickness of 200 nm to a final thickness of 50 nm.

Manev *et al.* calculate the capillary pressure from

$$P_c = \frac{2\sigma}{R_c} \tag{P10-12.2}$$

and they give a value of $P_c = 49.7$ mN/m$^2$ for this liquid film.

Manev *et al.* write the disjoining pressure in the form

$$\Pi_d = -\frac{A(h)}{6\pi h^3}. \tag{P10-12.3}$$

Note that their sign convention on $\Pi_d$ is opposite ours (see Eq. 10-2.2).

Predict $t_d$ vs. $R$ using the hydrodynamic model, with a constant value of $A = 2 \times 10^{-20}$ J. Compare your prediction to these data.

P10-13 Derive Eq. 10-2.11.

## References


Barber, A. D., and S. Hartland. (1976). The effects of surface viscosity on the axisymmetric drainage of planar liquid films. *Can J. Chem. Eng.* **54**, 279.

Bart, E.. 1968). The slow unsteady settling of a fluid sphere toward a flat fluid interface. *Chem. Eng. Sci.* **23**,193.

Bashforth, F., and J. C. Adams. (1883). An attempt to test the theories of capillary attraction. Cambridge University Press.

Blanchard, D. C., and L. D. Syzdek. (1972). Concentration of bacteria in jet drops from bursting bubbles. *J. Geophys. Res.* **77**, 5087.

Boulton-Stone, J. M., and J. R. Blake. (1993). Gas bubbles at a free surface. *J. Fluid Mech*, **254**, 437.

Brenner, H. (1961). The slow motion of a sphere through a viscous fluid towards a plane surface. *Chem. Eng. Sci.* **16**, 242.

Buevich, Yu. A., and E. Kh. Lipkina. (1975). Draining of liquid from thin axially symmetric films. *J. Appl. Mech. Tech. Phys.* **2**, 217.

Chen, J.-D. (1984). Effects of London-van der Waals and electric double layer forces on the thinning of a dimpled film between a small drop or bubble and a horizontal solid plane. *J. Coll. Int. Sci.* **98**, 329.

Chen, J.D., and J. C. Slattery. (1982). Effects of London-van der Waals forces on the thinning of a dimpled liquid film as a small drop or bubble approaches a horizontal solid plane. *AIChE J.* **28**, 955.

Chesters, A. K. (1975). The applicability of dynamic similarity criteria to isothermal, liquid–gas, two-phase flows without mass transfer. *Int. J. Multiphase Flow* **2**, 191.

Chesters, A.K., and G. Hofman, "Bubble Coalescence in Pure Liquids," Appl. Sci. Res., **38**, 353 (1982).

Derjaguin, B. V., and M. Kussakov. (1939). Anomalous properties of thin polymolecular films. *Acta Physicochim. U.R.S.S.* **10**, 25.

Drogaris, G., and P. Weiland. (1983). Studies of coalescence of bubble pairs. *Chem. Eng. Commun.*, **23**, 11.

Erneux, T., and S. H. Davis. (1993). Nonlinear rupture of free films. *Phys. Fluids* **A5**, 1117.

Flumerfelt, R. W., A. B. Catalano, and C.-H. Tong. (1981). On the coalescence characteristics of low tension oil–water–surfactant systems. In *Surface Phenomena in Enhanced Oil Recovery*, D. O. Sha (ed.). Plenum, New York, p. 571.

Frankel, S. P., and K. J. Mysels. (1962). On the dimpling during the approach of two interfaces. *J. Phys. Chem.* **66**, 190.

Gumerman, R. J., and G. M. Homsy. (1975). The stability of radially bounded thin films. *Chem. Eng. Commun.* **2**, 27.

Hartland, S. (1967). The approach of a liquid drop to a flat plate. *Chem. Eng. Sci.* **22**,1675.

Hartland, S., and J. D. Robinson. (1977). A model for an axisymmetric dimpled draining film. *J. Coll. Interface Sci.* **60**, 72.

Jain, R. K., and I. B. Ivanov. (1980). Thinning and rupture of ring-shaped films. *J. Chem. Soc., Faraday II* **76**, 250.

Jones, F. A., and S. D. R. Wilson. (1978). The film drainage problem in droplet coalescence. *J. Fluid Mech.* **87**, 263.

Kheshgi, H. S., and L. E. Scriven. (1991). Dewetting. Nucleation and growth of dry regions. *Chem. Eng. Sci.* **46**, 519.

Kientzler, C. F., A. B. Arons, D. C. Blanchard, and A. H. Woodcock. (1954). Photographic investigation of the projection of droplets by bubbles bursting at water surface. *Tellus* **6**, 1.

Lee, J. C., and T. D. Hodgson. (1968). Film flow and coalescence—I. Basic relations, film shape and criteria for interface mobility. *Chem. Eng. Sci.* **23**, 1375.

Lin, C.-Y., and J. C. Slattery. (1982). Thinning of a liquid film as a small drop or bubble approaches a solid plane. *AIChE J.* **28**, 147.

MacIntyre, F. (1972). Flow patterns in breaking bubbles. *J. Geophys. Res.* **77**, 5211.

MacKay, G. D. M., M. Suzuki and S. G. Mason. (1963). Approach of a solid sphere to a rigid plane interface, Part 2. *J. Coll. Sci.*, **18**, 103.

Manev, E. D., S. V. Sazdanova, and D. T. Wasan. (1984). Emulsion and foam stability—the effect of film size on film drainage. *J. Coll. Int. Sci.* **97**, 591.

Marrucci, G. (1969). A theory of coalescence. *Chem. Eng. Sci.* **24**, 975.

Oguz, H. N., and A. Prosperetti. (1990). Bubble entrainment by the impact of drops on liquid surfaces. *J. Fluid Mech.* **219**, 143.

Oolman, T. O., and H. W. Blanch. (1986). Bubble coalescence in stagnant liquids. *Chem. Eng. Commun.* **43**, 237.

Platikanov, D. (1964). Experimental investigation on the 'dimpling' of thin liquid films. *J. Phys. Chem.* **68**, 3619.

Pozrikidis, C. (1990). The deformation of a liquid drop moving normal to a plane wall. *J. Fluid Mech.* **215**, 331.

Pumphrey, H. C., and P. E. Elmore. (1990). The entrainment of bubbles by drop impacts. *J. Fluid Mech.* **220**, 539.

Radoev, B. P., A. D. Sheludko, and E. D. Manev. (1983). Critical thickness of thin liquid films. theory and experiment. *J. Coll. Int. Sci.* **95**, 254.

Sharma, A., and E. Ruckenstein. (1986). An analytical nonlinear theory of thin film rupture and its application to wetting films. *J. Coll. Int. Sci.* **113**, 456.

Sheludko, A. (1967). Thin liquid films. *Adv. Coll. Int. Sci.* **1**, 391.

Tsekov, R., and E. Ruckenstein. (1994). Dimple formation and its effect on the rate of drainage in thin liquid films. *Coll. Surf. A* **82**, 255.

Vrij, A. (1966). Possible mechanism for the spontaneous rupture of thin, free liquid films. *Disc. Farad. Soc.* **42**, 23.

Williams, M. B., and S. H. Davis. (1982). Nonlinear theory of film rupture. *J. Coll. Int. Sci.* **90**, 220.

Wu, R., and S. Weinbaum. (1982). On the development of fluid trapping beneath deformable fluid-cell membranes. *J. Fluid Mech.* **121**, 315.

Yang, Y. M., and J. R. Maa. (1984). Bubble coalescence in dilute surfactant solutions. *J. Coll. Int. Sci.* **98**, 120.

A liquid jet, just beyond the point of having broken down into a train of drops, impacts upon a rigid planar surface. The ensuing flow moves out radially, a hydraulic jump is formed, and part of each impacting drop gives rise to a spray of smaller droplets. Photo by Mario Errico.

# *Afterword*

It would not be difficult to add a few more chapters to this book. A number of interesting topics have not been included here, but one must make a choice and I have decided to keep this book small. Hopefully the interested reader will find additional topics of interest while pursuing the extensive references that accompany most chapters. My goal has been to demonstrate the success, and the failure, of simple models of the flows described throughout. There is, to me, a certain excitement that accompanies the observation that with some relatively simple ideas and a few tools one can predict the dynamics of some fairly complex flows.

At the same time there is a curiosity regarding what the future will bring. Should we expect, in the next decade, to be able to simulate the behavior of an unstable jet as it impacts on a rigid planar surface? The figure on the left shows a high speed photograph of such an event. Although some progress has been made in modeling the drop breakup dynamics prior to impact, the ensuing dynamics—the spread of the fluid across the lower surface, the onset of a hydraulic jump, and the conversion of some of the flow into a spray subsequent to impact—seems at the time of writing to be well beyond the capabilities of this generation of computational fluid dynamicists. My hope is that this book will arouse those who will ultimately develop the tools to model such a complex flow.

# Index

Printed in Dunstable, United Kingdom

91336861R00183